Emerging Technology in Food Processing

Yogesh Kumar • Swarnima Dey

Emerging Technology in Food Processing

Springer

Yogesh Kumar
Department of Food Technology
Faculty of Science and Humanities
SRM University
Delhi-NCR, Sonepat, Haryana, India

Swarnima Dey
SRM University
Delhi-NCR, Sonepat, Haryana, India

ISBN 978-3-032-24504-5 ISBN 978-3-032-24505-2 (eBook)
https://doi.org/10.1007/978-3-032-24505-2

This Springer imprint is published by the registered company Springer Nature Switzerland AG
The registered company address is: Gewerbestrasse 11, 6330 Cham, Switzerland

This book is dedicated to beloved food technologists who seek to explore the new era of food science. May it provide them with sufficient knowledge to pursue further research and contribute to the future of food science.

Preface

The global food system is undergoing a rapid transformation driven by advances in science, engineering, and digital technologies. This book presents an insightful and comprehensive exploration of emerging technologies that are revolutionizing food processing and food systems, with a focused perspective for academia, industry professionals, researchers, and students.

The book aims to provide a structured understanding of cutting-edge developments such as plant-based and cell-based foods, space foods, three-dimensional (3D) food printing, personalized nutrition, precision agriculture, nanotechnology in food systems, artificial intelligence in food processing, smart food packaging, and food waste reduction technologies. These topics are discussed to bridge the gap between theoretical knowledge and practical application, enabling faculty members and students to stay updated with current and future trends in food technology.

In addition, this book serves as a guide for industry stakeholders by outlining the implementation of precision agriculture, smart farming systems, and personalized nutrition approaches, supporting data-driven decision-making and sustainable practices. Special emphasis is placed on underexplored and emerging research areas, including food waste reduction technologies and nutraceutical applications, with the objective of stimulating further innovation and research.

Finally, the book highlights the emerging skill sets and competencies required for future roles in food technology and food processing, preparing readers to adapt to evolving professional demands in this multidisciplinary field.

The book is intended to act as a valuable reference for advancing knowledge, fostering innovation, and supporting sustainable growth in the future of food science and technology.

Delhi-NCR, Sonepat, Haryana, India Yogesh Kumar
Delhi-NCR, Sonepat, Haryana, India Swarnima Dey

Acknowledgments

The authors (Dr. Yogesh Kumar and Dr. Swarnima Dey) gratefully acknowledge the contributions and support of all individuals and institutions who assisted in the successful completion of this book.

Dr. Yogesh Kumar sincerely thanks Dr. Swarnima Dey, co-author of this book, for her dedication and timely completion of the assigned chapters. Her careful review and thorough rechecking of the manuscript significantly improved the accuracy, clarity, and overall quality of the content.

We are thankful to the reviewers for their constructive feedback and insightful suggestions, which helped strengthen the technical content of the manuscript.

Our sincere thanks are extended to Daniel Falatko, Senior Editor, for showing interest in this book and for providing the opportunity to bring this work to publication. We also acknowledge Sofia Priya Dharshini V, Production Editor, for her professional guidance and support throughout the manuscript development process.

The authors gratefully acknowledge the Springer Nature team for providing a reputable platform for this work and for accepting the proposal for publication.

Finally, we sincerely thank SRM University Delhi–NCR, Sonepat, Haryana, India, for providing the institutional support and time required to complete this book within the stipulated schedule.

Competing Interests The authors have no competing interests to declare that are relevant to the content of this manuscript.

Contents

Abbreviations

≈	Approximately
°C	Degree Celsius
2D	Two Dimensional
3D	Three Dimensional
3D FP	Three-Dimensional Food Printing
AA	Acetic Acid
ABTS	2,2′-Azinobis-(3-Ethylbenzothiazoline-6-Sulfonic Acid)
Ag	Silver
AGI	Artificial General Intelligence
AgNPs	Silver Nanoparticles
AI	Artificial Intelligence
Akt	Protein Kinase B
Al_2O_3	Aluminum Oxide
ALD	Atomic Layer Deposition
AM	Additive Manufacturing
ANN	Artificial Neural Network
APACHE	Advanced Plant Habitat and Controlled Environment (simulation environment used by NASA)
API	Application Programming Interface
APOA2	Apolipoprotein A2
APOE	Apolipoprotein E
ARGOS	Active Response Gravity Offload System
ASTP	Apollo Soyuz Test Project
ATP	Adenosine Triphosphate
Au	Gold
AuNPs	Gold Nanoparticles
B Vitamins	B complex Vitamins
B12	Vitamin B12
B_2	Vitamin B_2 (Riboflavin)
BD	Big Data
BES	Bioelectrochemical System

BET	Brunauer Emmett Teller (Surface Area Analysis)
BLSS	Bioregenerative Life Support System
BMD	Bone Mineral Density
BMI	Body Mass Index
BOD	Biological Oxygen Demand
BRD	Bovine Respiratory Disease
BSE	Bovine Spongiform Encephalopathy
CAD	Computer-Aided Design
$CaSO_4$	Calcium Sulfate
CEA	Controlled Environment Agriculture
CHP	Crew Health and Performance
CNES	Centre National d'Études Spatiales (French Space Agency)
CNN	Convolutional Neural Network
CNTs	Carbon Nanotubes
CO	Carbon Monoxide
CO_2	Carbon Dioxide
COD	Chemical Oxygen Demand
CRP	C-Reactive Protein
CSRM	Crew State and Risk Model
CV	Computer Vision
CVD	Chemical Vapor Deposition
CVDs	Cardiovascular Diseases
DASH	Dietary Approaches to Stop Hypertension
DBD	Dielectric Barrier Discharge
DCS	Decompression Sickness
DDGS	Dried Distiller's Grains with Solubles
DL	Deep learning
DLR	German Aerospace Center (Deutsches Zentrum für Luft- und Raumfahrt)
DNA	Deoxyribonucleic Acid
DNC	Differentiable Neural Computer
DPPH	2,2-Diphenyl-1-Picrylhydrazyl
DSC	Differential Scanning Calorimetry
DSMs	Digital Surface Models
E. coli	*Escherichia coli*
EAB	Electroactive Bacteria
EAE	Enzyme-Assisted Extraction
ECa	Soil Electrical Conductivity
EET	Extracellular Electron Transfer
EGF	Epidermal Growth Factor
EIP	Emerging Infections Program
ERK	Extracellular Signal-Regulated kinase
ES	Eggshell
ESM	Eggshell Membrane

ESM	Equivalent System Mass
EVA	Extravehicular Activity
EVOH	Ethylene Vinyl Alcohol
FAO	Food and Agriculture Organization
FBS	Fetal Bovine Serum
FDA	Food and Drug Administration
FDR	Frequency Domain Reflectometry
FFF	Fused Filament Fabrication
FGF	Fibroblast Growth Factor
FSC	Food Supply Chain
FSCW	Food Supply Chain Waste
FTIR	Fourier Transform Infrared Spectroscopy
FTO	Fat Mass and Obesity-Associated Gene
FW	Food Waste
g	Gram
G′	Storage Modulus (Elastic Modulus)
G″	Loss Modulus (Viscous Modulus)
GC	Group Specific Component (Vitamin D-Binding Protein Gene)
GCMS	Gas Chromatography Mass Spectrometry
G-code	Geometric Code (Machine Instruction Code)
GDL	Glucono Delta Lactone
GHG	Greenhouse Gas
GI	Gastrointestinal
GI	Glycemic Index
GIS	Geographic Information System
GMOs	Genetically Modified Organisms
GMOT	Grass Measurement Optimization Tool
GNSS	Global Navigation Satellite Systems
GPR	Ground Penetrating Radar
GPS	Global Positioning System
GPx	Glutathione Peroxidase
GRAS	Generally Recognized as Safe
GRS	Gamma Ray Spectrometry
GRU	Gated Recurrent Unit
h	Hour
HACCP	Hazard Analysis and Critical Control Point
HbA1c	Glycated Hemoglobin
HDL-C	High Density Lipoprotein Cholesterol
HDPE	High Density Polyethylene
HFE	High Iron Fe Gene (Homeostatic Iron Regulator)
HHP	High Hydrostatic Pressure
HM	High Moisture
HMEC	High Moisture Extrusion Cooking
HPLC	High Performance Liquid Chromatography

HTC	Hydrothermal Carbonization
IDDSI	International Dysphagia Diet Standardisation Initiative
IGF	Insulin-Like Growth Factor
IL-6	Interleukin-6
ILE	Ionic Liquid Extraction
IMEA	Injury Modes and Effects Analysis
IMF	Intermediate Moisture Food
IoT	Internet of Things
ISRO	Indian Space Research Organisation
ISS	International Space Station
K_2	Vitamin K_2 (Menaquinone)
Kg	Kilogram
KPI	Key Performance Indicator
KV/cm	Kilovolt per Centimeter
LbL	Layer-by-Layer (Deposition)
LCA	Life Cycle Assessment
LC-MS	Liquid Chromatography Mass Spectrometry
LCT	Lactase Gene
LDL	Low-Density Lipoprotein
LDL-C	Low-Density Lipoprotein Cholesterol
LDLR	Low-Density Lipoprotein Receptor
LED	Light-Emitting Diode
LEO	Low Earth Orbit
LiDAR	Light Detection and Ranging
LLDPE	Linear Low Density Polyethylene
LSS	Life Support Systems
LSTM	Long Short-Term Memory
MAE	Microwave-Assisted Extraction
MANN	Memory Augmented Neural Network
MAP	Modified Atmosphere Packaging
MAPK	Mitogen-Activated Protein Kinase
MATS	Microwave-Assisted Thermal Sterilization
MBE	Molecular Beam Epitaxy
MELiSSA	Micro Ecological Life Support System Alternative
MES	Microbial Electrosynthesis
$MgCl_2$	Magnesium Chloride
Min	Minute
MIT	Massachusetts Institute of Technology
ML	Machine Learning
MMT	Montmorillonite
MOCVD	Metal Organic Chemical Vapor Deposition
MPa	Megapascal
MSPC	Multivariate Statistical Process Control
MTHFR	Methylenetetrahydrofolate Reductase
mTOR	Mechanistic (mammalian) Target of Rapamycin

Na	Sodium
NASA	National Aeronautics and Space Administration
NCDs	Non-Communicable Diseases
NFC	Near Field Communication
NF-κB	Nuclear Factor Kappa B
NIR	Near Infrared
NIRS	Near Infrared Spectroscopy
NLP	Natural Language Processing
NMR	Nuclear Magnetic Resonance
NN	Neural Network
NPs	Nanoparticles
O_2	Oxygen
PA	Precision Agriculture
PACES	Physical and Cognitive Exploration Simulations
PATS	Pressure-Assisted Thermal Sterilization
PBM	Plant-Based Meat
PBMs	Plant-Based Meat Analogs
PECVD	Plasma-Enhanced Chemical Vapor Deposition
PEF	Pulsed Electric Field
pH	Potential of Hydrogen
PI3K	Phosphoinositide 3-Kinase
PID	Proportional Integral Derivative
PLA	Polylactic Acid
PLE	Pressurized Liquid Extraction
PLF	Precision Livestock Farming
PP	Polypropylene
PPARs	Peroxisome Proliferator-Activated Receptors
PPARα	Peroxisome Proliferator-Activated Receptor Alpha
PRYSM	Precision Robotic Sprayer
PUFA	Polyunsaturated Fatty Acids
PVD	Physical Vapor Deposition
QA	Quality Assurance
QC	Quality Control
QR	Quick Response
R&D	Research and Development
RAG	Retrieval-Augmented Generation
RF	Random Forest
RFID	Radio Frequency Identification
RHM	Rank Hovis McDougall
RL	Reinforcement Learning
RLHF	Reinforcement Learning from Human Feedback
RMS	Root Mean Square

RNA	Ribonucleic Acid
RNN	Recurrent Neural Network
ROS	Reactive Oxygen Species
rpm	Revolutions per Minute
SCFAs	Short-Chain Fatty Acids
SCP	Single Cell Protein
SDGs	Sustainable Development Goals
SEM	Scanning Electron Microscopy
SERMs	Selective Estrogen Receptor Modulators
SFE	Supercritical Fluid Extraction
SFF	Solid Freeform Fabrication
SFP	Smart Food Packaging
SiO_2	Silicon Dioxide (Silica)
SiO_x	Silicon Oxide (Non-stoichiometric)
SLNs	Solid Lipid Nanoparticles
SLS	Selective Laser Sintering
SMEAT	Skylab Medical Experiments Altitude Test
SnO_2	Tin(IV) Oxide
SNP	Single Nucleotide Polymorphism
SOD	Superoxide Dismutase
spp.	Species
STE	Stirred Tank Extraction
STE/UP	Stirred Tank Extraction with Ultrasonication Pretreatment
STL	Standard Tessellation Language
STRs	Stirred-tank Bioreactors
SVM	Support Vector Machine
SVR	Support Vector Regression
TAS2R38	Taste Receptor Type 2 Member 38
TCR	Total Carotenoid Recovery
TDR	Time-domain Reflectometry
TFC	Total Flavonoid Content
TFP	Traditional Food Packaging
TGF-β	Transforming Growth Factor-Beta
TiO_2	Titanium Dioxide
TNO	Netherlands Organisation for Applied Scientific Research
TPC	Total Phenolic Content
TTI	Time-Temperature Indicator
TVP	Textured Vegetable Protein
UAE	Ultrasound-Assisted Extraction
UAHE	Ultrasound Assisted Heating Extraction
UAV	Unmanned Aerial Vehicle
UNEP	United Nations Environment Programme
USA	United States of America
USD	United States Dollar

USE	Universal Sentence Encoder
UV	Ultraviolet
UVVis	Ultraviolet Visible Spectroscopy
VFAs	Volatile Fatty Acids
VOCs	Volatile organic compounds
VRA	Variable Rate Application
VRT	Variable-rate technology
w/w	Weight per Weight
WHO	World Health Organization
WSN	Wireless Sensor Network
Zn^{2+}	Zinc Ion
ZnO	Zinc Oxide
β	Beta
β-Conglycinin	Beta-Conglycinin
β-Glucans	Beta-Glucans
ω-3	Omega-3 Fatty Acids

Chapter 1
Overview of Emerging Technology in Food Processing

Abbreviations

AI	Artificial Intelligence
ALD	Atomic Layer Deposition
ANN	Artificial Neural Network
APOE	Apolipoprotein E
BLSS	Bioregenerative Life Support System
BMI	Body Mass Index
CAD	Computer-Aided Design
CNN	Convolutional Neural Network
CVD	Chemical Vapor Deposition
DNA	Deoxyribonucleic Acid
EAE	Enzyme-Assisted Extraction
FTO	Fat Mass and Obesity-Associated Gene
GHG	Greenhouse Gas
GMOs	Genetically Modified Organisms
GRU	Gated Recurrent Unit
HDPE	High-Density Polyethylene
HHP	High Hydrostatic Pressure
IMF	Intermediate Moisture Food
IoT	Internet of Things
ISS	International Space Station
LCA	Life Cycle Assessment
LCT	Lactase Gene
LSTM	Long Short-Term Memory
MAE	Microwave-Assisted Extraction
MAP	Modified Atmosphere Packaging
MATS	Microwave-Assisted Thermal Sterilization
ML	Machine Learning

Y. Kumar, S. Dey, *Emerging Technology in Food Processing*,
https://doi.org/10.1007/978-3-032-24505-2_1

MTHFR	Methylenetetrahydrofolate Reductase
NFC	Near Field Communication
PATS	Pressure-Assisted Thermal Sterilization
PEF	Pulsed Electric Field
PLA	Polylactic Acid
PLE	Pressurized Liquid Extraction
QR	Quick Response
RFID	Radio Frequency Identification
RNN	Recurrent Neural Network
SFE	Supercritical Fluid Extraction
SFF	Solid Freeform Fabrication
SNP	Single Nucleotide Polymorphism
TiO_2	Titanium Dioxide
TTI	Time-Temperature Indicator
TVP	Textured Vegetable Protein
UAE	Ultrasound-Assisted Extraction
UAV	Unmanned Aerial Vehicle
UV	Ultraviolet
VRA	Variable Rate Application
ZnO	Zinc Oxide

1.1 Introduction

The global food system is undergoing a profound transformation driven by population growth, climate change, resource constraints, shifting consumer preferences, and rapid technological innovation. Conventional food processing and production systems, which historically emphasized large-scale efficiency, standardization, and cost reduction, are increasingly challenged by demands for sustainability, safety, nutrition, transparency, and ethical responsibility. In response, a range of emerging technologies is reshaping food processing across the entire value chain from primary production and raw material utilization to processing, packaging, distribution, and consumption. These technologies collectively support the transition toward resilient, sustainable, and intelligent food systems capable of meeting future global food demands (Poore & Nemecek, 2018).

One of the most visible shifts in food processing is the development of alternative protein sources, particularly plant-based meat analogs and cell-based (cultivated) meat. Plant-based meat analogs are designed to mimic the sensory, structural, and nutritional properties of conventional meat while relying primarily on plant-derived ingredients such as soy, wheat gluten, legumes, and novel proteins. Although such products have existed for centuries, advances in food structuring, extrusion, flavor science, and protein functionality have propelled them into mainstream markets (Joshi & Kumar, 2015). Environmental concerns, including greenhouse gas emissions, land and water use, and inefficiencies associated with livestock production, have significantly accelerated their adoption (Poore & Nemecek, 2018). Alongside environmental benefits, plant-based meat alternatives address ethical concerns related to animal welfare and public health issues such as antibiotic resistance and zoonotic diseases.

Complementing plant-based innovations, cell-based meat represents a paradigm shift in protein production. Cultivated meat is produced by growing animal cells under controlled conditions, eliminating the need to raise and slaughter animals. Conceptualized as early as 1930 through the idea of "self-reproducing steaks," cell-based meat has progressed from theory to experimental and pilot-scale production. NASA-supported development of lab-grown fish for space missions (Benjaminson et al., 2002) represents one of the earliest applications of cellular agriculture, later followed by the public unveiling of the first cultured beef burger in 2013, a landmark achievement in the field (Post et al., 2020). More recent advances, such as cultivated steaks, cell-based seafood, fermentation-derived dairy proteins, and successful meat bioprinting aboard the International Space Station, demonstrate the scalability and versatility of cellular agriculture (Stephens et al., 2018). Together, plant-based and cell-based proteins exemplify how emerging technologies are redefining food processing to align with sustainability and ethical imperatives.

Emerging food technologies are also closely linked to the future of space food systems, which serve as testbeds for extreme condition food processing and preservation. Early space missions relied on highly simplified, shelf-stable foods designed for safety and functionality, but longer missions revealed the importance of palatability, nutrition, and psychological well-being (Perchonok & Bourland, 2002; Douglas et al., 2016). In the era of the International Space Station, food processing technologies matured to include advanced preservation, packaging, and cultural dietary accommodation (Abercromby et al., 2022). Looking ahead to long-duration lunar and Martian missions, reliance on Earth-based resupply is no longer feasible. Consequently, regenerative food systems incorporating hydroponics, cellular agriculture, microbial protein production, and three-dimensional food printing are gaining prominence (Zhang et al., 2022). These innovations not only sustain human life in space but also generate transferable knowledge for improving food resilience and sustainability on Earth.

Precision agriculture represents another critical pillar of emerging food processing technologies, addressing challenges at the primary production stage. Precision agriculture integrates sensors, Internet of Things (IoT) devices, unmanned aerial vehicles (UAVs), global navigation satellite systems, and data analytics to enable site-specific management of crops and livestock (Monteiro et al., 2021). In crop systems, precision tools support variable rate application of water, fertilizers, and pesticides, improving resource efficiency and reducing environmental impact (Praveen et al., 2024). In livestock systems, precision livestock farming technologies facilitate real-time monitoring of animal health, behavior, and productivity, enhancing welfare and disease management. The integration of artificial intelligence (AI), machine learning (ML), cloud computing, and blockchain further strengthens traceability, decision-making, and sustainability across agricultural and food processing systems (Polymeni et al., 2024).

Advances in digital manufacturing have also introduced three-dimensional (3D) food printing as a novel food processing approach. By applying additive manufacturing principles to edible materials, 3D food printing enables precise control over food structure, texture, composition, and nutritional profile (Sun et al., 2015a; Derossi et al., 2020). This technology supports personalized nutrition, elderly care, medical diets, and the utilization of alternative ingredients and food by-products, aligning with circular economy principles (Dankar et al., 2018). Despite challenges related to material properties, scalability, and regulation, 3D food printing illustrates how digital fabrication can redefine food design and processing.

Personalized nutrition further exemplifies the convergence of food processing with biotechnology and data science. Moving beyond generalized dietary guidelines, personalized nutrition tailors dietary recommendations based on genetic makeup, metabolic responses, gut microbiota, and lifestyle factors (Ordovas & Berciano, 2020). Advances in genomics, metabolomics, and microbiome research enable precision dietary interventions that support disease prevention and health optimization (Ferguson et al., 2016). These developments influence food processing by driving demand for customized formulations, functional foods, and nutritionally optimized products.

Nanotechnology has emerged as a powerful enabling technology in food processing, offering solutions for food safety, quality, shelf-life extension, and nutrient delivery. At the nanoscale, materials exhibit unique properties that enhance packaging performance, pathogen detection, and contaminant remediation (Singh et al., 2019). Nanocomposite and active packaging systems improve barrier properties and inhibit microbial growth, while intelligent packaging incorporating nanosensors enables real-time monitoring of food freshness and spoilage (Patel et al., 2021). Nano-encapsulation technologies enhance the stability and bioavailability of bioactive compounds, supporting functional food development. However, safety, regulatory oversight, and consumer acceptance remain critical considerations.

Closely related to nanotechnology is the development of bio-based sensors and intelligent packaging materials. Natural pigments and plant-derived compounds integrated into biodegradable polymers can act as pH-sensitive indicators of food spoilage, offering safer and more sustainable alternatives to synthetic dyes (Kuswandi & Moradi, 2018; Priyadarshi et al., 2026). These systems align with global sustainability policies and contribute to reduced food waste while enhancing consumer trust.

Finally, emerging technologies play a vital role in addressing the growing challenge of food waste. Food waste arises across the entire supply chain and represents both an environmental burden and a lost resource (Amasuomo & Baird, 2016b). Advanced extraction technologies, bioprocessing, and thermochemical conversion enable the valorization of food waste into bioactive compounds, bioenergy, and value-added products. Digital tools such as AI, IoT, smart packaging, and blockchain enhance monitoring, traceability, and waste reduction across food systems (da Costa et al., 2022).

This chapter provides an overview on emerging technologies in food processing, including alternative proteins, cellular agriculture, space food systems, precision agriculture, digital manufacturing, personalized nutrition, nanotechnology, intelligent packaging, and waste valorization, which are collectively reshaping the global food landscape. Their integration supports sustainability, resilience, efficiency, and ethical responsibility, positioning food processing at the forefront of technological innovation in the twenty-first century.

1.2 Emerging Technology in Food Processing

The major emerging technologies in food processing are highlighted in Table 1.1 along with their core principles, and applications, demonstrating their role in supporting sustainability, innovation, and resilience in modern food systems.

1.2.1 Plant-Based Foods

Plant-based foods have gained significant attention as an emerging technological domain in food processing due to increasing concerns about sustainability, health, and food system resilience. Advances in processing technologies have enabled the transformation of plant raw materials into nutritionally adequate, safe, and sensory appealing foods that can serve as alternatives to animal-derived products (Craig & Fresán, 2021).

Plant-based foods can be broadly classified into whole plant foods, minimally processed foods, protein ingredients, and advanced analogues. Whole plant foods include legumes, cereals and grains, nuts and seeds, and fruits and vegetables. Legumes such as soybeans, lentils, and chickpeas are particularly important due to their high protein content, dietary fiber, and bioactive compounds, while being low in saturated fat (Akhtar et al., 2022). Cereals like wheat, rice, and oats provide energy and B vitamins, whereas nuts, seeds, fruits, and vegetables contribute essential fatty acids, antioxidants, vitamins, minerals, and phytochemicals necessary for human health (Öncü Öner, 2023).

Minimally processed plant foods retain much of their original nutritional quality while improving safety and digestibility. Traditional soy foods such as tofu and tempeh are produced using basic processing techniques, including soaking, grinding, heating, and coagulation, resulting in high-quality protein products (Craig & Fresán, 2021). Fermented plant foods, including tempeh, kinema, natto, and miso, utilize microbial activity to enhance flavor, mineral bioavailability, protein digestibility, and shelf life, while also providing probiotic and functional benefits (Samtiya et al., 2021).

Table 1.1 Emerging technologies in food processing and their applications

Technology area	Description	Applications	References
Plant-based foods	Utilization of plant raw materials and advanced processing (extrusion, fermentation, precision fermentation) to create meat and dairy alternatives	Sustainable protein production, reduced environmental footprint, functional and fortified foods	Craig and Fresán (2021); Akhtar et al. (2022); Dupuis et al. (2023)
Cell-based foods	In vitro cultivation of animal cells using tissue engineering and cellular agriculture	Cultivated meat, seafood, dairy, ethical protein production, reduced land and water use	Post et al. (2020); Stephens et al. (2018); Rubio et al. (2020)
Space food systems	Advanced preservation, packaging, and regenerative food production for extreme environments	Long shelf-life foods, hydroponics, cellular agriculture, 3D food printing for space missions	Perchonok and Bourland (2002); Douglas et al. (2021); Bell et al. (2022)
Precision agriculture	Site-specific farm management using sensors, IoT, AI, UAVs, and geospatial analytics	Optimized input use, yield enhancement, reduced environmental impact, climate resilience	Bwambale et al. (2022); Júnior et al. (2024); Khan et al. (2023)
3D food printing	Additive manufacturing of foods using digital designs and printable food inks	Customized food shapes, personalized nutrition, functional and medical foods	Sun et al. (2015b); Godoi et al. (2016); Pant et al. (2021)
Personalized nutrition	Tailoring diets using phenotypic, genetic, microbiome, metabolomic, and behavioral data	Precision nutrition, functional foods, disease prevention, consumer-centric food design	Ordovas et al. (2018); Barrow et al. (2020); Bush et al. (2020)
Nanotechnology in food science	Engineering materials at nanoscale to enhance food quality, safety, and packaging performance	Antimicrobial packaging, nano-encapsulation, smart sensors, shelf-life extension	Jafari and McClements (2018); Chaudhry et al. (2008); Baskaran et al. (2025)
Smart food packaging	Integration of indicators, sensors, data carriers, and nano-enabled systems	Real-time quality monitoring, traceability, reduced food waste, consumer trust	Kuswandi et al. (2011); Ahari and Soufiani (2021); Motelica et al. (2020)
Artificial intelligence in food processing	Data-driven learning and intelligent automation using ML, ANN, CNN, and computer vision	Quality inspection, process optimization, shelf-life prediction, food safety	Abiodun et al. (2019); Nayak et al. (2020); Liakos et al. (2025)
Food waste reduction technologies	Valorization of food by-products via green extraction, bioprocessing, and thermochemical methods	Bioactive recovery, functional ingredients, bioenergy, circular economy	Amasuomo & Baird (2016a, b); da Costa et al. (2022); Baskaran et al. (2025)

Technological advancements have led to the development of plant-based protein ingredients such as protein concentrates, protein isolates, and textured vegetable protein (TVP). Protein concentrates and isolates derived from soy, pea, and wheat offer enhanced protein content and functional properties such as emulsification and gelation, making them suitable for diverse food applications (Švarc et al., 2022). TVP, produced mainly through extrusion cooking of defatted soy flour, exhibits fibrous, meat-like textures that closely mimic animal muscle structure (Maningat et al., 2022).

Plant-based meat analogues represent one of the most innovative areas of food processing. These products may be soy-based, wheat-based (such as seitan), mycoprotein-based, or formulated from mixed protein systems to improve amino acid balance and texture (Yusuf, 2023). Similarly, plant-based dairy alternatives derived from soy, almond, oat, and coconut employ emulsification, fermentation, enzymatic treatment, and fortification to replicate the sensory and nutritional characteristics of conventional dairy products (Munekata et al., 2020).

Next-generation technologies such as mycelium-based foods, precision fermentation, recombinant proteins, and 3D food printing represent the future of plant-based food processing. These approaches enable sustainable production of functional proteins, customized textures, and enhanced sensory attributes, positioning plant-based foods as a central component of future food systems (Dupuis et al., 2023).

1.2.2 Cell-Based Foods

Cell-based foods, also referred to as cultured or cultivated foods, represent a transformative innovation in food processing that aims to decouple animal-derived food production from conventional livestock farming. By cultivating animal cells under controlled in vitro conditions, cell-based technologies seek to deliver products with comparable sensory and nutritional properties to conventional animal-based foods while addressing sustainability, ethical, and food safety concerns (Post et al., 2020).

Cell-based foods can be classified based on biological source, product structure, cellular composition, and processing approach. Based on biological origin, products include cell-based meat, derived from livestock muscle and fat cells; cell-based seafood, produced from aquatic organism cells; cell-based dairy, where milk proteins and fats are synthesized using mammary cells or precision cultures; and cell-based eggs, which focus on producing egg proteins without whole egg formation (Stephens et al., 2018).

From a structural perspective, products range from unstructured forms such as minced meat and patties, to semi-structured products like sausages, and finally to structured or whole cut products that closely mimic natural muscle architecture, including steaks and fillets. These distinctions are closely linked to the complexity of tissue engineering involved (Ben-Arye & Levenberg, 2019).

The cellular makeup of cell-based foods significantly influences their quality and functionality. Muscle-based products emphasize myocytes for lean meat analogues, while fat-enhanced products incorporate cultured adipocytes to improve flavor and mouthfeel. Co-cultured systems, which integrate muscle, fat, and connective tissue cells, are essential for producing whole cut products with realistic texture and marbling (Kim et al., 2024).

Processing approaches include scaffold-based systems, where edible scaffolds such as collagen or alginate guide tissue formation, and scaffold-free systems, which rely on cell self-aggregation and are suitable for unstructured products. Hybrid products, combining cultured cells with plant-based matrices, offer a pragmatic pathway to reduce cost and improve scalability while maintaining sensory appeal (O'Neill et al., 2021).

Compared to conventional animal-based foods, cell-based foods require significantly less land and water and have the potential for lower greenhouse gas emissions, particularly when renewable energy is integrated into production systems. They also eliminate the need for routine antibiotic use and reduce the risk of zoonotic pathogens due to highly controlled production environments (Rubio et al., 2020).

Nutritionally, cell-based foods offer complete proteins identical to conventional meat and allow customization of fat composition and micronutrients. While plant-based foods are highly scalable and environmentally efficient, they often lack complete protein profiles and require extensive formulation to mimic meat-like texture and flavor (O'Neill et al., 2021).

Cell-based foods represent a disruptive yet complementary technology within modern food processing. By integrating principles of cell biology, tissue engineering, and food science, they offer a pathway toward sustainable, ethical, and nutritionally optimized food systems.

1.2.3 Space Foods Systems and Emerging Technologies in Human Spaceflight

Human space exploration has continuously driven innovation in food processing, preservation, and packaging to ensure astronaut health, safety, and mission success. From the earliest Mercury missions to long-duration habitation aboard the International Space Station (ISS), space food systems have evolved in response to increasing mission durations, crew size, and the need for nutritional adequacy and psychological well-being.

During the Mercury missions (1958–1963), food systems were rudimentary, consisting mainly of cubes, powders, and semi-liquid pastes packed in aluminum tubes and plastic laminates. Processing technologies, such as freezing and freeze-drying, ensured microbial safety and stability. These missions primarily demonstrated that eating and digestion were feasible in microgravity, establishing foundational principles of food containment and safety (Perchonok & Bourland, 2002).

The Gemini program (1961–1966) marked a significant advancement with extended missions lasting up to 2 weeks. Vacuum-dried and freeze-dried foods such as soups, eggs, and vegetables were introduced, along with rehydratable pouches featuring one-way valves. These innovations improved menu variety, ease of consumption, and in-package rehydration, reducing crew fatigue and enhancing nutritional intake.

With the Apollo missions (1961–1969), including lunar landings, food systems had to balance mass constraints with improved palatability and energy density. Dehydrated foods, intermediate moisture foods (IMF), and thermostabilized products were packaged in flexible laminates, aluminum cans, and collapsible tubes. Thermal sterilization and IMF stabilization enabled shelf stability suitable for lunar missions while enhancing crew acceptance.

The Skylab space station (1973) represented a paradigm shift toward crew comfort and long-duration living. Foods such as steak, ice cream, yogurt, bread, and even espresso were made available. Advanced technologies, including irradiation, microwave-assisted thermal sterilization (MATS), and pressure-assisted thermal sterilization (PATS), achieved shelf lives of up to 5 years. The inclusion of familiar foods significantly improved crew morale and psychological well-being (Douglas et al., 2021).

The Apollo-Soyuz Test Project (1975) demonstrated international interoperability of space food systems. Standardized processing methods such as dehydration and thermal processing, combined with flexible laminate pouches and aluminum cans, allowed shared meals between US and Soviet crews, reinforcing the importance of harmonized food standards.

In the Space Shuttle and ISS era (1998–2011), space food systems matured into highly reliable, long-duration solutions. Thermostabilized, irradiated, and dehydrated foods packaged in flexible retort pouches, HDPE containers, and multi-layer co-extruded films supported missions lasting several months. Emphasis shifted toward waste reduction, crew autonomy, and nutritional optimization.

Looking ahead, emerging technologies are shaping future space food systems. Hydroponics and plant growth systems provide fresh food, oxygen generation, and psychological benefits with high water efficiency, making them suitable for Mars habitats (Oluwafemi et al., 2018). Cellular agriculture and microbial food production offer compact, sustainable protein sources with minimal resource inputs, integrating well with bioregenerative life support systems (Stephens & Ellis, 2020; Bell et al., 2022). 3D food printing enables personalized nutrition and menu flexibility, while bioregenerative life support systems (BLSS) aim to close the loop between food, air, water, and waste recycling (Häder et al., 2018).

Collectively, these advancements demonstrate how space food technologies not only support human survival beyond Earth but also inspire sustainable food processing innovations applicable to terrestrial food systems.

1.2.4 Precision Agriculture

Precision agriculture represents a transformative approach to modern farming that leverages emerging technologies to optimize resource use, enhance productivity, and promote environmental sustainability. Unlike conventional farming, which relies on uniform field management, precision agriculture is based on site-specific decision-making using real-time data, advanced sensors, automation, and analytics. The integration of digital tools such as remote sensing, Internet of Things (IoT), artificial intelligence (AI), machine learning (ML), and geospatial technologies has enabled farmers to manage spatial and temporal variability more effectively across crops, soil, water, and livestock systems.

One of the core application areas of precision agriculture is soil management, where site-specific soil analysis supports optimized fertilizer and water application. Technologies such as soil sensors, geostatistical mapping, and variable rate application (VRA) systems allow precise nutrient delivery based on soil heterogeneity, improving crop yield while reducing nutrient losses and environmental pollution (Júnior et al., 2024; Surendran et al., 2024). Closely linked is water management, where smart irrigation systems use soil moisture sensors, weather data, and automated controllers to improve water-use efficiency, a critical factor under increasing climate variability (Bwambale et al., 2022; Lakhiar et al., 2024).

Crop monitoring has been revolutionized through the use of unmanned aerial vehicles (UAVs), multispectral sensors, and satellite imagery. These technologies enable early detection of crop stress, nutrient deficiencies, and growth variability using vegetation indices and remote analytics (Neupane & Baysal-Gurel, 2021; Zidan & Febriyanti, 2024). In parallel, nutrient management systems employ VRA fertilization and sensor-based nutrient stress detection to deliver precise nutrient doses, improving nutrient use efficiency and reducing fertilizer overuse (Nawar et al., 2017; Verma et al., 2020).

Precision agriculture also plays a significant role in pest and disease management. Remote sensing data combined with predictive models and AI-driven decision support systems enable early warning and targeted pesticide application, minimizing chemical inputs and enhancing integrated pest management strategies (Zhang et al., 2019; Pawase et al., 2023). Similarly, GPS-guided planting and harvesting systems improve operational accuracy, reduce overlap, and support yield monitoring, contributing to higher efficiency and reduced fuel consumption (Weiss et al., 2020).

Emerging technologies further support climate adaptation by integrating forecasting models and resilient farming practices that help farmers anticipate extreme weather events and adjust management strategies accordingly (Altieri & Nicholls, 2017). Data management platforms, including cloud-based systems and ML-driven analytics, consolidate large datasets from multiple sources, providing actionable insights and predictive capabilities for farm decision-making (Khan et al., 2023).

Beyond crop systems, precision agriculture extends to livestock management, where wearable sensors, imaging systems, and RFID technologies monitor animal health, nutrition, behavior, and reproduction. These tools improve welfare, productivity, and traceability while reducing labor requirements (Parsons et al., 2023). Additionally, environmental impact monitoring technologies track greenhouse gas emissions, manure composition, and energy use, supporting carbon management and sustainable land use practices (Shorten & Welten, 2022). Therefore, precision agriculture integrates emerging technologies to create intelligent, data-driven farming systems. By enhancing efficiency, sustainability, and resilience, it plays a crucial role in meeting future food security challenges while minimizing environmental impacts.

1.2.5 3D Food Printing

Three-dimensional (3D) food printing is an emerging additive manufacturing technology that enables the fabrication of customized food products through the layer-by-layer deposition of edible materials based on digital designs. Adapted from solid freeform fabrication (SFF) principles, this technology eliminates the need for molds or dies and allows unprecedented flexibility in food shape, texture, internal structure, and nutritional composition (Sun et al., 2015b, c). As consumer demand grows for personalized nutrition, functional foods, and novel gastronomic experiences, 3D food printing is gaining increasing attention from researchers and the food industry.

At the core of 3D food printing lies digital 3D model creation using computer-aided design (CAD) software. CAD enables precise control over geometry, porosity, infill patterns, portion size, and ingredient distribution, which directly influences texture and mouthfeel (Sher & Tutó, 2015). The digital model is subsequently converted into printable instructions through slicing and toolpath generation, where the model is divided into multiple two-dimensional layers and translated into machine-readable code. This step governs critical parameters such as layer thickness, extrusion rate, and deposition sequence, thereby determining printing accuracy and product fidelity (Pitayachaval et al., 2018).

A decisive factor in successful 3D food printing is the formulation of suitable "food inks." Food inks are printable edible materials whose rheological and viscoelastic properties must be carefully tailored to ensure smooth extrusion and structural stability after deposition. Ideal food inks exhibit shear-thinning behavior, appropriate yield stress, and rapid shape recovery to support layer stacking and interlayer adhesion (Phuhongsung et al., 2020). Common food inks include chocolate, dough, cheese, starch pastes, fruit and vegetable purees, and protein-rich formulations, enabling applications across bakery, confectionery, and functional foods (Pant et al., 2021; Sher & Tutó, 2015). The use of additives such as hydrocolloids, emulsifiers, and gelling agents further enhances printability and mechanical strength (Pant et al., 2021).

Several 3D food printing techniques have been developed, each with distinct working principles and applications. Extrusion-based 3D food printing is the most widely used approach, relying on mechanical, pneumatic, or screw-driven extrusion of semi-solid food inks. It offers high material versatility and control over internal structure but suffers from slow printing speeds and often requires post-processing such as baking or cooling (Godoi et al., 2016). Inkjet-based printing, by contrast, deposits micro-droplets of low-viscosity inks and excels in high-resolution decoration and precise nutrient dosing, though it is limited in structural strength (Severini & Derossi, 2016). Binder jetting enables complex geometries by selectively bonding powder beds but requires extensive post-processing and poses hygiene challenges (Sun et al., 2015b). Heat-fused or selective sintering methods provide excellent surface finish and immediate solidification but are constrained by high energy consumption and potential nutrient degradation (Yang et al., 2017).

Beyond structural considerations, nutritional retention and microbiological safety are critical challenges for 3D printed foods. Temperature sensitivity during printing and post-processing can affect nutrient stability, particularly in thermoreversible materials such as chocolate and fat-based inks (Chua et al., 2018). Nevertheless, with continued advances in material science, process optimization, and digital design, 3D food printing holds significant promise as a transformative technology for future food manufacturing.

1.2.6 Personalized Nutrition

Personalized nutrition has emerged as a transformative approach at the intersection of nutrition science, food processing, and digital health technologies. Unlike conventional dietary guidelines that adopt a "one-size-fits-all" paradigm, personalized nutrition tailors dietary recommendations based on individual biological, metabolic, genetic, microbiome, and behavioral characteristics. Advances in omics technologies, data analytics, and food processing innovations have accelerated its application, positioning personalized nutrition as a key emerging technology for improving health outcomes and optimizing food systems (Verma et al., 2018).

Phenotype-based personalized nutrition represents the most established and clinically accessible approach. It relies on observable physical and biochemical traits such as age, sex, body mass index (BMI), body composition, lipid profile, blood pressure, and glycemic markers. These parameters are widely used in clinical nutrition therapy, weight management, and chronic disease prevention due to their cost-effectiveness and strong evidence base (Laddu & Hauser, 2019). However, phenotype-based strategies primarily reflect downstream metabolic outcomes and have limited predictive capacity for early metabolic disturbances.

Genotype-based personalized nutrition, grounded in nutrigenetics, uses inherited genetic variations such as single nucleotide polymorphisms (SNPs) in genes like APOE, FTO, MTHFR, and LCT to inform dietary advice. This approach enables predictive and preventive strategies by identifying individual susceptibilities to nutrient metabolism and disease risk (Barrow et al., 2020). Despite its high personalization potential, challenges remain, including high analytical costs, ethical concerns, and limited translation of genetic data into clear dietary guidance.

Microbiome-based personalized nutrition focuses on the composition and functional activity of the gut microbiota. Dietary modulation of microbial diversity and short-chain fatty acid production has been shown to influence metabolic health, immune function, and nutrient bioavailability (Yeşilyurt et al., 2022). This approach is highly dynamic and responsive to diet, making it particularly relevant for food processing innovations such as personalized fiber formulations and probiotic-enriched foods. However, high inter-individual variability and the lack of standardized clinical guidelines limit its widespread adoption.

Metabolome-based personalized nutrition provides real-time insights into individual metabolic responses by analyzing metabolites in blood or urine. This high sensitivity approach integrates genetic and environmental influences, offering precise dietary interventions for metabolic disorders (Barrow et al., 2020). Its application in routine practice is constrained by expensive analytical techniques and complex data interpretation.

Behavior-based personalized nutrition complements biological data by incorporating lifestyle factors such as food preferences, eating patterns, physical activity, sleep, and stress. While biologically less specific, it plays a critical role in enhancing dietary adherence and long-term sustainability of personalized interventions (Ronteltap et al., 2013).

The most advanced model is integrated or multi-omics personalized nutrition, which combines phenotypic, genomic, microbiome, metabolomic, and behavioral data. Supported by digital platforms, artificial intelligence, and wearable technologies, this approach enables dynamic and highly accurate dietary recommendations (Ordovas et al., 2018). In food processing, this integration facilitates the development of functional foods, nutraceuticals, and customized dietary products aligned with individual health profiles.

The personalized nutrition represents a paradigm shift in food processing and dietary management, moving toward precision, prevention, and consumer-centric health solutions. While challenges related to cost, scalability, and data integration persist, ongoing technological advancements are expected to position personalized nutrition as a cornerstone of future food and health systems (Bush et al., 2020).

1.2.7 Nanotechnology in Food Science

Nanotechnology has emerged as a transformative field in food science, offering innovative solutions for food processing, packaging, safety, and quality enhancement. By manipulating materials at the nanoscale (1–100 nm), it is possible to tailor physicochemical properties such as surface area, reactivity, mechanical strength, and antimicrobial activity. In food systems, nanotechnology is primarily applied through the synthesis of nanomaterials using top-down, bottom-up, biological (green), and hybrid approaches, each with distinct mechanisms, advantages, and limitations (Jafari & McClements, 2018).

Top-down approaches involve breaking down bulk materials into nanoscale structures using physical or chemical means. Mechanical milling, including high-energy ball milling, relies on impact and shear forces to fracture materials, making it suitable for producing nanoclays and mineral fillers used in food packaging nanocomposites to improve mechanical and barrier properties (Biswas et al., 2012). Other top-down methods such as chemical etching, vapor deposition, and laser ablation allow precise surface modification and high-purity nanocoatings but are often limited by high costs, low throughput, and environmental concerns (Jafari & McClements, 2018; Mandracci, 2019).

In contrast, bottom-up approaches build nanomaterials atom by atom or molecule by molecule through chemical or physical assembly. Chemical precipitation and sol-gel processing are widely used due to their simplicity, scalability, and applicability to metal and metal oxide nanoparticles such as silver (Ag), zinc oxide (ZnO), and titanium dioxide (TiO_2), which are valued for their antimicrobial and UV-protective properties in food packaging (Yu et al., 2018). Advanced techniques such as chemical vapor deposition (CVD) and atomic layer deposition (ALD) provide atomic-scale control over coating thickness, enabling high-barrier nanolayers for flexible packaging, though their industrial adoption is constrained by cost and substrate limitations (Mandracci, 2019). Hydrothermal and solvothermal methods further enable the synthesis of highly crystalline nanoparticles with controlled morphology for sensing and antimicrobial applications (Qu et al., 2025).

Growing concerns over sustainability and safety have accelerated the development of biological or green synthesis methods. Microbial and plant-extract-mediated synthesis uses natural reducing and stabilizing agents, resulting in biocompatible nanomaterials with reduced toxicity (Ghosh et al., 2021). Although these approaches are environmentally friendly, challenges such as slow kinetics, batch variability, and limited size control remain. Biomimetic and biofabrication strategies, inspired by natural mineralization pathways, offer exceptional structural precision and are increasingly explored for smart packaging and sensor applications (Kulkarni et al., 2023).

A wide range of nanomaterial types are utilized in food science. Metallic and metal oxide nanomaterials exhibit strong antimicrobial and photocatalytic activity, making them effective for shelf-life extension and food safety (Chaudhry et al., 2008). Carbon-based nanomaterials such as graphene and carbon nanotubes provide superior mechanical strength and gas barrier performance for high-performance packaging and smart sensing systems (Abdelmonem, 2020). Natural biopolymeric nanomaterials, including nanocellulose, chitosan, and protein nanoparticles, are particularly attractive due to their biodegradability, biocompatibility, and suitability for edible coatings and bioactive delivery (Baskaran et al., 2025). Hybrid and composite nanomaterials further enhance functionality by combining the advantages of organic and inorganic components (Abd-Elsalam, 2020).

The nanotechnology holds immense promise in advancing food science by enabling safer, smarter, and more sustainable food systems. Continued research focused on scalability, regulatory compliance, and toxicological assessment will be essential for its responsible industrial implementation.

1.2.8 Smart Food Packaging

Smart food packaging has emerged as a key technological advancement in modern food processing systems, responding to growing demands for food safety, quality assurance, sustainability, and supply chain transparency. Unlike conventional packaging, which primarily serves as a passive barrier, smart packaging integrates indicators, sensors, and data carriers to actively monitor, record, and communicate information about food quality and environmental conditions throughout processing, storage, and distribution.

Indicator-based smart packaging represents one of the most widely adopted technologies due to its simplicity and cost-effectiveness. Time temperature indicators (TTIs) are designed to monitor cumulative thermal exposure and detect temperature abuse during storage and transportation. These systems, based on enzymatic, chemical, or diffusion reactions, provide irreversible visual color changes and are extensively used in cold chain monitoring of meat, dairy, seafood, and frozen foods (Ahari & Soufiani, 2021). Freshness indicators further enhance food safety by detecting spoilage-related metabolites and pH changes caused by microbial activity, commonly using natural dyes such as anthocyanins and curcumin (Kuswandi et al., 2011). Gas indicators, including oxygen, carbon dioxide, and hydrogen sulfide sensors, are particularly important in modified atmosphere packaging (MAP) systems to assess package integrity and microbial spoilage (Rodrigues et al., 2021).

Sensor-based packaging systems offer more precise and quantitative monitoring capabilities. Chemical sensors measure changes in pH, gas composition, or volatile organic compounds, making them suitable for high-value and highly perishable foods (Kerry et al., 2006). Biosensor-based packaging incorporates biological recognition elements such as enzymes, antibodies, or DNA probes to detect specific pathogens including *E. coli*, *Salmonella*, and *Listeria*, thereby significantly improving food safety assurance (Salgado et al., 2021). Physical sensors, such as temperature, humidity, and shock sensors, play a critical role in monitoring environmental and mechanical stresses during transportation and storage (Kuswandi et al., 2011).

Data carrier-based smart packaging enhances traceability and information exchange across the food supply chain. Barcodes and QR codes enable product identification and provide static information to consumers, while RFID and NFC technologies allow real-time tracking, inventory management, and anti-counterfeiting measures, particularly in global food distribution systems (Regattieri et al., 2007). Integrated smart packaging systems combine sensing, data carriers, and digital communication, supporting consumer engagement and transparency (Realini & Marcos, 2014).

Recent advances in food processing have enabled the development of advanced smart packaging technologies, including printed electronics and nano-enabled systems. Printed smart packaging uses conductive inks and flexible circuits to fabricate low-cost sensors suitable for large-scale applications. Nanotechnology further enhances sensor sensitivity and response speed through the use of nanoparticles and nanocomposites, although concerns related to migration, toxicity, and regulatory approval remain (Motelica et al., 2020).

Sustainability has become a central focus in smart packaging development. Bio-based smart packaging systems utilize biodegradable polymers such as starch, chitosan, cellulose, and polylactic acid (PLA), often combined with natural indicators, to reduce environmental impact while maintaining functional performance (Ahari & Soufiani, 2021). These systems align with circular economy principles, although trade-offs between functionality, stability, and cost must be carefully managed.

The smart food packaging represents a significant emerging technology in food processing, offering real-time quality monitoring, enhanced food safety, reduced food waste, and improved consumer trust. Continued advances in materials science, nanotechnology, and digital integration are expected to further expand its industrial adoption, positioning smart packaging as a cornerstone of future intelligent food systems.

1.2.9 Artificial Intelligence in Food Processing

Artificial intelligence (AI) has emerged as a transformative technology in food processing, marking a shift from conventional rule-based automation toward intelligent, adaptive, and data-driven systems. AI enables machines to mimic human cognitive abilities such as learning, reasoning, perception, and decision-making, allowing food processing systems to manage complex, variable, and large-scale datasets more efficiently than traditional approaches (Abiodun et al., 2019; Dodiya & Shah, 2021). By integrating symbolic reasoning with data-driven learning, AI supports intelligent decision-making, adaptive control, and advanced automation across food manufacturing operations.

An AI system typically comprises several core components, including a data or knowledge base, learning mechanisms, reasoning and inference engines, decision-making modules, and supporting hardware and software infrastructure (Abiodun et al., 2019). These components enable autonomous process control, real-time monitoring, and intelligent manufacturing. Based on capability, AI can be classified into Narrow AI, General AI, and Super AI. Current food industry applications predominantly rely on Narrow AI, which is designed for specific tasks such as inspection, prediction, and optimization (Hintze, 2016; Dodiya & Shah, 2021).

Compared with conventional automation, which operates on fixed, predefined rules, AI systems continuously learn from data and adapt to changing conditions. This distinction is particularly important in food processing, where raw material variability, environmental fluctuations, and complex biochemical reactions introduce uncertainty (Frankó et al., 2020). Machine learning (ML), a main subset of AI, plays a central role by enabling systems to model nonlinear and multidimensional relationships between processing inputs and quality outputs without requiring complete mechanistic understanding (Khan et al., 2022; Menichetti et al., 2023).

ML approaches used in food processing include supervised, unsupervised, and reinforcement learning. Supervised learning relies on labeled datasets for tasks such as quality grading, defect detection, and shelf-life prediction, with performance strongly dependent on data quality (Khan et al., 2022). Unsupervised learning identifies hidden patterns in unlabeled data and is particularly useful for raw material clustering and anomaly detection (Sharma et al., 2026). Reinforcement learning enables systems to learn optimal actions through trial-and-error interactions with the processing environment, supporting dynamic process control and adaptive parameter optimization (Wu et al., 2024).

Advanced AI models such as artificial neural networks (ANNs) are widely used to capture complex nonlinear relationships in food systems, supporting quality prediction, process modeling, and shelf-life estimation (Nayak et al., 2020). Deep learning architectures further enhance performance, with convolutional neural networks (CNNs) excelling in image-based inspection tasks such as surface defect detection, contamination identification, and packaging inspection (Ding et al., 2023). Recurrent neural networks (RNNs), including long short-term memory (LSTM) and gated recurrent unit (GRU) models, are effective for time-series analysis in spoilage prediction, microbial risk forecasting, and cold-chain monitoring (Ding et al., 2023).

Computer vision systems, combining imaging technologies with AI algorithms, have become indispensable in modern food processing. Using RGB, hyperspectral, and multispectral imaging, these systems enable non-destructive, high-speed quality inspection, grading, defect detection, and automated sorting, significantly reducing reliance on manual inspection while improving consistency and efficiency (Liakos et al., 2025).

Across the food value chain, AI supports raw material quality assessment, process optimization and control, predictive maintenance, shelf-life prediction, food safety and hazard detection, intelligent packaging, and supply chain traceability (Liakos et al., 2025). Emerging applications such as machine learning oriented nanotechnology, personalized nutrition, and AI-driven food waste reduction further demonstrate the expanding scope of intelligent food systems (Malefors et al., 2022).

Despite its promise, challenges remain, including data availability and quality, model interpretability, cybersecurity risks, high implementation costs, and the need for skilled interdisciplinary professionals. Addressing these challenges through standardization, regulatory alignment, and collaborative research will be critical. The AI and ML are poised to become foundational technologies in food processing, enabling safer, more efficient, and more sustainable food systems for the future.

1.2.10 Food Waste Reduction Technology

Food waste reduction technologies have emerged as a critical component of sustainable food systems, driven by increasing concerns over resource depletion, environmental impact, and food security. Large volumes of food waste generated across agricultural production, food processing, distribution, and consumption stages contain valuable bioactive compounds, nutrients, and energy-rich components. Modern food waste reduction technologies focus not only on minimizing waste generation but also on valorizing food by-products into high-value ingredients, functional materials, bioenergy, and nutraceuticals, thereby supporting circular economy principles.

Food processing wastes such as fruit and vegetable peels, pomace, seeds, shells, and cereal by-products are rich sources of bioactive compounds. For example, coffee shells contain caffeic acid with antibacterial and antioxidant properties, while fruit and vegetable peels are abundant in polyphenols, flavonoids, dietary fibers, carotenoids, and pectins that exhibit antimicrobial, antioxidant, anticancer, and digestive health benefits. Broccoli seeds are a notable source of sulforaphane, a compound with anticarcinogenic activity, whereas tomato seeds and pistachio pomace provide unsaturated fatty acids with antioxidant properties. These bioactives can be recovered and utilized in functional foods, dietary supplements, edible coatings, and active food packaging.

Cereal-based food wastes such as wheat bran, rice bran, maize pericarp, barley spent grains, and oat hulls are among the most significant agro-industrial by-products. They are rich in dietary fibers (arabinoxylans, β-glucans), phenolic acids (ferulic acid), essential fatty acids, proteins, and antioxidants such as γ-oryzanol and tocopherols. Through advanced processing techniques including enzymatic hydrolysis, microfluidization, air classification, fermentation, and micronization, these components can be transformed into value-added ingredients for bakery products, nutraceutical formulations, prebiotics, and biodegradable packaging materials.

Animal and seafood-based food wastes also represent valuable resources. By-products such as feathers, blood, offal, bones, skins, viscera, and shells contain proteins, amino acids, collagen, gelatin, chitin, chitosan, omega-3 fatty acids, calcium carbonate, and bioactive peptides. These materials can be recovered using rendering, enzymatic hydrolysis, membrane filtration, acid–alkali treatment, and supercritical fluid extraction (SFE). Applications include animal feed, functional food ingredients, biomedical materials, probiotics, and bio-based packaging. Notably, seafood shell wastes are increasingly utilized for chitin and chitosan production, which possess antimicrobial and film-forming properties suitable for food preservation.

Emerging extraction and conversion technologies play a pivotal role in efficient food waste valorization. Green and non-thermal techniques such as microwave-assisted extraction (MAE), ultrasound-assisted extraction (UAE), supercritical fluid extraction (SFE), pressurized liquid extraction (PLE), pulsed electric field (PEF), and high hydrostatic pressure (HHP) enhance extraction yields while preserving thermolabile compounds and reducing solvent use. Biological processes such as enzyme-assisted extraction (EAE), fermentation, and anaerobic digestion enable the production of bioactive peptides, organic acids, prebiotics, ethanol, and biogas. Thermochemical methods, including pyrolysis, gasification, and hydrothermal carbonization, further allow conversion of heterogeneous food waste into biofuels, biochar, and carbon-based functional materials.

The food waste reduction technologies represent a transformative and emerging approach in food science and technology. By integrating advanced extraction methods, bioprocessing, and material recovery strategies, food waste can be converted from an environmental burden into a valuable resource. Continued innovation, supported by policy frameworks and industrial adoption, will be essential to scale these technologies and achieve sustainable, low-waste food systems.

1.3 Ethical, Social, and Environmental Considerations

The rapid advancement of emerging technologies in food science ranging from space food systems and precision agriculture to 3D food printing, plant/cell-based food, precision nutrition, nanotechnology, artificial intelligence (AI), and smart food packaging, food waste reduction technologies has significantly transformed food production, processing, distribution, and consumption. While these innovations offer substantial benefits in terms of efficiency, sustainability, food security, and personalization, they also raise complex ethical, social, and environmental concerns. Addressing these considerations is essential to ensure that technological progress aligns with societal values, environmental protection, public trust, and equitable access to food systems (Table 1.2).

1.3.1 Ethical Considerations

Ethical concerns in emerging food technologies primarily relate to safety, transparency, data governance, equity, and long-term societal impacts. Technologies such as nanotechnology, AI, and cellular agriculture introduce novel materials, algorithms, and production systems whose risks may not be fully understood at the time of deployment. The ethical principle of precaution, therefore, becomes critical, particularly in food systems where public health is directly involved (Chaudhry et al., 2008).

Nanotechnology in food science raises ethical questions related to human exposure, toxicity, and long-term health effects. Engineered nanomaterials such as silver, zinc oxide, and titanium dioxide nanoparticles exhibit antimicrobial properties and are widely investigated for food packaging and preservation applications. However, their small size enables interaction with biological systems at the cellular and molecular levels, raising concerns regarding bioaccumulation, cytotoxicity, and unintended ecological effects (Jafari & McClements, 2018). Ethical deployment requires rigorous toxicological evaluation, transparent risk communication, and regulatory oversight to prevent harm while enabling innovation.

Artificial intelligence introduces ethical challenges related to algorithmic transparency, accountability, and bias. AI-driven decision-making systems in food processing, quality inspection, and supply chain management rely heavily on large

Table 1.2 Ethical, social, and environmental considerations associated with emerging technologies in food science

Dimension	Technology area	Considerations	Implications	References
Ethical	Nanotechnology	Potential toxicity, bioaccumulation, long-term health and ecological risks	Need for precautionary principle, rigorous risk assessment, transparent communication	Chaudhry et al. (2008); Jafari and McClements (2018)
	Artificial intelligence (AI)	Algorithmic bias, lack of transparency, data ownership and privacy	Ethical data governance, accountability, explainable AI systems	Abiodun et al. (2019); Liakos et al. (2025)
	3D food printing & personalized nutrition	Food authenticity, cultural identity, consumer manipulation	Balance innovation with informed choice and cultural respect	Sun et al. (2015b); Pant et al. (2021)
	Plant-based foods	Animal welfare, environmental responsibility	Reduced ethical concerns related to livestock farming	Craig and Fresán (2021)
	Cell-based foods	Moral status of cultured cells, technology access	Elimination of animal slaughter, ethical protein production	Post et al. (2020); Rubio et al. (2020)
	Smart packaging	Food transparency, waste reduction	Ethical consumption, reduced food loss	Realini and Marcos (2014)
Social	Precision agriculture	Labor displacement, digital divide	Risk of marginalizing smallholder farmers	Weiss et al. (2020); Surendran et al. (2024)
	AI & automation	Workforce transformation, skill gaps	Need for reskilling and inclusive policies	Parsons et al. (2023)
	Consumer acceptance	Perceived unnaturalness of novel foods	Importance of labeling, public engagement	Chaudhry et al. (2008); Severini and Derossi (2016)
	Food waste valorization	Reintroduction of waste-derived ingredients	Safety assurance and consumer trust required	Malefors et al. (2022)
	Space food technologies	Psychological well-being, food familiarity	Improved morale in confined environments	Douglas et al. (2021)

(continued)

Table 1.2 (continued)

Dimension	Technology area	Considerations	Implications	References
Environmental	Precision agriculture	Reduced water, fertilizer, and GHG emissions	Improved resource efficiency, but e-waste concerns	Bwambale et al. (2022); Lakhiar et al. (2024)
	Nanotechnology-based packaging	Reduced spoilage, nanoparticle release risks	Need for green synthesis and LCA evaluation	Abdelmonem (2020); Ghosh et al. (2021)
	AI in food systems	Energy optimization vs. data-center emissions	Demand for energy-efficient algorithms	Frankó et al. (2020); Menichetti et al. (2023)
	Food waste reduction technologies	Methane reduction, circular bioeconomy	Resource recovery with environmental safeguards	Mirabella et al. (2014)
	Space & BLSS technologies	Closed-loop sustainability models	High energy demand limits scalability	Häder et al. (2018)
Governance	Cross-cutting technologies	Regulatory gaps, global harmonization	Adaptive regulation and stakeholder participation	Mandracci (2019); Liakos et al. (2025)

datasets. If these datasets are incomplete, biased, or poorly curated, AI models may reinforce inequities or produce unreliable decisions affecting food safety and quality (Abiodun et al., 2019; Liakos et al., 2025). Moreover, the use of AI-based surveillance systems in agriculture and food manufacturing raises concerns about data ownership, privacy, and corporate control over farmer and consumer data.

3D food printing and personalized nutrition technologies present ethical dilemmas related to food authenticity, cultural identity, and consumer manipulation. While personalized food designs can improve nutrition for vulnerable populations such as the elderly or astronauts, excessive customization driven by commercial algorithms may influence consumer behavior in ways that prioritize profit over health (Sun et al., 2015b; Pant et al., 2021). Ethical frameworks must therefore balance innovation with respect for cultural food traditions and informed consumer choice.

In space food systems, ethical considerations extend to issues of resource allocation and technology transfer. Technologies developed for human spaceflight, such as advanced preservation methods, bioregenerative life support systems (BLSS), and cellular food production, have significant potential for addressing food insecurity on Earth. Ethical responsibility lies in ensuring that such innovations are adapted for terrestrial benefit rather than remaining exclusive to high-cost space programs (Douglas et al., 2021; Stephens & Ellis, 2020). Plant-based foods are widely recognized for their lower environmental footprint compared to conventional animal-based foods. They generally require less land, water, and energy and generate fewer greenhouse gas emissions, contributing positively to climate change mitigation and biodiversity conservation (Craig & Fresán, 2021). Ethically, plant-based foods reduce reliance on animal agriculture, aligning with concerns about animal welfare. Cell-based foods introduce a novel ethical paradigm by producing animal proteins without slaughter. This technology addresses ethical concerns related to animal welfare while potentially reducing the environmental impacts associated with livestock farming, such as deforestation and antibiotic overuse (Post et al., 2020; Rubio et al., 2020). Smart food packaging contributes to ethical and environmental sustainability by improving food safety, reducing food waste, and enhancing supply chain transparency. By enabling real-time monitoring of freshness, temperature, and spoilage, smart packaging technologies help prevent unnecessary disposal of safe food and support responsible consumption (Realini & Marcos, 2014).

1.3.2 Social Considerations

Emerging food technologies profoundly influence social structures, labor dynamics, consumer trust, and food accessibility. Precision agriculture, AI-driven automation, and smart food processing systems improve productivity and efficiency but may also displace traditional labor roles, particularly in small-scale and developing agricultural communities (Weiss et al., 2020; Parsons et al., 2023). The digital divide characterized by unequal access to technology, infrastructure, and technical knowledge can exacerbate existing socioeconomic inequalities if not addressed through inclusive policies and capacity-building initiatives.

Precision agriculture relies heavily on advanced sensors, satellite imagery, UAVs, and data analytics to optimize resource use and crop management. While these technologies enhance yields and environmental performance, their high initial costs may limit adoption among smallholder farmers, reinforcing the dominance of large agribusinesses (Surendran et al., 2024; Khan et al., 2023). Social equity considerations, therefore, require supportive frameworks, subsidies, and open-access platforms to democratize technological benefits.

Consumer perception and acceptance represent another major social dimension. Technologies such as nanotechnology, AI-generated foods, and 3D printed foods often face consumer hesitancy due to perceived unnaturalness or safety concerns. Transparent labelling, science-based communication, and public engagement are essential to build trust and promote informed decision-making (Chaudhry et al., 2008; Severini & Derossi, 2016). Failure to address public concerns may lead to resistance similar to that observed with genetically modified organisms (GMOs).

Food waste reduction technologies offer significant social benefits by enhancing food availability, reducing costs, and improving nutrition through the valorization of by-products. The recovery of bioactive compounds from food waste for functional foods and nutraceuticals has the potential to support public health and circular economies (Malefors et al., 2022). However, ethical questions arise regarding the reintroduction of waste-derived ingredients into the food chain, emphasizing the need for strict safety standards and consumer transparency.

Space food technologies also provide unique social insights, particularly in relation to psychological well-being. Studies from Skylab and ISS missions demonstrate that familiar foods and improved food variety significantly enhance crew morale and mental health during long-duration missions (Douglas et al., 2021). These findings have broader implications for institutional food systems on Earth, such as hospitals, military facilities, and disaster.

1.3.3 Environmental Considerations

Environmental sustainability is a central driver behind many emerging food technologies, yet their environmental footprints must be critically assessed across their life cycles. Precision agriculture significantly reduces water use, fertilizer runoff, and greenhouse gas emissions through site-specific management and smart irrigation systems (Bwambale et al., 2022; Lakhiar et al., 2024). However, the production, maintenance, and disposal of electronic sensors, drones, and digital infrastructure contribute to electronic waste and energy consumption, necessitating life cycle assessment (LCA)-based evaluation.

Nanotechnology-based packaging offers improved barrier properties, antimicrobial protection, and shelf-life extension, thereby reducing food spoilage and waste. Nevertheless, environmental risks arise from nanoparticle release during production, use, or disposal, potentially affecting soil, water, and microbial ecosystems (Abdelmonem, 2020; Baskaran et al., 2025). Environmentally benign synthesis methods, such as green and biological approaches, are therefore critical for minimizing ecological impact (Ghosh et al., 2021).

Artificial intelligence contributes to environmental sustainability by optimizing processing parameters, reducing energy consumption, minimizing product losses, and enabling predictive maintenance (Frankó et al., 2020; Menichetti et al., 2023). AI-driven supply chain optimization further reduces transportation emissions and food waste. However, the growing computational demand of AI systems, including deep learning models, increases energy use and carbon emissions associated with data centers, highlighting the need for energy-efficient algorithms and renewable energy integration.

Food waste reduction technologies offer the most direct environmental benefits by diverting organic waste from landfills, reducing methane emissions, and recovering valuable resources. Technologies such as anaerobic digestion, fermentation, and thermochemical conversion transform waste into bioenergy, biochar, and bio-based materials, contributing to circular bioeconomies (Mirabella et al., 2014). Nonetheless, large-scale implementation must consider emissions, water use, and potential contamination risks to ensure net environmental gains.

Space food systems and bioregenerative life support technologies exemplify closed-loop sustainability models, integrating food production with air, water, and waste recycling (Häder et al., 2018). These systems provide valuable blueprints for sustainable food production in extreme environments and urban settings on Earth. However, their high energy requirements and technological complexity must be addressed before widespread terrestrial adoption.

Effective governance is essential to address ethical, social, and environmental challenges associated with emerging food technologies. Regulatory frameworks must evolve alongside technological innovation to ensure food safety, consumer protection, environmental stewardship, and fair market practices. International harmonization of standards is particularly important for nanotechnology, AI, and space-derived food technologies, which operate across global supply chains (Mandracci, 2019; Liakos et al., 2025).

Risk assessment methodologies must be updated to account for novel materials, data-driven decision systems, and hybrid technologies. Post-market surveillance, adaptive regulation, and stakeholder participation are increasingly recognized as essential components of responsible innovation. Furthermore, interdisciplinary collaboration among food scientists, ethicists, policymakers, engineers, and social scientists is critical for anticipating unintended consequences and aligning technological development with societal goals.

1.4 Health Conscious and Functional Food Innovations

The growing global emphasis on health, wellness, and disease prevention has significantly reshaped modern food systems, leading to the rapid expansion of health-conscious and functional food innovations. Functional foods are defined as foods that provide benefits beyond basic nutrition by improving physiological functions, reducing disease risk, or enhancing overall well-being (Fig. 1.1). These innovations are driven by rising lifestyle-related disorders, aging populations, increased consumer awareness, and advances in food science and emerging technologies (Jafari & McClements, 2018).

Health-conscious food innovation increasingly integrates emerging technologies such as nanotechnology, precision agriculture, 3D food printing, artificial intelligence (AI), and food waste valorization. Together, these technologies enable the development of safer, personalized, sustainable, and nutritionally optimized food products.

Plant-based foods are strongly associated with health benefits due to their high content of dietary fiber, antioxidants, phytochemicals, and unsaturated fats. Advances in processing have enabled the fortification of plant-based products with essential micronutrients such as calcium, vitamin B12, iron, and omega-3 fatty acids, addressing common nutritional gaps associated with plant-based diets (Munekata et al., 2020). Functional plant-based foods, including fermented products and

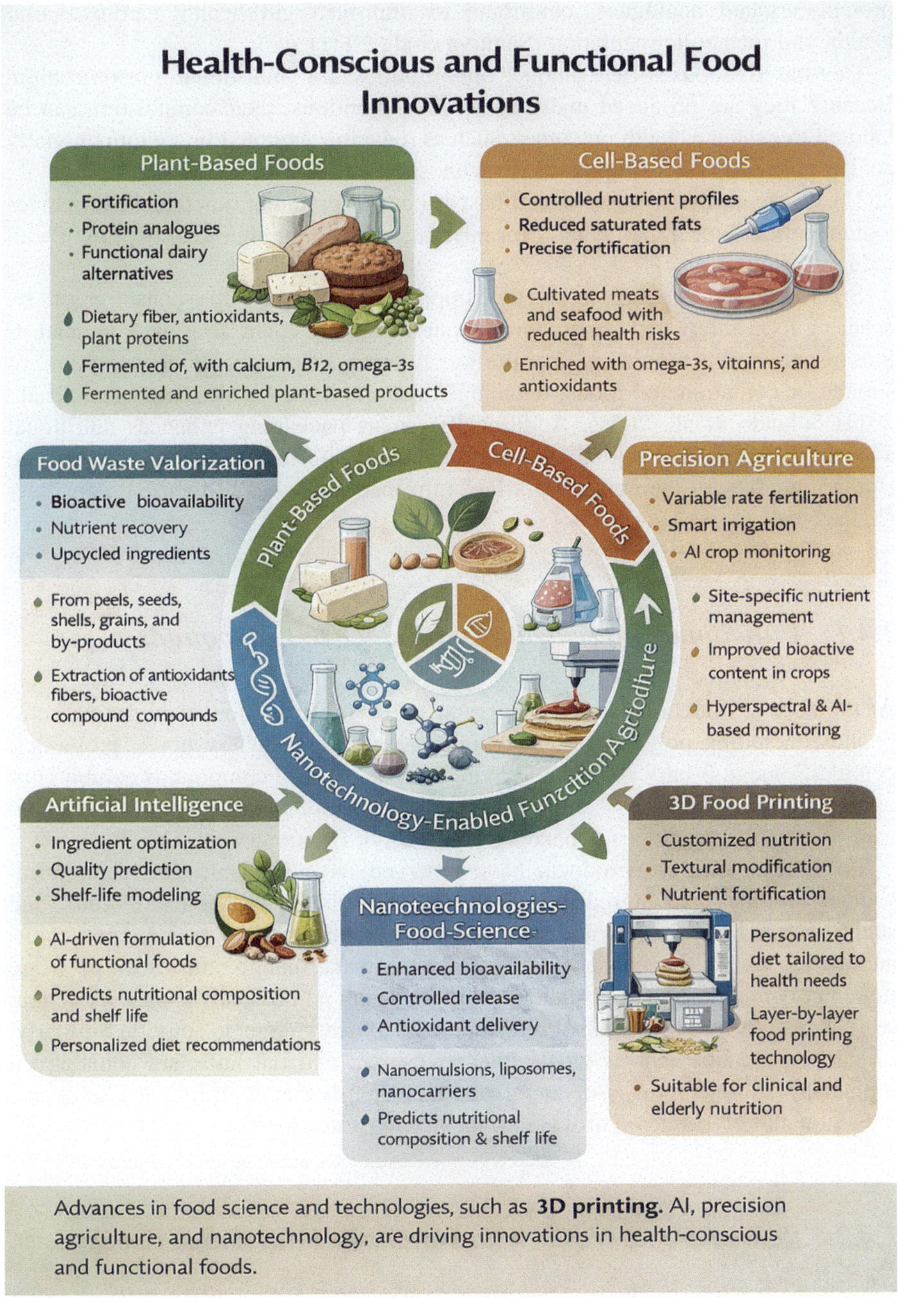

Fig. 1.1 Food innovations for health and sustainability

protein-enriched analogues, contribute to improved gut health, cardiovascular health, and metabolic regulation (Samtiya et al., 2021).

Cell-based foods offer unique opportunities for nutritional customization. Because they are produced under controlled conditions, their composition can be tailored to enhance health outcomes, such as reducing saturated fat, optimizing fatty acid profiles, or enriching products with specific micronutrients (O'Neill et al., 2021). These attributes position cell-based foods as potential functional alternatives to conventional meat, particularly for populations seeking high-quality protein with reduced health risks.

Smart food packaging indirectly supports health-conscious consumption by ensuring food safety and quality throughout the supply chain. Freshness indicators, biosensors, and temperature monitoring systems help prevent consumption of spoiled or contaminated foods, reducing foodborne illness risks (Kuswandi et al., 2011; Salgado et al., 2021). Additionally, smart packaging enhances nutritional awareness by enabling digital access to ingredient information, allergen alerts, and storage guidance, thereby supporting informed and health-focused consumer choices.

1.4.1 Functional Ingredients and Bioactive Compounds

At the core of functional food innovation is the incorporation of bioactive compounds, including polyphenols, flavonoids, carotenoids, dietary fibers, probiotics, peptides, omega-3 fatty acids, and phytochemicals. These compounds exhibit antioxidant, antimicrobial, anti-inflammatory, antidiabetic, anticancer, and cardioprotective properties (Chaudhry et al., 2008; Baskaran et al., 2025).

Food processing by-products have emerged as sustainable and economical sources of functional ingredients. Fruit and vegetable peels, pomace, seeds, and shells contain high concentrations of polyphenols, pectins, flavonoids, carotenoids, and dietary fibers. For instance, apple, citrus, banana, mango, and carrot peels are rich in polyphenols and fibers that support gut health and oxidative stress reduction, while broccoli seeds provide sulforaphane with anticarcinogenic activity. Similarly, cereal by-products such as wheat bran, rice bran, and oat hulls are abundant in β-glucans, arabinoxylans, γ-oryzanol, and tocopherols that contribute to cholesterol reduction and glycemic control (Jafari & McClements, 2018).

1.4.2 Role of Food Waste Valorization in Functional Foods

Food waste reduction technologies play a crucial role in health-oriented food innovation by enabling the recovery and reutilization of valuable nutrients and bioactives. Advanced extraction techniques such as microwave-assisted extraction (MAE), ultrasound-assisted extraction (UAE), supercritical fluid extraction (SFE), pressurized liquid extraction (PLE), pulsed electric field (PEF), and enzyme-assisted extraction (EAE) allow efficient recovery of thermolabile and sensitive compounds while minimizing solvent use and environmental impact (Malefors et al., 2022).

Recovered bioactives are increasingly incorporated into functional beverages, bakery products, dairy alternatives, edible coatings, and nutraceuticals. Animal and seafood by-products further contribute collagen, gelatin, chitin, chitosan, bioactive peptides, and omega-3 fatty acids, which are widely used in joint-health supplements, protein-enriched foods, and antimicrobial packaging (Abd-Elsalam, 2020). Thus, waste valorization not only supports circular economy principles but also enhances the nutritional density of health-conscious foods.

1.4.3 Nanotechnology-Enabled Functional Foods

Nanotechnology has become a major enabler in functional food development by improving bioavailability, stability, controlled release, and efficacy of health-promoting compounds. Nanoscale delivery systems such as nanoemulsions, liposomes, polymeric nanoparticles, and biopolymeric nanocarriers enhance the solubility and absorption of poorly bioavailable compounds such as carotenoids, polyphenols, curcumin, and omega-3 fatty acids (Jafari & McClements, 2018).

Natural biopolymeric nanomaterials, including nanocellulose, chitosan nanoparticles, and protein-based nanocarriers, are particularly attractive for health-conscious applications due to their biodegradability and safety. Metallic and metal oxide nanoparticles, such as ZnO and TiO_2, are used primarily in antimicrobial and UV-protective packaging rather than direct food incorporation, supporting food safety and shelf-life extension (Chaudhry et al., 2008).

Despite their benefits, consumer skepticism (or consumer hesitancy) toward nanotechnology in foods highlights the importance of transparent risk assessment, regulatory oversight, and clear communication regarding safety and benefits.

1.4.4 Precision Agriculture and Health-Oriented Food Production

Precision agriculture contributes indirectly but significantly to functional food innovation by improving nutrient quality, consistency, and sustainability at the production level. Site-specific nutrient and water management ensures optimal accumulation of bioactives such as polyphenols, antioxidants, and vitamins in crops. Smart irrigation and variable rate fertilization reduce excessive chemical inputs, resulting in cleaner raw materials with improved nutritional profiles (Surendran et al., 2024; Lakhiar et al., 2024).

Furthermore, AI-enabled crop monitoring and disease prediction reduce pesticide residues, supporting the production of safer and health-conscious foods. Precision livestock technologies also enhance animal health and product quality, influencing the nutritional composition of meat, milk, and eggs (Parsons et al., 2023).

1.4.5 Personalization Through 3D Food Printing

Three-dimensional food printing represents a breakthrough innovation in personalized nutrition, a central theme of health-conscious food systems. By combining digital design, tailored food inks, and controlled deposition, 3D food printing enables customization of macronutrient composition, micronutrient fortification, texture, and portion size based on individual dietary needs (Sun et al., 2015b).

This technology is particularly promising for elderly nutrition, clinical diets, sports nutrition, and space foods, where precise nutrient delivery and texture modification are critical. Functional ingredients such as proteins, fibers, probiotics, and antioxidants can be incorporated into printed foods, while rheological control using hydrocolloids ensures structural stability and palatability (Pant et al., 2021). Although scalability and regulatory challenges remain, 3D food printing aligns strongly with future health-focused food manufacturing.

1.4.6 Artificial Intelligence in Functional Food Development

Artificial intelligence accelerates innovation in health-conscious foods by supporting ingredient selection, formulation optimization, quality prediction, and shelf-life modeling. Machine learning models analyze complex interactions between ingredients, processing conditions, and nutritional outcomes, enabling the design of functional foods with targeted health benefits (Khan et al., 2022).

AI-driven consumer analytics further supports product personalization by linking dietary patterns, health data, and preferences to tailored food solutions. Computer vision systems ensure quality and safety in functional food production, while predictive models optimize bioactive retention during processing and storage (Liakos et al., 2025).

1.5 Conclusion

The global food system is undergoing a fundamental transformation as emerging technologies redefine how food is produced, processed, packaged, and consumed. Driven by population growth, environmental pressures, resource limitations, and evolving consumer expectations, conventional food processing models are no longer sufficient to ensure long-term sustainability, safety, and nutritional adequacy. In response, a diverse range of innovative technologies is reshaping the food value chain and enabling the transition toward resilient and intelligent food systems.

Alternative protein technologies, particularly plant-based and cell-based foods, illustrate how food processing can align with sustainability, ethical responsibility, and public health goals. Plant-based meat analogues reduce environmental impacts and reliance on animal agriculture, while cell-based meat offers a transformative pathway to produce animal proteins without livestock farming. Together, these approaches demonstrate the potential of food processing to address climate change, animal welfare, and food security challenges simultaneously.

Advances in precision agriculture, digital manufacturing, and personalized nutrition further highlight the integration of data science, biotechnology, and food engineering. Technologies such as IoT, artificial intelligence, 3D food printing, and omics-based nutrition enable resource-efficient production, customized diets, and improved health outcomes. Meanwhile, innovations in nanotechnology, intelligent and smart packaging, and bio-based sensors enhance food safety, shelf life, traceability, and consumer trust while supporting waste reduction and sustainability goals.

Emerging technologies also play a critical role in addressing food waste through valorization strategies and digital monitoring systems, transforming waste streams into valuable resources. Collectively, these innovations position food processing at the forefront of twenty-first-century technological advancement. Their successful implementation will depend on interdisciplinary collaboration, robust regulatory frameworks, ethical governance, and public acceptance, ultimately shaping a sustainable, resilient, and equitable global food system for the future.

References

Abdelmonem, A. M. (2020). Application of carbon-based nanomaterials in food preservation area. In *Carbon nanomaterials for agri-food and environmental applications* (pp. 583–593). Elsevier.

Abd-Elsalam, K. A. (2020). Multifunctional hybrid nanomaterials for sustainable agri-food and ecosystems: A note from the editor. In *Multifunctional hybrid nanomaterials for sustainable agri-food and ecosystems* (pp. 1–19). Elsevier.

Abercromby, A., Douglas, G., Kalogera, K., Somers, J., Suresh, R., Thompson, M., et al. (2022, July). NASA crew health & performance capability development for exploration: 2021 to 2022 overview. In *51st International conference on environmental systems*.

Abiodun, O. I., Jantan, A., Omolara, A. E., Dada, K. V., Umar, A. M., Linus, O. U., et al. (2019). Comprehensive review of artificial neural network applications to pattern recognition. *IEEE Access, 7*, 158820–158846.

Ahari, H., & Soufiani, S. P. (2021). Smart and active food packaging: Insights in novel food packaging. *Frontiers in Microbiology, 12*, 657233.

Akhtar, S., Ismail, T., Layla, A., Hussain, M., & Qamar, M. (2022). An overview of plant-based protein rich products. In *Plant protein foods* (pp. 27–60). Springer.

Altieri, M. A., & Nicholls, C. I. (2017). The adaptation and mitigation potential of traditional agriculture in a changing climate. *Climatic Change, 140*(1), 33–45. https://doi.org/10.1007/s10584-013-0909-z

Amasuomo, E., & Baird, J. (2016a). Solid waste management trends in Nigeria. *Journal of Management and Sustainability, 6*, 35.

Amasuomo, E., & Baird, J. (2016b). The concept of waste and waste management. *Journal of Management and Sustainability, 6*, 88.

Barrow, M., Bell, L., & Bell, C. (2020). Transforming personalized nutrition practice. *Nutrition Reviews, 78*(12), 1046–1051.

Baskaran, R., Chandradass, J., & Thirugnanasambandham, T. (2025). Advances in sustainable polymeric materials derived from renewable biopolymers integrated with nanotechnology for active food preservation: A review. *Polymer-Plastics Technology and Materials, 65*, 1–40.

Bell, P. J., Paras, F. E., Mandarakas, S., Arcenal, P., Robinson-Cast, S., Grobler, A. S., & Attfield, P. V. (2022). An electro–microbial process to uncouple food production from photosynthesis for application in space exploration. *Life, 12*(7), 1002.

Ben-Arye, T., & Levenberg, S. (2019). Tissue engineering for clean meat production. *Frontiers in Sustainable Food Systems, 3*, 46.

Benjaminson, M. A., Gilchriest, J. A., & Lorenz, M. J. A. A. (2002). In vitro edible muscle protein production system (MPPS): Stage 1, fish. *Acta Astronautica, 51*(12), 879–889.

Biswas, A., Bayer, I. S., Biris, A. S., Wang, T., Dervishi, E., & Faupel, F. (2012). Advances in top–down and bottom–up surface nanofabrication: Techniques, applications & future prospects. *Advances in Colloid and Interface Science, 170*(1–2), 2–27.

Bush, C. L., Blumberg, J. B., El-Sohemy, A., Minich, D. M., Ordovás, J. M., Reed, D. G., & Behm, V. A. Y. (2020). Toward the definition of personalized nutrition: A proposal by the American Nutrition Association. *Journal of the American College of Nutrition, 39*(1), 5–15.

Bwambale, E., Abagale, F. K., & Anornu, G. K. (2022). Smart irrigation monitoring and control strategies for improving water use efficiency in precision agriculture: A review. *Agricultural Water Management, 260*, 107324. https://doi.org/10.1016/j.agwat.2021.107324

Chaudhry, Q., Scotter, M., Blackburn, J., Ross, B., Boxall, A., Castle, L., et al. (2008). Applications and implications of nanotechnologies for the food sector. *Food Additives and Contaminants, 25*(3), 241–258.

Chua, C. K., Tan, C., Li, L., & Wong, G. (2018). Enhancing 3D printability of pureed food by addition of hydrocolloids. In *Proceedings of the 3rd international conference on progress in additive manufacturing (Pro-AM 2018)* (pp. 662–666). Nanyang Technological University. https://doi.org/10.25341/D42C7V

Craig, W. J., & Fresán, U. (2021). International analysis of the nutritional content and a review of health benefits of non-dairy plant-based beverages. *Nutrients, 13*(3), 842.

da Costa, T. P., Gillespie, J., Cama-Moncunill, X., Ward, S., Condell, J., Ramanathan, R., & Murphy, F. (2022). A systematic review of real-time monitoring technologies and its potential application to reduce food loss and waste: Key elements of food supply chains and IoT technologies. *Sustainability, 15*(1), 614.

Dankar, I., Haddarah, A., Omar, F. E., Sepulcre, F., & Pujolà, M. (2018). 3D printing technology: The new era for food customization and elaboration. *Trends in Food Science & Technology, 75*, 231–242.

Derossi, A., Husain, A., Caporizzi, R., & Severini, C. (2020). Manufacturing personalized food for people uniqueness. An overview from traditional to emerging technologies. *Critical Reviews in Food Science and Nutrition, 60*(7), 1141–1159.

Ding, H., Tian, J., Yu, W., Wilson, D. I., Young, B. R., Cui, X., et al. (2023). The application of artificial intelligence and big data in the food industry. *Foods, 12*(24), 4511.

Dodiya, M., & Shah, M. (2021). A systematic study on shaping the future of solar prosumage using deep learning. *International Journal of Energy and Water Resources, 5*(4), 477–487.

Douglas, G. L., Cooper, M., Bermudez-Aguirre, D., & Sirmons, T. (2016). *Risk of performance decrement and crew illness due to an inadequate food system* (No. JSC-CN-37577).

Douglas, G. L., Wheeler, R. M., & Fritsche, R. F. (2021). Sustaining astronauts: Resource limitations, technology needs, and parallels between spaceflight food systems and those on Earth. *Sustainability, 13*(16), 9424.

Dupuis, J. H., Cheung, L. K., Newman, L., Dee, D. R., & Yada, R. Y. (2023). Precision cellular agriculture: The future role of recombinantly expressed protein as food. *Comprehensive Reviews in Food Science and Food Safety, 22*(2), 882–912.

Ferguson, L. R., De Caterina, R., Görman, U., Allayee, H., Kohlmeier, M., Prasad, C., et al. (2016). Guide and position of the international society of nutrigenetics/nutrigenomics on personalised nutrition: Part 1-fields of precision nutrition. *Lifestyle Genomics, 9*(1), 12–27.

Frankó, A., Vida, G., & Varga, P. (2020). Reliable identification schemes for asset and production tracking in industry 4.0. *Sensors, 20*(13), 3709.

Ghosh, S., Ahmad, R., Zeyaullah, M., & Khare, S. K. (2021). Microbial nano-factories: Synthesis and biomedical applications. *Frontiers in Chemistry, 9*, 626834.

Godoi, F. C., Prakash, S., & Bhandari, B. R. (2016). 3D printing technologies applied for food design: Status and prospects. *Journal of Food Engineering, 179*, 44–54.

Häder, D. P., Braun, M., & Hemmersbach, R. (2018). Bioregenerative life support systems in space research. In *Gravitational biology I: Gravity sensing and graviorientation in microorganisms and plants* (pp. 113–122). Springer International Publishing.

Hintze, A. (2016). Understanding the four types of artificial intelligence. *Consulté le, 10*(11), 2021.

Jafari, S. M., & McClements, D. J. (Eds.). (2018). *Nanoemulsions: Formulation, applications, and characterization*. Academic Press.

Joshi, V. K., & Kumar, S. (2015). Meat analogues: Plant based alternatives to meat products—A review. *International Journal of Food and Fermentation Technology, 5*(2), 107–119.

Júnior, M. R. B., de Almeida Moreira, B. R., dos Santos Carreira, V., de Brito Filho, A. L., Trentin, C., & de Souza, F. L. P. (2024). Precision agriculture in the United States: A comprehensive meta-review inspiring further research, innovation, and adoption. *Computers and Electronics in Agriculture, 221*, 108993. https://doi.org/10.1016/j.compag.2024.108993

Kerry, J. P., O'Grady, M. N., & Hogan, S. A. (2006). Past, current and potential utilisation of active and intelligent packaging systems for meat and muscle-based products. *Meat Science, 74*(1), 113–130.

Khan, M. I. H., Sablani, S. S., Nayak, R., & Gu, Y. (2022). Machine learning-based modeling in food processing applications: State of the art. *Comprehensive Reviews in Food Science and Food Safety, 21*(2), 1409–1438.

Khan, A., Hassan, M., & Shahriyar, A. K. (2023). Optimizing onion crop management: A smart agriculture framework with IoT sensors and cloud technology. *Applied Research in Artificial Intelligence and Cloud Computing, 6*, 49–67.

Kim, W. J., Kim, Y., Ovissipour, R., & Nitin, N. (2024). Plant-based biomaterials as scaffolds for cellular agriculture. *Future Foods, 10*, 100468.

Kulkarni, D., Sherkar, R., Shirsathe, C., Sonwane, R., Varpe, N., Shelke, S., et al. (2023). Biofabrication of nanoparticles: Sources, synthesis, and biomedical applications. *Frontiers in Bioengineering and Biotechnology, 11*, 1159193.

Kuswandi, B., & Moradi, M. (2018). Improvement of food packaging based on functional nanomaterial. In *Nanotechnology: Applications in energy, drug and food* (pp. 309–344). Springer International Publishing.

Kuswandi, B., Jayus, Restyana, A., Abdullah, A., Heng, L. Y., & Ahmad, M. (2011). A novel colorimetric food package label for fish spoilage based on polyaniline film. *Food Control, 22*(3–4), 596–602.

Laddu, D., & Hauser, M. (2019). Addressing the nutritional phenotype through personalized nutrition for chronic disease prevention and management. *Progress in Cardiovascular Diseases, 62*(1), 9–14.

Lakhiar, I. A., Yan, H., Zhang, C., et al. (2024). A review of precision irrigation water-saving technology under changing climate for enhancing water use efficiency, crop yield, and environmental footprints. *Agriculture, 14*(6), 1141. https://doi.org/10.3390/agriculture14061141

Liakos, K. G., Athanasiadis, V., Bozinou, E., & Lalas, S. I. (2025). Machine learning for quality control in the food industry: A review. *Foods, 14*(19), 3424.

Malefors, C., Secondi, L., Marchetti, S., & Eriksson, M. (2022). Food waste reduction and economic savings in times of crisis: The potential of machine learning methods to plan guest attendance in Swedish public catering during the Covid-19 pandemic. *Socio-Economic Planning Sciences, 82*, 101041.

Mandracci, P. (Ed.). (2019). *Chemical vapor deposition for nanotechnology*. BoD–Books on Demand.

Maningat, C. C., Jeradechachai, T., & Buttshaw, M. R. (2022). Textured wheat and pea proteins for meat alternative applications. *Cereal Chemistry, 99*(1), 37–66.

Mirabella, N., Valentina C., & Serenella, S. (2014). Current options for the valorization of food manufacturing waste: A review. *Journal of Cleaner Production, 65*, 28–41.

Menichetti, G., Ravandi, B., Mozaffarian, D., & Barabási, A. L. (2023). Machine learning prediction of the degree of food processing. *Nature Communications, 14*(1), 2312.

Monteiro, A., Santos, S., & Gonçalves, P. (2021). Precision agriculture for crop and livestock farming—Brief review. *Animals, 11*(8), 2345.

Motelica, L., Ficai, D., Oprea, O. C., Ficai, A., & Andronescu, E. (2020). Smart food packaging designed by nanotechnological and drug delivery approaches. *Coatings, 10*(9), 806.

Munekata, P. E., Domínguez, R., Budaraju, S., Roselló-Soto, E., Barba, F. J., Mallikarjunan, K., et al. (2020). Effect of innovative food processing technologies on the physicochemical and nutritional properties and quality of non-dairy plant-based beverages. *Foods, 9*(3), 288.

Nawar, S., Corstanje, R., Halcro, G., Mulla, D., & Mouazen, A. M. (2017). Delineation of soil management zones for variable-rate fertilization: A review. *Advances in Agronomy, 143*, 175–245. https://doi.org/10.1016/bs.agron.2017.01.003

Nayak, J., Vakula, K., Dinesh, P., Naik, B., & Pelusi, D. (2020). Intelligent food processing: Journey from artificial neural network to deep learning. *Computer Science Review, 38*, 100297.

Neupane, K., & Baysal-Gurel, F. (2021). Automatic identification and monitoring of plant diseases using unmanned aerial vehicles: A review. *Remote Sensing, 13*(19), 3841. https://doi.org/10.3390/rs13193841

O'Neill, E. N., Cosenza, Z. A., Baar, K., & Block, D. E. (2021). Considerations for the development of cost-effective cell culture media for cultivated meat production. *Comprehensive Reviews in Food Science and Food Safety, 20*(1), 686–709.

Oluwafemi, F. A., De La Torre, A., Afolayan, E. M., Olalekan-Ajayi, B. M., Dhital, B., Mora-Almanza, J. G., et al. (2018). Space food and nutrition in a long term manned mission. *Advances in Astronautics Science and Technology, 1*(1), 1–21.

Öncü Öner, T. (2023). Health effects of plant-based foods and their components. In *Plant-based foods: Ingredients, technology and health aspects* (pp. 137–178). Springer International Publishing.

Ordovas, J. M., & Berciano, S. (2020). Personalized nutrition and healthy aging. *Nutrition Reviews, 78*(Supplement_3), 58–65.

Ordovas, J. M., Ferguson, L. R., Tai, E. S., & Mathers, J. C. (2018). Personalized nutrition and health. *BMJ, 361*, bmj.k2173.

Pant, A., Lee, A. Y., Karyappa, R., Lee, C. P., An, J., Hashimoto, M., et al. (2021). 3D food printing of fresh vegetables using food hydrocolloids for dysphagic patients. *Food Hydrocolloids, 114*, 106546.

Parsons, I. L., Norman, D. A., Karisch, B. B., Webb, S. L., Stone, A. E., Proctor, M. D., & Street, G. M. (2023). Automated walk-over-weigh system to track daily body mass and growth in grazing steers. *Computers and Electronics in Agriculture, 212*, 108113. https://doi.org/10.1016/j.compag.2023.108113

Patel, J. K., Patel, A., & Bhatia, D. (2021). Introduction to nanomaterials and nanotechnology. In *Emerging technologies for nanoparticle manufacturing* (pp. 3–23). Springer International Publishing.

Pawase, P. P., Nalawade, S. M., Bhanage, G. B., Wagh, D. P., & Borkar, D. S. (2023). Variable rate fertilizer application technology for nutrient management: A review. *International Journal of Agricultural and Biological Engineering, 16*(3), 11–19. https://doi.org/10.25165/j.ijabe.20231603.7483

Perchonok, M., & Bourland, C. (2002). NASA food systems: Past, present, and future. *Nutrition, 18*(10), 913–920.

Phuhongsung, P., Zhang, M., & Devahastin, S. (2020). Investigation on 3D printing ability of soybean protein isolate gels and correlations with their rheological and textural properties via LF-NMR spectroscopic characteristics. *LWT, 122*, 109019.

Pitayachaval, P., Sanklong, N., & Thongrak, A. (2018). A review of 3D food printing technology. In *MATEC web of conferences* (Vol. 213, p. 01012). EDP Sciences.

Polymeni, S., Skoutas, D. N., Sarigiannidis, P., Kormentzas, G., & Skianis, C. (2024). Smart agriculture and greenhouse gas emission mitigation: A 6G-IoT perspective. *Electronics, 13*(8), 1480.

Poore, J., & Nemecek, T. (2018). Reducing food's environmental impacts through producers and consumers. *Science, 360*(6392), 987–992.

Post, M. J., Levenberg, S., Kaplan, D. L., Genovese, N., Fu, J., Bryant, C. J., et al. (2020). Scientific, sustainability and regulatory challenges of cultured meat. *Nature Food, 1*(7), 403–415.

Praveen, B. V. S., Verma, R. K., Deepika, B., & Bhavani, P. (2024). A comprehensive review of sustainable technologies and environmental impacts for waste-to-energy conversion. In *International conference on novel materials and technologies for energy and environment* (pp. 69–95). Springer Nature Singapore.

Priyadarshi, R., Khan, A., Riahi, Z., & Rhim, J. W. (2026). Toxicity and safety of nanomaterials in food packaging applications: A dimension-based review. *Comprehensive Reviews in Food Science and Food Safety, 25*(1), e70374.

Qu, B., Xiao, Z., & Luo, Y. (2025). Sustainable nanotechnology for food preservation: Synthesis, mechanisms, and applications of zinc oxide nanoparticles. *Journal of Agriculture and Food Research, 19*, 101743.

Realini, C. E., & Marcos, B. (2014). Active and intelligent packaging systems for a modern society. *Meat Science, 98*(3), 404–419.

Regattieri, A., Gamberi, M., & Manzini, R. (2007). Traceability of food products: General framework and experimental evidence. *Journal of Food Engineering, 81*(2), 347–356.

Rodrigues, C., Souza, V. G. L., Coelhoso, I., & Fernando, A. L. (2021). Bio-based sensors for smart food packaging—Current applications and future trends. *Sensors, 21*(6), 2148.

Ronteltap, A., van Trijp, J. C. M., Berezowska, A., & Goossens, J. (2013). Nutrigenomics-based personalised nutritional advice: In search of a business model? *Genes & Nutrition, 8*, 153–163.

Rubio, N. R., Xiang, N., & Kaplan, D. L. (2020). Plant-based and cell-based approaches to meat production. *Nature Communications, 11*(1), 6276.

Salgado, P. R., Di Giorgio, L., Musso, Y. S., & Mauri, A. N. (2021). Recent developments in smart food packaging focused on biobased and biodegradable polymers. *Frontiers in Sustainable Food Systems, 5*, 630393.

Samtiya, M., Aluko, R. E., Dhewa, T., & Moreno-Rojas, J. M. (2021). Potential health benefits of plant food-derived bioactive components: An overview. *Foods, 10*(4), 839.

Severini, C., & Derossi, A. (2016). Could the 3D printing technology be a useful strategy to obtain customized nutrition? *Journal of Clinical Gastroenterology, 50*, S175–S178.

Sharma, P., Bisht, A., Agarwal, G., & Rana, M. (2026). Unsupervised learning techniques in food science: Clustering and dimensionality reduction. In *Artificial intelligence in food science* (pp. 81–95). Academic Press.

Sher, D., & Tutó, X. (2015). Review of 3D food printing. *Temes de Disseny, 31*, 104–117.

Shorten, P. R., & Welten, B. G. (2022). An acoustic sensor technology to detect urine excretion. *Biosystems Engineering, 214*, 90–106. https://doi.org/10.1016/j.biosystemseng.2021.12.004

Singh, R., Mann, B., Sharma, R., & Singh, S. (2019). Application of nanotechnology in functional foods. In *Nanoscience for sustainable agriculture* (pp. 547–579). Springer International Publishing.

Stephens, N., & Ellis, M. (2020). Cellular agriculture in the UK: A review. *Wellcome Open Research, 5*, 12.

Stephens, N., Di Silvio, L., Dunsford, I., Ellis, M., Glencross, A., & Sexton, A. (2018). Bringing cultured meat to market: Technical, socio-political, and regulatory challenges in cellular agriculture. *Trends in Food Science & Technology, 78*, 155–166.

Sun, J., Peng, Z., Yan, L., Fuh, J. H., & Hong, G. S. (2015a). 3D food printing—An innovative way of mass customization in food fabrication. *International Journal of Bioprinting, 1*(1), 27–38.

Sun, J., Peng, Z., Zhou, W., Fuh, J. Y., Hong, G. S., & Chiu, A. (2015b). A review on 3D printing for customized food fabrication. *Procedia Manufacturing, 1*, 308–319.

Sun, J., Zhou, W., Huang, D., Fuh, J. Y., & Hong, G. S. (2015c). An overview of 3D printing technologies for food fabrication. *Food and Bioprocess Technology, 8*(8), 1605–1615.

Surendran, U., Nagakumar, K. C. V., & Samuel, M. P. (2024). Remote sensing in precision agriculture. In *Digital agriculture: A solution for sustainable food and nutritional security* (pp. 201–223). Springer.

Švarc, P. L., Jensen, M. B., Langwagen, M., Poulsen, A., Trolle, E., & Jakobsen, J. (2022). Nutrient content in plant-based protein products intended for food composition databases. *Journal of Food Composition and Analysis, 106*, 104332.

Verma, M., Hontecillas, R., Tubau-Juni, N., Abedi, V., & Bassaganya-Riera, J. (2018). Challenges in personalized nutrition and health. *Frontiers in Nutrition, 5*, 117.

Verma, P., Chauhan, A., & Ladon, T. (2020). Site specific nutrient management: A review. *Journal of Pharmacognosy and Phytochemistry, 9*(1), 233–236.

Weiss, M., Jacob, F., & Duveiller, G. (2020). Remote sensing for agricultural applications: A meta-review. *Remote Sensing of Environment, 236*, 111402. https://doi.org/10.1016/j.rse.2019.111402

Wu, M., Liu, W., & Zheng, S. (2024). Intelligent food safety: A prediction model based on attention mechanism and reinforcement learning. *Applied Artificial Intelligence, 38*(1), 2379731.

Yang, F., Zhang, M., & Bhandari, B. (2017). Recent development in 3D food printing. *Critical Reviews in Food Science and Nutrition, 57*(14), 3145–3153.

Yeşilyurt, N., Yılmaz, B., Ağagündüz, D., & Capasso, R. (2022). Microbiome-based personalized nutrition as a result of the 4.0 technological revolution: A mini literature review. *Process Biochemistry, 121*, 257–262.

Yu, H., Park, J. Y., Kwon, C. W., Hong, S. C., Park, K. M., & Chang, P. S. (2018). An overview of nanotechnology in food science: Preparative methods, practical applications, and safety. *Journal of Chemistry, 2018*(1), 5427978.

Yusuf, E. H. (2023). Comparison of life cycle assessments and nutritional contents of soy protein and wheat protein (seitan) based vegan bacon products for human and environmental health. *Journal of the Science of Food and Agriculture, 103*(7), 3315–3321.

Zhang, J., Huang, Y., Pu, R., Gonzalez-Moreno, P., Yuan, L., Wu, K., & Huang, W. (2019). Monitoring plant diseases and pests through remote sensing technology: A review. *Computers and Electronics in Agriculture, 165*, 104943. https://doi.org/10.1016/j.compag.2019.104943

Zhang, L. Z., Dong, H. S., Yu, Y. B., Liu, L. Y., & Zang, P. (2022). Application and challenges of 3D food printing technology in manned spaceflight: A review. *International Journal of Food Science and Technology, 57*(8), 4906–4917.

Zidan, F., & Febriyanti, D. E. (2024). Optimizing agricultural yields with artificial intelligence-based climate adaptation strategies. *IAIC Transactions on Sustainable Digital Innovation, 5*, 136–147.

Chapter 2
Plant-Based Foods

Abbreviations

B vitamins	B complex Vitamins
B_2	Vitamin B_2 (Riboflavin)
BMD	Bone Mineral Density
BMI	Body Mass Index
BSE	Bovine Spongiform Encephalopathy
$CaSO_4$	Calcium Sulfate
CRP	C Reactive Protein
CVDs	Cardiovascular Diseases
DNA	Deoxyribonucleic Acid
E. coli	*Escherichia coli*
FAO	Food and Agriculture Organization
FDA	Food and Drug Administration
g	Gram
GDL	Glucono Delta Lactone
GI	Gastrointestinal
GI	Glycemic Index
GPx	Glutathione Peroxidase
h	Hour
HM	High Moisture
HMEC	High Moisture Extrusion Cooking
IL-6	Interleukin-6
K_2	Vitamin K_2 (Menaquinone)
Kg	Kilogram
LDL	Low-Density Lipoprotein
$MgCl_2$	Magnesium Chloride
Min	Minute
MPa	Megapascal

Y. Kumar, S. Dey, *Emerging Technology in Food Processing*,
https://doi.org/10.1007/978-3-032-24505-2_2

Na	Sodium
PBM	Plant-Based Meat
PBMs	Plant-Based Meat Analogues
pH	Potential of Hydrogen
PUFAs	Polyunsaturated Fatty Acids
RHM	Rank Hovis McDougall
RNA	Ribonucleic Acid
Rpm	Revolutions per Minute
SCP	Single Cell Protein
SERMs	Selective Estrogen Receptor Modulators
SOD	Superoxide Dismutase
spp.	Species
TVP	Textured Vegetable Protein
US	United States
USA	United States of America
w/w	Weight per Weight
≈	Approximately
°C	Degree Celsius
3D	Three Dimensional
β	Beta
β-conglycinin	Beta-conglycinin
β-glucans	Beta-glucans
ω-3	Omega-3 Fatty Acids

2.1 Introduction

In recent years, global attention toward sustainable and ethical food production has intensified, leading to the emergence and rapid development of plant-based foods (PBFs) (McClements & Grossmann, 2021a). Among these, meat analogues, also known as meat substitutes, mock meats, faux meats, or imitation meats, have attracted considerable interest from both consumers and researchers (Hu, 2003). These products are designed to approximate the aesthetic, textural, and chemical characteristics of conventional meat, allowing consumers to experience familiar sensory attributes while adhering to alternative dietary, environmental, or ethical preferences (Hu, 2003). The concept of plant-based foods, particularly meat substitutes, is not entirely new; however, modern technological advancements and shifting consumer priorities have transformed them from niche products into mainstream alternatives within the global food system (McClements & Grossmann, 2022).

The rise in the consumption of vegetable proteins in food products can be attributed to multiple intersecting factors. Over the past few decades, concerns over animal-borne diseases such as bovine spongiform encephalopathy (BSE), avian influenza, and swine flu have led many consumers to question the safety and sustainability of traditional meat production (McClements & Grossmann, 2022). Additionally, a global shortage of animal protein in certain regions has prompted food industries and policymakers to explore alternative protein sources that can supplement or replace conventional livestock production (Nolden & Forde, 2023). Religious and ethical considerations have also played a significant role: communities seeking halal, kosher, or vegetarian foods have demonstrated a consistent demand for products that align with their dietary restrictions while maintaining desirable taste and texture profiles (McClements & Grossmann, 2021a). Furthermore, economic considerations, including the high cost of animal feed, transportation, and environmental management, have made plant-based alternatives increasingly attractive for both producers and consumers (Hu, 2003).

From an environmental perspective, the shift toward plant-based foods is particularly compelling. A meat-based diet requires vastly greater environmental resources than a diet centered on grains, legumes, and vegetables. Estimates suggest that between 2 and 15 kilograms of plant-based feed are required to produce just 1 kilogram of meat, depending on the species and production system (Nolden & Forde, 2023). This inefficiency translates into high land, water, and energy usage, as well as elevated greenhouse gas emissions. Livestock production is known to be a major contributor to deforestation, methane emissions, and water pollution (Nolden & Forde, 2023). In contrast, plant-based diets are generally associated with a lower ecological footprint, offering a sustainable pathway to feed the growing global population without placing excessive strain on natural ecosystems (Langyan et al., 2022).

Despite the clear environmental advantages, advances in animal agriculture have led to increased cost efficiency and productivity in meat production. Through technological innovations, intensification of farming systems, and genetic selection of livestock, the industry has been able to produce meat at lower costs and higher volumes than ever before (McClements & Grossmann, 2021a, b). However, these improvements have not come without consequences. Intensive animal farming practices often raise concerns regarding animal welfare, antibiotic resistance, and environmental degradation (McClements & Grossmann, 2021a, b). In light of these challenges, both academic institutions and industry leaders have intensified efforts to improve the sensory characteristics, nutritional quality, and consumer acceptance of plant-based meat (PBM) alternatives. These efforts aim to bridge the gap between traditional meat products and plant-based substitutes, ensuring that consumers do not have to compromise on flavor, texture, or cultural familiarity (Alcorta et al., 2021).

Simultaneously, a new frontier known as cellular agriculture has emerged, introducing cell-based or cultured meat as a potential complement to plant-based proteins (McClements & Grossmann, 2022). Unlike traditional plant-based analogues, which rely on vegetable proteins such as soy, pea, or wheat gluten, cell-based meat is produced by cultivating animal cells directly in controlled environments. This innovative approach replicates the structure and composition of conventional meat without the need for animal slaughter. Although still in its early stages of commercialization, cellular agriculture represents a significant technological advancement that could further diversify the spectrum of sustainable protein sources available in the future (Aydar, 2023).

The global demand for meat continues to grow, underscoring the urgency of developing sustainable alternatives. According to the Food and Agriculture Organization (FAO) of the United Nations, the worldwide demand for meat was projected to reach 455 million metric tons by 2050, representing a 76% increase from 2005 levels (McClements & Grossmann, 2021a, b). Similarly, the global demand for fish is expected to rise to approximately 140 million metric tons by the same year. This surge is largely driven by population growth, urbanization, and rising incomes, particularly in middle-income countries such as China, where economic development has significantly altered dietary patterns. In contrast, consumption levels in high-income countries, including the United Kingdom, Germany, and the United States, have begun to plateau or even decline slightly, as health and environmental awareness influence consumer choices (McClements & Grossmann, 2021a, b). Meanwhile, low-income countries such as India exhibit relatively constant rates of meat consumption, constrained by cultural norms, affordability, and availability (McClements & Grossmann, 2022).

This global pattern supports the "inverted U-shaped" theory describing the relationship between income and meat consumption. According to this model, as a nation's income rises, meat consumption initially increases due to greater purchasing power and dietary diversification (McClements & Grossmann 2021a). However, beyond a certain threshold, consumption tends to stagnate or decline as wealthier populations become more conscious of health implications, ethical considerations, and environmental sustainability. As such, plant-based foods have found strong support in markets where consumers are motivated not by economic necessity but by values-based decisions seeking foods that are nutritious, sustainable, and aligned with broader global goals such as climate action and resource conservation (Nolden & Forde, 2023).

The modern plant-based food movement thus exists at the intersection of nutrition, technology, environmental science, and consumer behavior. Advances in food technology, such as extrusion cooking, emulsification, and flavor encapsulation, have enabled manufacturers to recreate the complex sensory properties of meat using plant proteins like soy, pea, lentil, chickpea, and mycoprotein (Hashempour-Baltork et al., 2020). Additionally, innovative food companies are now employing biotechnology, fermentation, and natural flavor engineering to enhance the realism of meat substitutes. Products such as plant-based burgers, sausages, nuggets, and dairy alternatives have rapidly entered mainstream markets, appealing to both vegetarian and flexitarian consumers who seek to reduce but not necessarily eliminate meat consumption (Hashempour-Baltork et al., 2020).

Beyond the consumer level, the adoption of plant-based foods carries profound implications for global food security. As the world population is projected to exceed 9 billion by 2050, ensuring an adequate supply of high-quality protein will be a major challenge (Ogutu et al., 2024). Shifting dietary habits toward plant-based proteins could alleviate pressure on natural resources, reduce dependence on industrial livestock systems, and contribute to a more equitable and resilient food distribution model. Governments and international organizations are increasingly recognizing the importance of supporting plant-based innovation through policy frameworks, research funding, and public awareness campaigns (Ogutu et al., 2024).

The evolution of plant-based foods represents a pivotal shift in the global food landscape driven by environmental urgency, technological innovation, health awareness, and ethical consciousness (Hashempour-Baltork et al., 2020). Meat analogues and other plant-derived alternatives offer a practical and scalable solution to the intertwined challenges of climate change, population growth, and sustainable resource management. While traditional animal agriculture continues to adapt through efficiency gains, the accelerating interest in plant-based and cell-based proteins signals a transformative rethinking of how humanity produces and consumes food (Langyan et al., 2022). As research and innovation progress, the line between conventional and alternative proteins may continue to blur, ushering in an era of diverse, sustainable, and inclusive food systems capable of nourishing both people and the planet (Nolden & Forde, 2023).

2.2 Types of Plant-Based Foods

2.2.1 Textured Vegetable Protein (TVP/Soya Meat)

Textured vegetable protein (TVP), commonly referred to as soya meat or soy chunks, is a plant-derived protein product developed to serve as a meat substitute (Riaz, 2011; Bakhsh et al., 2022). TVP is obtained primarily from defatted soy flour, which is a by-product of soybean oil extraction. It was first developed in the United States and introduced to the European market in the late 1960s as a cost-effective, protein-rich alternative to animal meat (Hong et al., 2022). Figure 2.1 illustrates a schematic representation of the different types of plant-based foods.

TVP is extremely high in protein, typically containing 50–70% protein by dry weight, though this percentage decreases after rehydration, since it absorbs large amounts of water during cooking (Bakhsh et al., 2022). It is cholesterol free, low in fat, and a complete protein source, containing all essential amino acids required for human nutrition. TVP can be manufactured in various forms, such as granules, chunks, flakes, or strips, making it highly versatile for use in diverse culinary applications (Hong et al., 2022). Essentially, TVP mimics the texture and appearance of meat, allowing it to be used as a full or partial meat replacement in recipes, making it popular among vegetarians, vegans, and flexitarians (Hong et al., 2022). Table 2.1 presents an integrated classification of the major types of plant-based foods.

2.2.1.1 Processing of Textured Vegetable Protein

The production of TVP involves several mechanical, thermal, and biochemical steps designed to extract, refine, and restructure plant proteins into a fibrous, meat-like

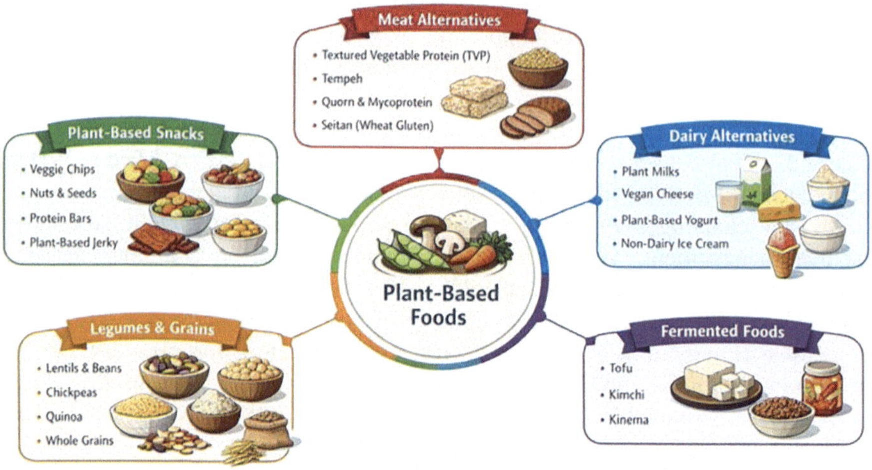

Fig. 2.1 Schematic representation of the different types of plant-based foods

texture (Riaz, 2011; Choi et al., 2022). The process can be divided into three key stages: (1) preparation of soybeans, (2) protein isolation and drying, and (3) texturization via extrusion (Zhang et al., 2019).

1. *Preparation of Soybeans*

The process begins with cleaning, washing, and soaking the soybeans. Typically, beans are soaked in water at around 30 °C for three hours (Choi et al., 2022). This step serves multiple purposes:

- It softens the seed coat (husk), facilitating removal.
- It activates endogenous enzymes, improving digestibility.
- It helps eliminate anti-nutritional factors such as trypsin inhibitors and oligosaccharides that can cause flatulence.
- After soaking, the softened beans are washed thoroughly to remove the husk completely, ensuring a smoother final texture. The dehulled beans are then dried at 70 °C for 5–8 hours, producing a moisture-stable substrate suitable for further processing (Choi et al., 2022).

2. *Extraction and Protein Concentration*

Next, oil extraction is carried out using solvent extraction (commonly hexane), leaving behind defatted soy flour, which becomes the base material for TVP. This flour contains approximately 50% protein and is rich in carbohydrates and dietary fiber. In some cases, soy protein concentrate or soy protein isolate is used instead of defatted flour to achieve higher protein purity. The proteins are then blended with controlled amounts of water, emulsifiers, and sometimes carbohydrates to create a dough-like mixture suitable for extrusion (Zhang et al., 2019).

Table 2.1 Classification of major types of plant-based foods

Category	Type of plant-based food	Primary source	Key Characteristics	Examples	Key References
Whole plant foods	Legumes	Soybean, lentils, peas	High protein, dietary fiber, bioactive compounds; low saturated fat	Soybean, chickpea, lentils	Akhtar et al. (2022); Samtiya et al. (2021)
	Cereals and grains	Wheat, rice, barley	Energy-rich; moderate protein; source of B vitamins	Wheat, rice, oats	Akhtar et al. (2022); Prakash et al. (2022)
	Nuts and seeds	Almond, peanut, flaxseed	Healthy fats, protein, antioxidants	Almonds, peanuts, chia	Samtiya et al. (2021)
	Fruits and vegetables	Diverse plant tissues	Vitamins, minerals, antioxidants, phytochemicals	Spinach, tomato, apple	Watzl (2008); Öncü Öner (2023)
Minimally processed plant foods	Traditional soy foods	Soybean	High-quality protein; minimal processing	Tofu, tempeh	Tamang (2012); Craig and Fresán (2021)
	Fermented plant foods	Soybean, legumes	Improved digestibility; probiotic and functional benefits	Kinema, natto, miso	Tamang (2015); Kharnaior and Tamang (2022)
Plant-based protein ingredients	Protein concentrates	Soy, pea, wheat	60–80% protein; functional properties	Soy protein concentrate	Švarc et al. (2022); McClements and Grossmann (2022)
	Protein isolates	Soy, pea	≥90% protein; high solubility and purity	Soy protein isolate	Ribas-Agustí et al. (2018); Rousseau et al. (2020)
	Textured vegetable protein (TVP)	Defatted soy flour	Fibrous, meat-like texture	TVP chunks, granules	Maningat et al. (2022); Zhang et al. (2019)
Plant-based meat analogues	Soy-based analogues	Soy protein	Complete protein; cholesterol-free	TVP-based meats	Yusuf (2023); Sridhar et al. (2023)
	Wheat-based analogues	Wheat gluten	Elastic, fibrous texture; high protein	Seitan	Muneer et al. (2025); Yusuf (2023)

(continued)

Table 2.1 (continued)

Category	Type of plant-based food	Primary source	Key Characteristics	Examples	Key References
	Mycoprotein-based foods	Fungal biomass	High protein and fiber; low fat	Quorn	McClements and Grossmann (2022)
	Mixed protein analogues	Soy, pea, wheat blends	Balanced amino acids; improved texture	Vegan burgers, sausages	Cornet et al. (2022); Zhao et al. (2022)
Plant-based dairy alternatives	Milk alternatives	Soy, almond, oat	Lactose-free; fortified with calcium and vitamins	Soy milk, oat milk	Craig and Fresán (2021); Munekata et al. (2020)
	Yogurt and cheese alternatives	Soy, nuts, coconut	Fermented or emulsified matrices	Soy yogurt, vegan cheese	McClements and Grossmann (2022)
Next-generation plant foods	Mycelium-based foods	Fungal cultures	Naturally fibrous; sustainable	Mycelium meat	Dupuis et al. (2023)
	Recombinant and precision-fermented proteins	Engineered microbes/plants	Enhanced flavor, nutrition, functionality	Recombinant heme proteins	Dupuis et al. (2023); Yano & Fu (2022)
	3D-printed plant foods	Plant protein pastes	Customized texture and structure	Printed plant-based meat	Wang et al. (2022)

3. *Texturization (Extrusion Processing)*

The core of TVP production lies in the extrusion process, which converts amorphous plant proteins into structured, fibrous textures that resemble meat. The process is typically carried out using a twin screw extruder under high temperature (120–180 °C) and pressure (3–10 MPa) conditions (Riaz, 2011; Hong et al., 2022).

Inside the Extruder

- The mechanical shear and heat cause proteins to denature and unfold.
- As the mixture passes through the screw barrel, the proteins align and cross-link, forming laminar, fibrous structures.
- When the mixture exits the die nozzle, the sudden drop in temperature and pressure causes steam expansion, puffing the product slightly and creating its porous texture.
- The extruded product is then dried and cooled to produce shelf stable dry TVP, which can later be rehydrated before cooking (Hong et al., 2022).

4. *Rehydration and Final Use*

Before consumption, TVP is typically rehydrated in warm water or broth for about 15–20 minutes. During this stage, it absorbs water up to three times its weight, becoming soft and spongy. This rehydrated product can then be seasoned, marinated, or cooked in a variety of dishes, offering a texture and mouthfeel similar to minced or chunked meat (Riaz, 2011).

2.2.1.2 Applications of Textured Vegetable Protein

TVP's versatility, high nutritional value, and neutral taste make it suitable for a wide range of culinary, industrial, and nutritional applications (Webb et al., 2023).

1. *Culinary Uses*

- Meat Substitutes: TVP is used to replace ground meat in dishes such as burgers, sausages, tacos, curries, stews, and pasta sauces.
- Meat Extender: In food service industries, it is often combined with minced meat to reduce cost while maintaining protein content and palatability.
- Ready to Eat Foods: TVP is a common ingredient in instant noodles, canned foods, and soups to boost protein levels.
- Vegetarian and Vegan Products: Forms the base for plant-based meat patties, nuggets, and kebabs (Maningat et al., 2022; Webb et al., 2023).

2. *Nutritional Applications*

Due to its complete amino acid profile, low fat content, and absence of cholesterol, TVP is widely used in nutritional programs, sports nutrition, and emergency food relief. It is often included in high-protein meal replacements, fortified cereals, and school feeding programs in developing nations (Guerrero et al., 2025).

3. *Industrial and Functional Uses*

- In food formulation, TVP acts as a texturizer and binder, improving product structure and water retention.
- It can be flavored and shaped into customized protein inclusions for snack bars, pet foods, and ready meals.
- Its high water absorption capacity and emulsifying ability enhance product juiciness and mouthfeel (Guerrero et al., 2025).

2.2.1.3 Quorn/Mycoprotein-Based Foods

Quorn is the brand name for a line of mycoprotein-based food products derived from the filamentous fungus *Fusarium venenatum*. Mycoprotein refers to protein-rich biomass produced through fungal fermentation, offering a sustainable, nutritious, and environmentally friendly alternative to animal-derived meat

(Hashempour-Baltork et al., 2020; Saeed et al., 2023). Owing to its favorable nutritional composition and lower environmental footprint, mycoprotein has gained significant attention as a next-generation protein source for sustainable food systems (Linder, 2024).

The concept of developing a fungus-based meat substitute was initiated by the British company Rank Hovis McDougall (RHM) in the 1960s, following growing concerns about global food shortages and protein deficiency. After more than 15 years of intensive research, development, and safety testing, Quorn was first commercialized in 1985 in the United Kingdom. It remains one of the most successful examples of single-cell protein (SCP) technology applied to the food industry (Whittaker et al., 2020).

Mycoprotein-based foods, such as Quorn, are characterized by a high protein content (approximately 45–50%), low saturated fat, zero cholesterol, and a favorable fatty acid profile. The fiber-rich cell walls of the fungal hyphae, composed mainly of chitin and β-glucans, contribute to digestive health (Hashempour-Baltork et al., 2020). Additionally, the polyunsaturated fatty acids (PUFAs) found in fungal cell membranes, and the high-quality proteins within the cytoplasm, make Quorn a nutritionally complete and functional food (Saeed et al., 2023). In short, Quorn is a cholesterol-free, low-fat, high-protein food designed to replicate the sensory and nutritional characteristics of meat, while providing superior sustainability and dietary benefits (Shahid et al., 2024).

2.2.1.4 Processing of Mycoprotein (Quorn Production)

The production of Quorn involves industrial scale fermentation using *Fusarium venenatum* under controlled conditions to maximize protein yield and maintain product safety. This biotechnological process is one of the most established applications of fungal single-cell protein technology and has been refined over several decades (Finnigan, 2011; Whittaker et al., 2020). The process can be divided into five main stages: (1) fermentation, (2) harvesting, (3) RNA reduction, (4) texturization, and (5) product formulation.

1. *Fermentation Stage*

   - The process begins with the preparation of a fermentation medium, typically composed of glucose, mineral salts, vitamins, and nitrogen sources such as ammonia or ammonium salts. The fungal spores of *Fusarium venenatum* are inoculated into this medium within large airlift or continuous stirred-tank bioreactors, each holding up to 150,000 liters (Finnigan, 2011; Whittaker et al., 2020).
   - The fermentation temperature is maintained around 30 °C, with pH near 6.0, and continuous aeration and agitation to promote optimal mycelial growth (Lonchamp et al., 2022).

- Under these aerobic conditions, the fungus grows as filamentous hyphae, forming a fibrous network that closely resembles the structure of animal muscle fibers (Lonchamp et al., 2022).
- The carbon source (usually glucose or molasses) is supplied continuously to sustain growth (Finnigan, 2011).
- The growth cycle lasts about four to six days, after which the fungal biomass reaches a high density, typically 15–20 g dry matter per liter of culture (Finnigan, 2011).

2. *Harvesting and Cell Separation*

At the end of the fermentation period, the mycelial biomass is separated from the liquid medium by filtration or centrifugation. The resulting wet biomass is washed repeatedly with water to remove residual nutrients, cell debris, and metabolic by-products. The harvested material typically has a moisture content of around 75% (Whittaker et al., 2020).

3. *RNA Reduction and Safety Control*

Fungal cells naturally contain relatively high levels of ribonucleic acid (RNA), which, if consumed in excess, can contribute to elevated uric acid levels in humans. Therefore, RNA reduction is a critical safety step in mycoprotein production (Hashempour-Baltork et al., 2020). To achieve this, the harvested biomass is heat-treated at temperatures between 64 and 68 °C for approximately 20–30 minutes. This treatment inactivates endogenous enzymes and promotes the degradation of RNA into smaller nucleotides, which are subsequently removed through washing and filtration. This step ensures that the final mycoprotein product complies with regulatory safety standards and is suitable for human consumption (Finnigan, 2011; Hashempour-Baltork et al., 2020).

4. *Texturization and Binding*

Following RNA reduction, the mycoprotein fibers are blended with binding agents to improve cohesiveness and texture. Traditionally, egg albumin has been used; however, plant-based binders such as potato protein or pea protein are increasingly employed to produce vegan formulations (Lonchamp et al., 2022). The mixture is then heat set, during which the binder proteins denature and form a gel network that binds the fungal hyphae together. This structured mass is frozen to stabilize the fibrous alignment and subsequently shaped into various formats, including fillets, patties, chunks, nuggets, or mince. The resulting texture closely mimics the chewiness and elasticity of meat, which is a key determinant of consumer acceptance (Whittaker et al., 2020).

5. *Product Formulation and Flavor Development*

To produce a diverse range of consumer products, the base mycoprotein is combined with flavoring agents, fats, colorants, and seasonings. Vegetable oils such as sunflower oil are commonly added to enhance juiciness and flavor release, while colorants, including caramel or beet extract, provide a cooked-meat appearance. Umami-rich ingredients, such as yeast extracts and hydrolyzed vegetable proteins, are incorporated to enhance savory flavor notes (Finnigan, 2011; Lonchamp et al., 2022).

Finally, the formulated products undergo cooking, sterilization, or freezing depending on the intended application, such as ready-to-cook or ready-to-eat formats. Owing to its efficient fermentation-based production and favorable environmental profile, mycoprotein has been recognized as a sustainable alternative to both conventional livestock products and many plant-based meat analogues (Shahid et al., 2024; Souza Filho et al., 2018).

2.2.1.5 Applications of Quorn/Mycoprotein-Based Foods

Quorn has become a central player in the alternative protein industry, offering versatile applications in food manufacturing, nutrition, and health-conscious diets. As one of the earliest and most commercially successful mycoprotein products, Quorn exemplifies the scalability and multifunctionality of filamentous fungal proteins in modern food systems (Whittaker et al., 2020; Huling, 2023). Recent reviews identify mycoprotein as a transformative protein source capable of supporting future food security, sustainability, and dietary diversification (Akinsemolu & Onyeaka, 2025; Hoxha & Taherzadeh, 2025).

1. *Culinary and Commercial Applications*
   - Meat Alternatives: Quorn products are designed to replace traditional meat items such as minced beef, chicken fillets, sausages, burgers, and nuggets.
   - Hybrid Foods: Used in combination with vegetables or legumes in ready meals, pastas, stir fries, and soups.
   - Processed Foods: Incorporated into convenience foods (e.g., frozen meals, sandwiches, salads) for a healthy, high-protein option.
   - Restaurant and Retail Markets: Widely sold in supermarkets and foodservice sectors worldwide, especially in Europe, the United States, and Australia (Whittaker et al., 2020; Huling, 2023; Akinsemolu & Onyeaka, 2025).
2. *Nutritional and Health Applications*

From a nutritional standpoint, Quorn is characterized by high-quality protein, substantial dietary fiber, low saturated fat content, and the absence of cholesterol, making it suitable for health-conscious diets. The fungal cell walls contain β-glucans and chitin, which function as prebiotic fibers that support gut health by modulating the intestinal microbiota and improving digestive function (Hashempour-Baltork et al., 2020; Akinsemolu & Onyeaka, 2025). Clinical and nutritional studies have shown that regular mycoprotein consumption may contribute to reductions in low-density lipoprotein (LDL) cholesterol, enhanced satiety, and improved glycemic control. These properties make Quorn a promising ingredient for functional foods and dietary strategies targeting cardiovascular health, weight management, and diabetes prevention (Hashempour-Baltork et al., 2020; Hoxha & Taherzadeh, 2025).

3. *Environmental and Sustainability Benefits*

The environmental advantages of mycoprotein production further strengthen Quorn's role as a sustainable protein source. Fermentation-based production systems require substantially less land and water and generate significantly lower greenhouse gas emissions compared with conventional livestock farming. Additionally, fungal fermentation can utilize renewable carbon sources and agro industrial by-products, enhancing circular bioeconomy integration and resource efficiency (Akinsemolu & Onyeaka, 2025; Hoxha & Taherzadeh, 2025). These sustainability benefits position Quorn not only as an alternative to animal-derived proteins but also as a strategic component of future resilient and climate-smart food systems. Consequently, mycoprotein is increasingly highlighted as a key driver of transformative change in global protein production and consumption patterns (Hoxha & Taherzadeh, 2025).

2.2.2 Tempeh

Tempeh is a traditional Indonesian fermented food made primarily from soybeans, known for its rich nutritional value, characteristic nutty flavor, and firm texture. It originated on the island of Java, Indonesia, and has been a staple protein source in Southeast Asia for centuries (Ahnan-Winarno et al., 2021). Today, it is recognized globally as a plant-based meat alternative and a key component of vegetarian and vegan diets (Ahnan-Winarno et al., 2021). Tempeh is produced through solid-state fermentation of cooked soybeans using a filamentous fungus, most commonly *Rhizopus oligosporus* (also referred to as *Rhizopus oryzae* or *Rhizopus microsporus*). The fermentation binds the soybeans together into a compact cake-like form through the interweaving of fungal mycelia. This process enhances digestibility, flavor, texture, and nutritional profile, transforming simple soybeans into a high-quality protein source with improved functional and sensory properties (Aaslyng & Højer, 2021).

Nutritionally, tempeh is rich in protein (about 19–21%), fiber, vitamins (notably B-complex), and minerals such as calcium, iron, and magnesium. Unlike other soy products, tempeh retains the natural soybean fiber and has a lower fat content and zero cholesterol (Ahnan-Winarno et al., 2021; Wilson et al., 2024). The fermentation also increases levels of bioactive compounds, including isoflavones in their aglycone form, which are known for their antioxidant and estrogenic properties. Overall, tempeh represents a functional food that provides not only macronutrients but also beneficial bioactive peptides and probiotics, promoting overall health and well-being (Subali et al., 2023).

2.2.2.1 Processing of Tempeh

The preparation of tempeh is a microbial fermentation process that transforms soaked and cooked soybeans into a firm, textured, and flavorful food. The process involves five major steps: selection and preparation of soybeans, cooking, inoculation with the starter culture, incubation (fermentation), and post-processing or preservation (Ahnan-Winarno et al., 2021; Górska et al., 2025).

1. *Selection and Preparation of Soybeans*

High-quality yellow soybeans with uniform size and minimal damage are preferred. The beans are cleaned to remove dust, stones, and other impurities, and then soaked in water for 8–12 hours at ambient temperature (approximately 25–30 °C). This soaking process softens the beans, activates endogenous enzymes, and initiates the removal of anti-nutritional factors such as trypsin inhibitors and phytic acid (Ahnan-Winarno et al., 2021).

After soaking, the softened beans are dehulled either manually or mechanically. The removal of the outer husk is essential because it allows the fungal mycelia to penetrate the bean surface during fermentation. The dehulled beans are then thoroughly rinsed to remove soluble sugars and other components that may interfere with fermentation (Górska et al., 2025).

2. *Cooking*

The dehulled soybeans are boiled or steamed for 30–45 minutes to fully cook and soften the cotyledons. Cooking serves several purposes:

- It destroys undesirable microorganisms that may compete with *Rhizopus* during fermentation.
- It improves substrate digestibility, making nutrients more available to the fermenting fungus.
- It helps achieve the desired moisture content (about 60–65%), which is optimal for fungal growth.

After cooking, the beans are drained and cooled to approximately 30 °C before inoculation (Ahnan-Winarno et al., 2021).

3. *Inoculation*

Once cooled, the cooked soybeans are inoculated with a starter culture containing spores of *Rhizopus oligosporus* (approximately 10^6 spores per gram of beans). The inoculum may be in the form of a commercial powder or a traditional tempeh starter (fermented rice flour or previous batch residues). Proper and uniform mixing of the starter ensures consistent fungal growth. At this stage, the beans may be lightly acidified using vinegar or lactic acid to achieve a pH around 4.5–5.0, which helps suppress unwanted bacterial growth during early fermentation (Aaslyng & Højer, 2021).

4. *Fermentation (Incubation)*

The inoculated beans are then packed into perforated containers or wrapped in banana or teak leaves, which allow air circulation. The fermentation process is carried out at 30–35 °C for 24–36 hours in a humid, aerated environment (Ahnan-Winarno et al., 2021).

During fermentation, *Rhizopus oligosporus* grows rapidly over the surface of the beans, producing white cotton-like mycelia that bind the beans together into a solid, firm cake (Górska et al., 2025). The fungus secretes enzymes such as proteases, lipases, and amylases, which hydrolyze complex macromolecules into simpler, more digestible forms:

- Proteases break down proteins into amino acids and peptides, improving digestibility and flavor.
- Amylases degrade residual starch into sugars that contribute mild sweetness.
- Lipases hydrolyze fats, producing free fatty acids and aroma compounds.

As fermentation progresses, pH rises gradually (up to 7.0–8.0) due to ammonia production, which inhibits spoilage bacteria and helps develop the characteristic tempeh aroma (Górska et al., 2025).

5. *Post-processing and Preservation*

After the fermentation reaches completion, the tempeh is cooled and packaged. It can be consumed fresh, refrigerated (lasting up to a week), or frozen for long-term storage. Some tempeh is pasteurized to extend shelf life, though this may reduce probiotic activity. Tempeh may be sliced, marinated, fried, grilled, or added directly to stews and soups. The firm, cohesive texture allows it to be cut into various shapes, similar to meat (Aaslyng & Højer, 2021).

2.2.2.2 Applications of Tempeh

Tempeh's versatility, high protein content, and fermentation-derived flavor profile make it widely applicable across culinary, nutritional, and industrial contexts. Owing to its firm texture, rich umami taste, and enhanced digestibility, tempeh has transitioned from a traditional fermented food to a globally recognized plant-based meat alternative and functional protein source (Ahnan-Winarno et al., 2021; Aaslyng & Højer, 2021).

1. *Culinary Applications*

   - Meat Substitute: Tempeh is commonly used as a replacement for beef, chicken, or fish in dishes like burgers, stir fries, kebabs, curries, and sandwiches.
   - Traditional Dishes: In Indonesia, tempeh is fried or braised with spices in dishes such as tempeh goreng or tempeh bacem.
   - Global Cuisine: Western adaptations include tempeh bacon, tempeh tacos, and tempeh crumbles used in vegan recipes.

Because of its chewy texture and ability to absorb marinades, tempeh mimics the mouthfeel and flavor retention of meat, making it highly popular in plant-based cooking (Ahnan-Winarno et al., 2021).

2. *Nutritional and Functional Benefits*

   - High Quality Protein: With a protein content comparable to meat (≈20%), tempeh is a vital source of amino acids for vegetarians.
   - Enhanced Digestibility: Fermentation breaks down oligosaccharides, reducing bloating and improving protein utilization.
   - Rich in Probiotics and Bioactive Compounds: Contains beneficial microorganisms and peptides that support gut health and immunity.
   - Cholesterol Free and Low in Saturated Fat: Supports heart health and aids in weight management.
   - Isoflavone Activation: Fermentation converts glycosidic isoflavones into their aglycone forms (genistein, daidzein), enhancing antioxidant activity and potential hormone-balancing effects (Ahnan-Winarno et al., 2021; Subali et al., 2023).

3. *Industrial and Commercial Uses*

   - Used in ready to cook vegetarian meals, snack bars, and frozen entrees.
   - Increasingly incorporated into hybrid protein formulations combining legumes, grains, and fungi.
   - Serves as a functional ingredient in developing fermented protein beverages or powdered protein supplements (Subali et al., 2023).

2.2.3 Tofu

Tofu, also known as soybean curd, is a protein-rich plant-based food produced by coagulating soymilk and pressing the resulting curds into soft, white blocks (Qin et al., 2022). It is one of the oldest and most widely consumed soy products, originating in China over 2000 years ago and subsequently spreading throughout East and Southeast Asia before achieving global recognition as a staple food in vegetarian and vegan diets (Zheng et al., 2020). Today, tofu plays a central role in modern plant-based food systems due to its nutritional quality, technological versatility, and cultural acceptance (Das et al., 2023). From a nutritional perspective, tofu is considered a high-quality protein source, providing approximately 8% protein, 4–5% lipids, and 2% carbohydrates on a fresh weight basis. It contains all essential amino acids in favorable proportions, along with unsaturated fatty acids and important minerals such as calcium, iron, magnesium, and phosphorus. Tofu is also a notable source of soy-derived bioactive compounds, particularly isoflavones including daidzein, genistein, and glycitein, which are associated with antioxidant, anti-inflammatory, and cardioprotective effects (Qin et al., 2022). Importantly, tofu is cholesterol free, low in calories, and contains dietary fiber (approximately 1%), making it a heart-healthy and functional food suitable for diverse population groups (Zheng et al., 2020).

From a technological standpoint, tofu production is analogous to cheese making. In this process, soymilk serves as the protein substrate, while coagulants such as calcium sulphate ($CaSO_4$), magnesium chloride ($MgCl_2$), or glucono-δ-lactone (GDL) are used to induce protein coagulation and curd formation (Zheng et al., 2020). Variations in coagulant type, concentration, temperature, and pressing conditions give rise to different tofu textures, commonly classified as silken (soft), regular (medium firm), firm, and extra firm tofu (Zheng et al., 2020). These textural differences directly influence tofu's culinary performance and suitability for different food applications. The simplicity, versatility, and nutritional balance of tofu have positioned it as a cornerstone of plant-based food innovation (Carreno & Dolle, 2018). It is widely used in traditional dishes, minimally processed foods, and increasingly in modern meat analogue products, including plant-based burger patties, tofu steaks, and hybrid formulations combining tofu with tempeh or other protein sources (Thwe et al., 2025). Beyond its culinary value, tofu also exemplifies how traditional fermented and non-fermented soy foods can be integrated into engineered plant-based food systems aimed at improving sustainability, protein security, and consumer acceptance (Das et al., 2023).

2.2.3.1 Processing of Tofu

The preparation of tofu involves five primary stages: (1) preparation of soybeans, (2) extraction of soymilk, (3) coagulation, (4) curd formation and pressing, and (5) finishing and storage. Each step influences the texture, flavor, and nutritional composition of the final product (Zhang et al., 2019; Zheng et al., 2020; Chen et al., 2023).

1. *Preparation of Soybeans*

High-quality yellow soybeans with uniform size and high protein content are selected. The beans are cleaned to remove dust, stones, and impurities and then soaked in clean water (25–30 °C) for 8–12 hours (Zhang et al., 2019). Soaking hydrates the beans, increasing their moisture content to approximately 120–125% of the dry weight, which softens the beans and facilitates grinding. During soaking, some anti-nutritional compounds such as oligosaccharides, trypsin inhibitors, and phytic acid are partially leached out, improving the digestibility and palatability of tofu (Guan et al., 2021).

2. *Extraction of Soymilk*

After soaking, the beans are drained and ground with water at a ratio of 1:6 to 1:8 (beans to water) to produce a fine soy slurry (Zheng et al., 2020). The slurry is then heated to 95–100 °C for 5–10 minutes to:

- Inactivate trypsin inhibitors and lipoxygenase enzymes (which cause beany flavor).
- Denature proteins for better coagulation.
- Sterilize the slurry, reducing microbial load (Chen et al., 2023).

The heated slurry is filtered through a muslin cloth or fine screen to separate the soymilk (filtrate) from the okara (residual fiber pulp). The soymilk obtained typically contains 5–7% solids and serves as the base for tofu production. The okara by-product is rich in fiber and is often used in animal feed or as a food ingredient (Zhang et al., 2019).

3. *Coagulation*

The freshly prepared soymilk is cooled to 70–80 °C, and a coagulant is added to induce protein aggregation and gel formation. The type and concentration of coagulant determine the texture, flavor, and mineral composition of tofu (Chen et al., 2023; Guan et al., 2021).

The commonly used coagulants include the following:

- Calcium sulfate ($CaSO_4$) yields a firm, slightly grainy tofu with high calcium content.
- Magnesium chloride ($MgCl_2$) produces a smoother, softer tofu and enhances mineral bioavailability.
- Glucono delta lactone (GDL) provides a delicate, silken tofu texture with mild taste due to gradual acidification.

The coagulant is dissolved in warm water and gently mixed into the hot soymilk. Coagulation occurs as the soy proteins (primarily glycinin and β-conglycinin) denature and aggregate into a network structure that traps water and lipids. The process typically takes 10–15 minutes, resulting in a soft curd whey separation (Guan et al., 2021).

4. *Curd Formation and Pressing*

Once coagulation is complete, the curds are allowed to settle for several minutes. The curd mass is then ladled into perforated molds lined with cheese cloth. The molds are pressed at controlled pressure to expel excess whey and achieve the desired firmness (Zhang et al., 2019). The pressing time and pressure determine the tofu type as represented in Table 2.2.

After pressing, the tofu blocks are cut into desired shapes and immersed in cold water to improve texture and remove residual coagulant taste (Zheng et al., 2020).

5. *Finishing and Storage*

Tofu has high moisture content (70–80%) and is therefore highly perishable. It is typically stored in cold water at 4 °C and consumed within 2–3 days. Commercially, tofu may be pasteurized or vacuum packed to extend shelf life. Some variants are deep fried, smoked, fermented, or frozen to produce value-added products such as stinky tofu, dried tofu, or frozen tofu puffs (Zheng et al., 2020).

Table 2.2 Pressing conditions and corresponding tofu types

Type	Pressing time	Characteristics
Silken tofu	No pressing (set in container)	Very soft, custard-like texture
Soft tofu	5–10 min light pressing	Smooth and delicate
Firm tofu	15–30 min moderate pressing	Dense and versatile
Extra firm tofu	45–60 min high pressure	Chewy, meat-like texture

2.2.3.2 Applications of Tofu

Tofu's mild flavor, soft texture, and high protein content make it one of the most versatile plant-based foods, suitable for both traditional and modern culinary uses. Beyond its culinary appeal, tofu also holds nutritional, functional, and industrial significance (Das et al., 2023; Qin et al., 2022; Usman & Xu, 2024).

1. *Culinary Applications*

   - Meat and Dairy Substitute: Tofu is widely used as a protein-rich substitute for meat, eggs, and cheese in vegetarian and vegan diets. It can be grilled, stir fried, scrambled, baked, or blended into smoothies and desserts (Usman & Xu, 2024).
   - Traditional Asian Dishes: Common dishes include Mapo Tofu (China), Agedashi Tofu (Japan), Dubujeon (Korea), and Tauhu Goreng (Indonesia) (Das et al., 2023).
   - Western Adaptations: Used in tofu burgers, tofu nuggets, plant-based lasagna, and protein shakes (Usman & Xu, 2024).
   - Fermented Tofu Products: In East Asia, tofu is fermented to create "stinky tofu" or "fermented bean curd," which develops a strong flavor and extended shelf life (Das et al., 2023).

Tofu's neutral flavor profile allows it to absorb marinades and seasonings, mimicking the taste and texture of animal-based foods (Zheng et al., 2020).

2. *Nutritional and Functional Benefits*

   - High Quality Protein: Tofu provides a complete amino acid profile, making it comparable to animal proteins (Qin et al., 2022).
   - Low in Fat and Calories: Aids in weight management and cardiovascular health (Usman & Xu, 2024).
   - Rich in Isoflavones: Soy isoflavones such as genistein and daidzein act as phytoestrogens, linked to reduced risks of osteoporosis, breast cancer, and menopausal symptoms (Qin et al., 2022).
   - No Cholesterol and Gluten Free: Ideal for heart patients and those with gluten intolerance (Zheng et al., 2020).
   - Source of Calcium and Iron: Especially when calcium sulfate is used as a coagulant, contributing to bone and blood health (Chen et al., 2023).

3. *Industrial and Commercial Uses*
    - Used in the production of plant-based dairy alternatives, such as tofu-based yogurts, cheese analogues, and desserts (Usman & Xu, 2024).
    - Serves as a functional ingredient in meat analogues to improve texture and moisture retention (Zheng et al., 2020).
    - Dehydrated tofu powders are used in nutritional supplements and protein fortification (Qin et al., 2022).
    - Tofu by-product (okara) is utilized in fiber-enriched snacks, baked goods, and animal feed (Das et al., 2023).

2.2.4 Kinema

Kinema is a traditional fermented soybean food indigenous to the eastern Himalayan region, particularly consumed by the Nepali, Lepcha, and Bhutia communities of Nepal, India (Sikkim and Darjeeling hills), and Bhutan (Tamang, 2015, 2024). It is a non-salted, solid-state fermented, alkaline food characterized by a sticky texture, strong ammonia-like aroma, and savory umami taste (Tamang & Nikkuni, 1996; Tamang, 2024). Kinema serves as both a protein-rich staple food and a cultural delicacy, deeply rooted in the food traditions of Himalayan households. Nutritionally, kinema is a high-protein fermented product, containing approximately 48% protein, 17% fat, 28% carbohydrates, and 7% ash on a dry weight basis. Its moisture content averages around 62%, which supports microbial fermentation. The fermentation process is dominated by *Bacillus* species, mainly *Bacillus subtilis*, *Bacillus licheniformis*, and *Bacillus pumilus*, which contribute to its unique sticky texture, alkalinity, and flavor (Tamang, 2015).

The ammoniacal aroma typical of kinema results from the degradation of amino acids and the release of ammonia during fermentation. The production of ammonia increases pH, creating a mildly alkaline environment that inhibits spoilage organisms while supporting beneficial microbial activity (Tamang & Nikkuni, 1996). From a nutritional perspective, kinema provides highly digestible protein, essential fatty acids, B vitamins, and bioactive peptides. It is comparable in protein quality to animal-based foods such as chicken or beef, making it a vital source of affordable plant protein in mountain diets (Tamang, 2015).

2.2.4.1 Processing of Kinema

Kinema production is a traditional, small-scale fermentation process that uses natural microbial flora without added starter cultures. The method varies slightly across regions, but the core process involves soaking, cooking, fermenting, and drying soybeans. The following describes the traditional household process practiced in Sikkim and Nepal (Tamang, 2015).

1. *Selection and Preparation of Soybeans*

High-quality yellow soybeans are selected for uniformity and absence of insect damage. The beans are cleaned thoroughly to remove dirt, stones, and foreign materials. Clean beans ensure better fermentation and prevent contamination by undesirable microorganisms (Nout et al., 1998).

2. *Soaking*

The cleaned soybeans are soaked in water at ambient temperature (10–25 °C) for 12–20 hours. Soaking hydrates the beans, softens the seed coat, and activates endogenous enzymes. This process also helps in removing oligosaccharides and some anti-nutritional compounds, improving digestibility (Tamang & Nikkuni, 1996).

3. *Cooking*

The soaked beans are boiled in water at 90–95 °C for about 90 minutes until they become soft but not mushy. Cooking denatures proteins, gelatinizes starches, and sterilizes the substrate, creating a suitable environment for microbial fermentation (Tamang & Nikkuni, 1996). It also ensures partial destruction of trypsin inhibitors and phytates, which interfere with protein and mineral absorption. After boiling, the water is drained, and the beans are lightly crushed or cracked to produce grits, increasing the surface area for microbial colonization. The beans are then cooled to 30–35 °C, the optimal temperature range for *Bacillus* growth (Tamang, 2015).

4. *Fermentation*

The warm, cooked soybean grits are wrapped in fresh leaves (usually fern, banana, or fig leaves) and placed inside a jute or cloth bag, which provides insulation and aeration. The wrapped package is left to ferment naturally for 24–72 hours at ambient temperature (25–35 °C) (Kharnaior & Tamang, 2022). During this period, naturally occurring *Bacillus* spp. from the environment and leaves proliferate rapidly. These bacteria are aerobic and spore forming, producing proteolytic enzymes that hydrolyze soybean proteins into peptides and amino acids. The enzymatic breakdown of amino acids leads to ammonia production, raising the pH (to around 8.0–9.0) and giving kinema its characteristic alkaline flavor and stickiness (Tamang, 2015).

The sticky, mucilaginous surface observed in fermented kinema is due to the secretion of extracellular polysaccharides and peptides by *Bacillus* spp. This sticky matrix enhances the product's texture and mouthfeel. The fermentation is considered complete when the product develops a distinctive ammoniacal aroma, a sticky texture, and light brown coloration. The entire process is spontaneous and self-limiting, typically reaching the desired characteristics within two days in warmer climates or up to three days in cooler regions (Tamang, 2024).

5. *Post Fermentation Handling*

After fermentation, kinema is air dried or sun dried to reduce moisture and extend shelf life. The dried kinema can be stored for several months and later rehydrated before consumption. Fresh kinema, however, is perishable and is typically consumed within a few days. Before eating, fresh kinema is fried lightly in oil and cooked with onions, tomatoes, chili, and spices, often combined with vegetables or rice. This process enhances flavor and aroma while reducing excess alkalinity (Tamang, 2015).

2.2.4.2 Applications of Kinema

Kinema is a highly versatile food used in both traditional cuisines and modern plant-based diets, valued for its unique flavor, nutritional benefits, and functional properties (Kharnaior & Tamang, 2022).

1. *Culinary Applications*

   - Traditional Dishes: In Nepal and Sikkim, kinema is used to prepare "Kinema curry," "Kinema achar" (pickle), and "Kinema soup." It is often served with rice or flatbread (Tamang, 2012, 2024).
   - Flavor Enhancer: The strong umami flavor derived from glutamic acid and peptides makes kinema a natural flavor enhancer similar to soy sauce or miso (Kharnaior & Tamang, 2022).
   - Protein Substitute: Kinema can replace meat in curries or stews, providing comparable protein levels and texture (Tamang, 2012).
   - Processed Foods: Dried kinema can be powdered and incorporated into instant soups, snack bars, or savory seasonings (Tamang, 2024).

The product's sticky consistency allows it to bind ingredients in plant-based patties and dumplings, offering functional advantages in vegan recipes (Kharnaior & Tamang, 2022).

2. *Nutritional and Functional Benefits*

   - High Protein Content: With nearly 48% protein (dry weight), kinema rivals animal protein sources like chicken or beef (Tamang, 2012).
   - Improved Digestibility: *Bacillus* fermentation breaks down macromolecules, making proteins and nutrients more bioavailable (Kharnaior & Tamang, 2022).
   - Alkaline and Probiotic Nature: The *Bacillus* species present in kinema contribute to gut microbiome balance and may exert antimicrobial effects (Tamang, 2012).
   - Rich in Vitamins and Minerals: Fermentation increases the bioavailability of vitamin B_2 (riboflavin), vitamin K_2, and iron (Tamang, 2012).

- Antioxidant and Anti-inflammatory Properties: Fermented soy peptides exhibit bioactive effects, potentially supporting immune and cardiovascular health (Kharnaior & Tamang, 2022).

Kinema's fermentation process also reduces anti-nutritional compounds, including phytic acid, tannins, and trypsin inhibitors, enhancing its nutritional efficiency (Tamang, 2012).

3. *Industrial and Commercial Uses*

- Functional Ingredient: Kinema powder can be incorporated into instant meal formulations, protein bars, and nutraceutical products (Tamang, 2024).
- Starter Culture Source: Pure strains of *Bacillus subtilis* isolated from kinema are used in laboratory studies and controlled fermentations for similar foods like Natto (Japan) and Thua Nao (Thailand) (Kharnaior & Tamang, 2022).
- Flavoring Agent: Kinema-derived peptides and amino acids are explored in the food flavor industry as natural umami enhancers (Tamang, 2024).
- Regional Branding and Export: Kinema is gaining recognition as a geographically indicated (GI) product in the eastern Himalayas, promoting local entrepreneurship and sustainable food heritage (Tamang, 2024).

Kinema is more than just a fermented soybean food; it represents the intersection of traditional wisdom, microbiological science, and nutritional sustainability. Through simple, natural fermentation, ordinary soybeans are transformed into a high-protein, digestible, and flavorful product with significant health-promoting properties (Kharnaior & Tamang, 2022). The predominance of *Bacillus* fermentation imparts kinema's characteristic alkaline taste, stickiness, and ammonia aroma, making it distinct among global fermented foods. Today, kinema is increasingly recognized as a model for indigenous fermentation technology and a potential candidate for industrial scale functional food development. By preserving its traditional preparation methods while modernizing safety and packaging standards, kinema can bridge the gap between local heritage foods and global plant-based innovation (Tamang, 2024).

2.2.5 Seitan

Seitan, often called wheat meat, wheat gluten, or gluten steak, is a high-protein plant-based food made by isolating the gluten portion of wheat flour and processing it into a chewy, fibrous mass that closely resembles the texture of cooked meat (Yusuf, 2023). It has been consumed in East Asia for centuries, particularly within Buddhist vegetarian cuisines of China and Japan, and has recently become a prominent meat alternative in Western vegan diets (Švarc et al., 2022).

The name "seitan" was popularized in Japan in the 1960s by macrobiotic food advocate George Ohsawa, though the product itself has much historical roots (Muneer et al., 2025). Chemically, seitan is composed primarily of gluten proteins namely gliadin and glutenin which, when hydrated and kneaded, form an elastic protein network. This matrix gives seitan its distinctive stringy, meat-like texture. Nutritionally, seitan contains about 75–80% protein (dry weight), making it one of the most concentrated plant protein sources available (Aunon-Lopez et al., 2025). It is low in fat, cholesterol free, and provides minerals such as iron, phosphorus, and selenium. However, unlike soy-based proteins, seitan is low in lysine, an essential amino acid, and thus benefits from pairing with legumes or soy products for balanced nutrition. Because of its neutral flavor, seitan readily absorbs marinades, spices, and sauces, allowing it to mimic the taste of chicken, beef, pork, or seafood. Its textural versatility and high protein content make it an indispensable ingredient in the modern plant-based meat industry (Aunon-Lopez et al., 2025).

2.2.5.1 Processing of Seitan

Seitan production involves gluten extraction from wheat flour, followed by hydration, kneading, extrusion, and thermal treatment to achieve the desired fibrous structure (Yusuf, 2023; Muneer et al., 2025; Aunon-Lopez et al., 2025). The process can be carried out by traditional wet-washing methods or through modern mechanical and extrusion technologies (Švarc et al., 2022).

1. *Preparation of Dough*

The process begins with wheat flour (typically high-protein bread flour) mixed with water at a ratio of 2:1 (flour to water) to form a smooth, elastic dough (Yusuf, 2023). The dough is kneaded thoroughly for about 10–15 minutes to develop the gluten network, ensuring that gliadin and glutenin proteins are aligned and cross-linked through hydrogen and disulfide bonds. This dough serves as the substrate for gluten extraction. Adequate kneading is crucial because it directly influences the elasticity, chewiness, and fibrous characteristics of the final seitan (Yusuf, 2023).

2. *Washing and Gluten Extraction*

After resting for 20–30 minutes, the dough is repeatedly washed under running water to remove starch granules and bran components. During washing, the starch suspension is drained off while the sticky, elastic gluten mass remains. This step can be done manually or mechanically in industrial settings. The washing continues until the rinse water becomes clear, indicating that most starch and soluble components have been removed. The result is raw wheat gluten, which contains approximately 65–70% moisture and 25–30% protein on a wet basis. The washed gluten is then squeezed to remove excess water. The removed starch water can be sedimented, dried, and reused as a by-product for noodles, bread improvers, or sweeteners, making the process resource efficient (Muneer et al., 2025).

3. *Texturization and Shaping*

The raw gluten mass is kneaded, stretched, or extruded to orient the protein fibers. In modern processing, the moisture content is adjusted to 40–80% (w/w) before extrusion. The mixture is then passed through a twin-screw or single-screw extruder at controlled temperature (75–120 °C) and pressure to produce a fibrous, layered structure resembling animal muscle. The extrudate can be shaped into various forms such as sheets, cylinders, or chunks by extruding through nozzles or slits. In traditional methods, gluten is manually stretched and folded multiple times to enhance its fibrous texture (Aunon-Lopez et al., 2025).

4. *Heat Treatment*

Following shaping, the seitan is subjected to thermal processing in boiling water or steam for 30–60 minutes. This heat treatment:

- Stabilizes the protein structure, fixing the texture.
- Denatures residual enzymes.
- Improves water holding capacity.
- Enhances flavor development through mild Maillard reactions.

During industrial processing, baking, roasting, or autoclaving may also be applied to further modify flavor and texture. The product can then be seasoned, marinated, or flavored to imitate specific meats (e.g., chicken, beef, duck) (Švarc et al., 2022).

5. *Finishing and Storage*

Once cooked, seitan can be sliced, diced, or shredded depending on the intended use (Muneer et al., 2025). It is often marinated in soy sauce, spices, and plant oils to enhance taste and color. For commercial products, seitan is vacuum sealed or frozen to extend shelf life. Dried seitan can also be produced by dehydrating cooked gluten for long-term storage and rehydrated before use (Aunon-Lopez et al., 2025).

2.2.5.2 Applications of Seitan

Seitan's meat-like texture, high protein content, and neutral flavor have led to diverse applications in culinary and industrial settings (Muneer et al., 2025).

1. *Culinary Applications*

   - Meat Substitute: Seitan is used as a replacement for chicken, beef, or pork in dishes like stir-fries, kebabs, burgers, sausages, and deli slices.
   - Ethnic Cuisine: In traditional Asian cooking, seitan appears as "kaofu" in Chinese dishes, "fu" in Japanese Buddhist cuisine, and "mien-chin" in other East Asian preparations.
   - Western Vegan Foods: Popular in vegan roasts, sandwich fillings, tacos, and mock steaks, often marketed as "plant-based meat."
   - Texture Enhancement: Because seitan can absorb marinades and hold structure under heat, it provides a chewy, satisfying bite similar to animal muscle.

Seitan's ability to retain its form during grilling or frying distinguishes it from softer soy products like tofu or tempeh (Yusuf, 2023).

2. *Nutritional and Functional Benefits*

- High Protein Density: Provides up to 25 g of protein per 100 g (wet weight), supporting muscle development and satiety.
- Low Fat and Cholesterol Free: Contributes to heart-healthy diets.
- Mineral Source: Supplies iron, selenium, and phosphorus.
- Low Carbohydrate Content: Suitable for low-carb and high-protein diets.
- Functional Texture: Gluten's viscoelasticity allows formation of fibers that mimic meat, making seitan essential in hybrid and analogue formulations.

However, seitan lacks lysine and is unsuitable for individuals with celiac disease or gluten intolerance, which limits its universality as a protein source (Muneer et al., 2025.

3. *Industrial and Commercial Uses*

- Plant-based Meat Formulations: Seitan serves as a base ingredient for commercial vegan meat analogues, often combined with soy, pea, or mycoprotein to achieve desired nutritional and textural profiles.
- Texturizing Agent: Used in food extrusion technology to enhance chewiness in imitation meats, sausages, and jerky.
- Frozen and Ready to Eat Foods: Widely utilized in retail vegan meals, marinated strips, and canned mock meats.
- Hybrid Products: Integrated with legumes or cereals to improve amino acid balance in protein-enriched snacks and bars.

Because of its elasticity and thermal stability, seitan is particularly suited for 3D food printing and structured plant-protein products, where precise texture control is required (Maningat et al., 2022; Akhtar et al., 2022).

2.3 Processing of Plant-Based Foods

The production of plant-based meat (PBM) involves a series of carefully designed steps that aim to replicate the sensory, nutritional, and functional characteristics of animal-derived meat (Ribas-Agustí et al., 2018). The processing of PBM is a multidisciplinary process, drawing from food chemistry, protein engineering, material science, and biotechnology (Munekata et al., 2020). Generally, it can be divided into three key stages: (1) protein isolation and functionalization, (2) formulation, and (3) processing and structuring. Each stage plays a crucial role in transforming raw plant materials into a realistic, high-quality meat analogue that mimics the appearance, flavor, texture, and nutritional value of conventional meat (Prakash et al., 2022; Rousseau et al., 2020). Table 2.3 summarizes the major processing methods used in the production of plant-based foods, highlighting their objectives and representative examples.

Table 2.3 Processing methods used in the production of plant-based foods

Processing stage/method	Description	Examples of plant-based foods	Purpose/functional outcome	References
Primary processing (cleaning, soaking, milling)	Removal of impurities followed by hydration and size reduction of raw plant materials	Soybeans, wheat, legumes, cereals	Improves digestibility, reduces antinutritional factors, prepares raw material for further processing	Samtiya et al. (2021); Öncü Öner (2023)
Protein extraction and fractionation	Separation of protein fractions using wet or dry processing techniques	Soy protein isolate, pea protein, wheat gluten (seitan)	Enhances protein concentration and functional properties	Mariotti (2025); Sridhar et al. (2023)
Texturization (extrusion cooking)	High-temperature, high-shear processing to create fibrous, meat-like textures	TVP, plant-based meat analogues	Develops meat-like structure, chewiness, and water-holding capacity	Sridhar et al. (2023); Munialo and Andrei (2023)
Fermentation	Use of microorganisms to modify nutritional and sensory characteristics	Tempeh, kinema, fermented soy products	Improves flavor, mineral bioavailability, gut health, and shelf life	Samtiya et al. (2021); Ketnawa et al. (2022)
Mycoprotein fermentation	Controlled fungal fermentation using filamentous fungi	Quorn (*Fusarium venenatum*)	Produces high-quality protein with fiber and low-fat content	Mariotti (2025); Munialo and Andrei (2023)
Thermal processing (cooking, pasteurization)	Application of heat to ensure safety and stability	Tofu, plant-based beverages, meat analogues	Inactivates pathogens, improves texture and shelf stability	Craig and Fresán (2021); Öncü Öner (2023)
Enzymatic treatment	Use of enzymes to modify protein, starch, or fiber structure	Plant-based dairy alternatives, protein hydrolysates	Enhances digestibility, flavor development, and functionality	Ketnawa et al. (2022); Sridhar et al. (2023)
Formulation and fortification	Addition of micronutrients, flavors, and stabilizers	Fortified plant-based milks, meat analogues	Improves nutritional adequacy and consumer acceptance	Viroli et al. (2023); Craig and Fresán (2021)
Packaging and preservation	Use of modified atmosphere or aseptic packaging	Ready-to-eat plant-based foods	Extends shelf life and maintains product quality	Viroli et al. (2023)

Protein Isolation and Functionalization

The first stage in plant-based meat production focuses on extracting and enhancing plant proteins to make them suitable for use in meat analogue formulations. Plant sources such as soybean, pea, wheat, mung bean, and faba bean are among the most commonly used due to their high protein content, functional properties, and global availability (Sim et al., 2021).

Protein Isolation

Protein isolation involves separating proteins from the rest of the plant matrix, carbohydrates, fibers, oils, and phenolic compounds. The main goal is to obtain a concentrated protein fraction with minimal impurities, high solubility, and appropriate physicochemical characteristics (McClements & Grossmann, 2022; Yan et al., 2023). The common methods include the following:

- Dry fractionation, which relies on milling and air classification to separate protein-rich fractions based on particle size and density. This method is energy efficient and avoids chemical solvents, though it typically yields a lower protein concentration (40–60%) (McClements & Grossmann, 2022).
- Wet extraction, which involves solubilizing proteins in water or an alkaline solution (pH 8–11) followed by isoelectric precipitation (pH 4.5–5.0). This technique yields protein isolates with up to 90% purity but requires significant water and energy input (Yan et al., 2023).

Membrane filtration and ultrafiltration techniques are also used to refine protein extracts, improving purity and functional stability while minimizing denaturation (Yan et al., 2023).

Protein Functionalization

Once extracted, proteins may not naturally exhibit the same mechanical or emulsifying properties as animal proteins (Wang et al., 2022). Therefore, functionalization is performed to improve properties such as solubility, water holding capacity, gelation, emulsification, and cross-linking ability. A widely applied approach is enzymatic hydrolysis, where proteins are partially broken down using proteases to expose hydrophobic amino acids and reactive groups. This modification enhances solubility, emulsifying ability, and the capacity to form cohesive networks during structuring (Leonard et al., 2020).

Other Functionalization Methods Include the Following

- Chemical modification (e.g., acetylation, succinylation) to alter surface charge and improve emulsifying stability (Yan et al., 2023).
- Physical modification (e.g., heat treatment, high-pressure processing, or ultrasonication) to denature protein structures partially, enhancing texturization potential (Mulla et al., 2022).
- Blending of multiple protein sources (such as soy and pea, or wheat gluten and faba bean) to achieve synergistic improvements in elasticity and chewiness (Sim et al., 2021).

Through functionalization, plant proteins become more adaptable to subsequent texturization processes, where they are transformed into fibrous, muscle-like structures (Sim et al., 2021).

Formulation

In the second stage, formulation, the isolated and functionalized plant proteins are combined with various auxiliary ingredients to create a balanced, palatable, and nutritionally complete product. This step determines not only the taste and nutritional profile but also the mouthfeel, juiciness, and color of the final PBM (McClements & Grossmann, 2022).

Core Ingredients

- Plant Proteins: These form the structural backbone of PBM. Common sources include soy protein isolate, pea protein, wheat gluten, rice protein, and potato protein (Munekata et al., 2020).
- Fats and Oils: Plant-based fats such as coconut oil, canola oil, and sunflower oil are used to replicate the marbling and juiciness of meat. They provide lubricity and improve flavor release during cooking (Ribas-Agustí et al., 2018).
- Carbohydrates and Binders: Ingredients such as starches, methylcellulose, pectin, and plant-based gums act as binding agents and moisture retainers, contributing to product cohesiveness and bite (Ribas-Agustí et al., 2018).
- Colorants: Natural pigments like beet juice extract, annatto, and leghemoglobin (a heme-like protein produced via fermentation) are incorporated to mimic the red-to-brown color transition of real meat during cooking (Munekata et al., 2020).
- Flavoring Agents: These include yeast extracts, amino acids, and Maillard reaction products that generate savory "meaty" aromas (Munekata et al., 2020).
- Micronutrients: Fortification with vitamin B12, iron, zinc, and omega-3 fatty acids ensures that PBMs match or surpass the nutrient profile of animal-derived meat (Ribas-Agustí et al., 2018).

Formulation Strategies

To achieve a desirable sensory experience, ingredients must be combined in precise ratios to form an emulsion system where water, oil, and protein interact in a stable matrix. Emulsification and mixing processes are optimized to prevent phase separation, ensuring that the final mixture exhibits the same cohesiveness, elasticity, and juiciness as meat (McClements & Grossmann, 2022).

In addition, pH adjustment, ionic strength control, and thermal preconditioning may be applied to fine-tune the interactions among protein molecules. These adjustments influence the way proteins unfold and cross-link during thermal processing, directly affecting texture and appearance (McClements & Grossmann, 2022).

Processing and Structuring

The final and most critical stage in plant-based meat production is the processing and structuring of the protein mixture into fibrous, meat-like textures. This is where technological innovation plays a central role, using mechanical, thermal, and biological methods to replicate the complex structure of animal muscle tissue (Rousseau et al., 2020).

2.3.1 Extrusion Technology

High Moisture Extrusion Cooking (HMEC) is one of the most widely used methods for structuring PBM. In this process, the protein mixture (typically containing 50–70% moisture) is fed into a twin screw extruder, where it undergoes controlled heating, shearing, and mixing (Zhang et al., 2019).

Inside the Extruder

- The screw elements apply mechanical shear and pressure, causing proteins to denature, align, and aggregate.
- Heat and moisture promote gelation and cross-linking, forming continuous fibrous networks.
- As the material exits the extruder through a cooling die, the temperature drops rapidly, causing phase separation between protein and water layers, which results in fibrous meat-like textures.

This process is used to produce products such as plant-based chicken strips, nuggets, and beef-like chunks (Cornet et al., 2022).

2.3.2 Shear Cell Technology

Shear cell technology is a novel structuring method that subjects plant protein blends to controlled shear forces under heat, aligning protein molecules to mimic the layered structure of animal muscle fibers (Wang et al., 2022). Unlike extrusion, shear cell technology offers a more uniform and gentle structuring environment, resulting in laminated textures that replicate the natural orientation of muscle fibers. This method reduces energy consumption compared to extrusion and allows better control over texture anisotropy, making it suitable for whole cut meat analogues such as steaks or fillets (Wang et al., 2022).

2.3.3 Mycelium Cultivation

Another promising avenue in PBM development is mycelium-based meat, which leverages the root-like structures of fungi to create natural fibrous textures. Mycelium offers superior texture, nutritional value, and sustainability (McClements & Grossmann, 2022).

The process typically involves the following:

- Substrate Preparation: Food waste or agricultural by-products serve as nutrient-rich substrates.
- Inoculation: Fungal spores (e.g., *Fusarium venenatum* or *Rhizopus oligosporus*) are introduced into the substrate.

- Fermentation: The culture is incubated (e.g., 30 °C, 110 rpm, 48 h) to allow mycelial growth.
- Harvesting and Drying: Biomass is collected, washed, frozen, and dried to form a protein-rich powder.
- Post-processing: The dried mycelium is combined with binders (e.g., egg albumin or plant alternatives), flavors, and colorants. After heating and extrusion, the product attains a texture comparable to that of chicken or beef.

Mycelium cultivation provides a high-protein, fiber-rich, and low-carbon alternative, often requiring less land and water than crop or livestock-based systems (McClements & Grossmann, 2022).

2.3.4 Recombinant Protein Additives

Biotechnological innovations have introduced recombinant proteins lab-engineered proteins designed to enhance the texture, taste, or nutrition of PBM. These are produced through genetic engineering of microorganisms such as *E. coli, Saccharomyces cerevisiae*, or *Pichia pastoris* (Dupuis et al., 2023; Park & Wi, 2016; Zhao et al., 2022).

The mechanism involves the following:

- Insertion of a target gene encoding the desired protein into a vector using enzymes like endonucleases and DNA ligases (Yemets et al., 2014).
- Transformation of the vector into host cells, which are then cultured to express the protein.
- Purification and incorporation of the recombinant protein into PBM formulations, where it functions as a flavor enhancer, emulsifier, or heme analog (e.g., recombinant soy leghemoglobin used in Impossible™ Burger) (Yemets et al., 2014).

This approach allows for precise molecular control over food functionality and supports the creation of next-generation plant-based meats that rival conventional meat in sensory quality (Park & Wi, 2016). The mechanism of processing plant-based foods is a sophisticated integration of protein science, food technology, and biotechnology. From the initial extraction and functionalization of plant proteins to advanced structuring techniques such as extrusion, shear cell processing, mycelium cultivation, and recombinant protein enhancement, each step contributes to producing a product that meets consumer expectations for taste, texture, and nutrition (Zhao et al., 2022). These technologies not only offer a sustainable alternative to meat production but also exemplify the innovative potential of modern food engineering in addressing global challenges of food security, health, and environmental sustainability (Zhao et al., 2022).

2.4 Health Benefits

In recent decades, the growing awareness of the interconnection between diet, health, and environmental sustainability has led to a substantial shift toward plant-based foods (PBFs) (Viroli et al., 2023; Mariotti, 2025). These foods derived primarily from plant sources such as legumes, grains, fungi, and cereals are designed to provide the nutritional, sensory, and functional qualities of animal-derived foods while reducing health risks associated with high meat consumption (Munialo & Andrei, 2023). Plant-based meat analogues like textured vegetable protein (TVP), Quorn, tempeh, tofu, kinema, and seitan have emerged as essential components of this nutritional evolution (Öncü Öner, 2023).

Plant-based foods are not only rich in high-quality proteins, vitamins, minerals, and fibers, but also low in cholesterol, saturated fats, and caloric density, which collectively promote better metabolic health (Craig & Fresán, 2021). Moreover, many of these foods contain bioactive compounds such as isoflavones, saponins, antioxidants, and prebiotic fibers, which contribute to disease prevention and long-term wellness. Figure 2.2 provides an overview of the key health benefits associated with plant-based foods, highlighting their biochemical, physiological, and clinical impacts, which are discussed in detail in the following section (Sridhar et al., 2023).

2.4.1 Cardiovascular Health

One of the most well-documented benefits of plant-based foods is their ability to lower the risk of cardiovascular diseases (CVDs) (Hu, 2003; Salehin et al., 2023). Diets rich in plant proteins, such as those derived from soy (TVP, tofu, tempeh, and kinema), are naturally cholesterol free and low in saturated fatty acids. In contrast, animal-derived proteins are often accompanied by high levels of saturated fat and cholesterol, which contribute to elevated low-density lipoprotein (LDL) levels and arterial plaque formation (Hemler & Hu, 2019).

Soy-based products, for example, contain isoflavones (daidzein, genistein, and glycitein) that exhibit phytoestrogenic and antioxidant properties. These compounds modulate lipid metabolism by reducing serum LDL cholesterol and triglycerides while simultaneously raising high-density lipoprotein (HDL) levels (Hu, 2003). The US Food and Drug Administration (FDA) recognizes that consumption of 25 grams of soy protein daily may reduce the risk of heart disease when incorporated into a low-fat diet (Salehin et al., 2023).

Mycoprotein-based foods like Quorn further enhance cardiovascular health by providing dietary fibers (β-glucans and chitin) that reduce cholesterol absorption in the intestine and improve satiety. Studies have shown that mycoprotein consumption lowers total cholesterol and LDL levels without affecting beneficial HDL cholesterol (Mariotti, 2025). Similarly, seitan, although derived from wheat gluten, remains

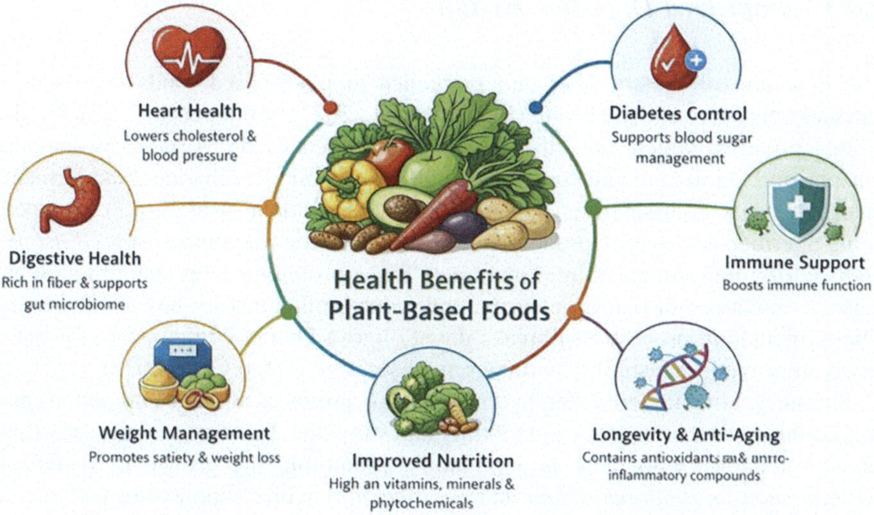

Fig. 2.2 Overview of the key health benefits associated with plant-based foods

cholesterol free and can support heart-healthy dietary patterns when combined with legumes for lysine balance (Salehin et al., 2023).

2.4.2 Weight Management and Obesity Prevention

Plant-based proteins generally possess a high satiety index, meaning they promote a feeling of fullness and reduce total calorie intake (Ivanova et al., 2021). Foods like TVP, tofu, and tempeh are protein dense yet low in fat, offering an ideal balance for individuals seeking to manage body weight. Protein and dietary fiber both contribute to delay gastric emptying and reduced postprandial hunger, which assist in appetite regulation (Bassin et al., 2024). Mycoprotein from Quorn has been clinically shown to lower energy intake in subsequent meals due to its slow digestibility and high fiber content. The fibrous texture of seitan and tempeh also encourages slower chewing, which increases sensory satisfaction and helps prevent overeating (Munialo & Andrei, 2023).

Moreover, plant-based foods typically have lower energy density, meaning fewer calories per gram allowing individuals to consume larger, more satisfying meals without excessive caloric intake. Long-term epidemiological studies have shown that populations adhering to vegetarian or flexitarian diets tend to have lower body mass indices (BMIs) compared to omnivores (Bassin et al., 2024).

2.4.3 Improved Digestive Health

The inclusion of dietary fiber and prebiotics in plant-based foods significantly enhances digestive and gut health (Ketnawa et al., 2022; Sridhar et al., 2023). Unlike animal proteins, which lack fiber, many plant-based foods, particularly tempeh, kinema, and Quorn, contain complex carbohydrates, oligosaccharides, and fermentable fibers that nourish beneficial gut microbiota (Sridhar et al., 2023). Tempeh, being a fermented soybean product, introduces probiotic microorganisms (*Rhizopus oligosporus*) into the gastrointestinal tract. These probiotics help restore microbial balance, enhance gut barrier integrity, and reduce inflammatory bowel conditions. The fermentation process also breaks down oligosaccharides responsible for flatulence, improving digestibility and nutrient absorption (Öncü Öner, 2023).

Similarly, kinema, fermented by *Bacillus* spp., produces alkaline compounds and free amino acids that improve gut motility and stimulate digestive enzyme secretion. Mycoprotein in Quorn acts as a prebiotic, promoting the growth of beneficial bacteria such as *Bifidobacterium* and *Lactobacillus* while suppressing pathogenic strains. Collectively, these properties contribute to enhanced gut microbiome diversity, reduced constipation, and improved metabolic health (Mariotti, 2025).

2.4.4 Antioxidant and Anti-inflammatory Effects

Plant-based foods are abundant in antioxidant substances that neutralize harmful free radicals and prevent oxidative stress-related damage to cells (Watzl, 2008; Samtiya et al., 2021). Soy-based products, in particular, are rich in isoflavones and phenolic acids, which exhibit anti-inflammatory and free radical scavenging activities. Isoflavones have been found to upregulate antioxidant enzymes such as superoxide dismutase (SOD) and glutathione peroxidase (GPx), thereby reducing oxidative stress in the cardiovascular and nervous systems. Regular consumption of soy-derived products like TVP, tofu, and tempeh has been associated with reduced markers of inflammation, including C-reactive protein (CRP) and interleukin-6 (IL-6) (Sridhar et al., 2023).

Quorn's mycoprotein also contains natural antioxidants, including ergosterol derivatives and β-glucans, which have been shown to modulate immune responses and reduce systemic inflammation. By lowering oxidative damage and inflammation, plant-based foods play a preventive role in chronic diseases such as atherosclerosis, diabetes, and cancer (Mariotti, 2025).

2.4.5 Cancer Prevention

Several epidemiological studies have linked plant-based diets with lower incidences of certain cancers, including breast, colon, and prostate cancers (DeClercq et al., 2022). This protective effect is primarily attributed to phytochemicals, dietary fiber, and lower intake of carcinogenic compounds (such as heme iron and nitrosamines) typically found in red and processed meats. Soy isoflavones act as selective estrogen receptor modulators (SERMs), mimicking the effect of estrogen in a protective rather than stimulatory manner (Rahman et al., 2020). These compounds help regulate hormonal activity, particularly in reducing the risk of hormone-dependent cancers like breast and prostate cancer (Rahman et al., 2020).

Tempeh and tofu, as minimally processed soy foods, retain these bioactive components in their natural forms, offering chemopreventive benefits. Additionally, fermented foods such as kinema and tempeh generate secondary metabolites like peptides, vitamins, and phenolics that possess anti-mutagenic and anti-carcinogenic properties (Rahman et al., 2020). Mycoprotein from Quorn provides dietary fiber and chitin, both of which contribute to colon health by reducing bile acid reabsorption and diluting carcinogenic metabolites in the colon, thereby reducing colorectal cancer risk (DeClercq et al., 2022).

2.4.6 Bone Health and Hormonal Balance

Soy-based products like tofu and tempeh are excellent sources of plant-based calcium, particularly when prepared with calcium sulfate ($CaSO_4$) as a coagulant (Kelly et al., 2024). They also contain isoflavones that enhance bone mineral density (BMD) by mimicking the protective effects of estrogen on bone tissue, especially beneficial for postmenopausal women at risk of osteoporosis (Zheng et al., 2020). Fermented soy foods further improve calcium absorption and bioavailability by reducing phytate content, an anti-nutritional factor that binds minerals. Moreover, mycoprotein in Quorn contains essential amino acids and phosphorus, both crucial for bone formation and maintenance (Ghadiri et al., 2023).

2.4.7 Metabolic Health and Diabetes Control

Plant-based diets are increasingly recognized for their role in improving insulin sensitivity and glycemic control. Foods such as tofu, tempeh, and TVP possess low glycemic indices (GI), meaning they produce a slower, more controlled rise in blood glucose levels (Johannesen et al., 2020; Mokhtari et al., 2022). The fiber content in tempeh and Quorn helps delay glucose absorption, stabilize blood sugar, and reduce insulin spikes. Isoflavones in soy also enhance pancreatic β-cell function and modulate insulin signaling pathways. Regular consumption of these foods can thus aid in type 2 diabetes prevention and management (Mokhtari et al., 2022).

2.4.8 Immune Function and Microbial Protection

Fermented plant-based foods like kinema and tempeh contain probiotic organisms that stimulate immune function (Jiang et al., 2021). These microbes promote the production of short-chain fatty acids (SCFAs) such as butyrate and acetate, which strengthen the intestinal barrier and modulate immune cell activity. Mycoprotein and soy peptides have also shown antimicrobial properties against harmful pathogens, contributing to improved resistance against infections (Di Sotto et al., 2020; London, 2010).

2.5 Conclusion

The growing global interest in plant-based foods (PBFs) represents a significant transformation in both nutrition science and sustainable food systems. Driven by health concerns, environmental awareness, ethical values, and economic considerations, the development and consumption of plant-based meat analogues (PBMs) and protein-rich substitutes have expanded remarkably in recent years. The shift away from traditional meat products is supported by a convergence of factors ranging from the rising incidence of animal-borne diseases and the global shortage of animal protein to the recognition of the environmental burden of intensive livestock farming. Consequently, plant-based foods such as textured vegetable protein (TVP), Quorn, tempeh, tofu, kinema, and seitan have emerged as key contributors to future food security and human health.

The production of plant-based meat analogues involves intricate biochemical and mechanical processes, beginning with protein isolation and functionalization, followed by formulation and processing. These steps enable plant-derived proteins to mimic the texture, flavor, and sensory attributes of animal meat. Technologies such as high moisture extrusion, shear cell technology, 3D printing, and mycelium cultivation have enhanced the organoleptic and structural properties of PBMs. Moreover, innovative methods like recombinant protein additives and fermentation techniques further improve the nutritional quality, digestibility, and appearance of plant-based foods. Through these technological advancements, PBMs can effectively simulate meat in taste, texture, and nutrient profile, thus providing a realistic and satisfying alternative for consumers.

The variety of plant-based foods available demonstrates the adaptability of this approach. TVP, derived from soybeans, serves as one of the earliest and most widely used meat substitutes, providing over 50% protein in its dry form. Quorn, a mycoprotein-based food from *Fusarium venenatum*, is rich in fiber, low in saturated fats, and free from cholesterol. Tempeh, a traditional Indonesian fermented product, offers high-quality protein, probiotics, and bioactive compounds, while tofu, produced from soy milk coagulation, provides a soft-textured, calcium-rich protein source. Similarly, kinema, a fermented soybean product indigenous to the Himalayas, exemplifies traditional biotechnology with its high protein and unique microbial properties. Seitan, derived from wheat gluten, offers exceptional texture and protein concentration, making it a staple in many modern vegan diets. Collectively, these foods represent a diverse and nutritious range of meat alternatives rooted in both cultural tradition and scientific innovation.

The health benefits of plant-based foods are extensive and well documented. These foods are naturally low in cholesterol and saturated fat and rich in dietary fiber, antioxidants, vitamins, and minerals, making them effective in reducing the risk of cardiovascular diseases, obesity, type 2 diabetes, and certain cancers. Fermented products like tempeh and kinema promote gut health and immune function through probiotic activity, while soy-based foods contribute isoflavones that support hormonal balance and bone health. The consumption of mycoprotein (Quorn) and gluten-based protein (seitan) further aids in satiety regulation, muscle maintenance, and metabolic balance. Beyond human health, the adoption of plant-based foods contributes to environmental sustainability by lowering greenhouse gas emissions, conserving land and water resources, and reducing reliance on animal agriculture.

In conclusion, the integration of plant-based foods into global diets is not merely a dietary trend but a scientific and ethical evolution toward a healthier and more sustainable food future. The combination of technological innovation, nutritional value, and environmental responsibility positions plant-based proteins as the foundation of twenty-first century nutrition. As consumer awareness continues to rise and production technologies advance, plant-based foods will play a pivotal role in addressing the intertwined challenges of public health, food security, and environmental conservation. Thus, embracing these foods represents both a personal and planetary investment in a balanced, resilient, and sustainable future.

References

Aaslyng, M. D., & Højer, R. (2021). Introducing tempeh as a new plant-based protein food item on the danish market. *Food, 10*(11), 2865.

Ahnan-Winarno, A. D., Cordeiro, L., Winarno, F. G., Gibbons, J., & Xiao, H. (2021). Tempeh: A semicentennial review on its health benefits, fermentation, safety, processing, sustainability, and affordability. *Comprehensive Reviews in Food Science and Food Safety, 20*(2), 1717–1767.

Akhtar, S., Ismail, T., Layla, A., Hussain, M., & Qamar, M. (2022). An overview of plant-based protein rich products. In *Plant Protein Foods* (pp. 27–60).

Akinsemolu, A. A., & Onyeaka, H. N. (2025). Mycoproteins as sustainable food sources: Current applications and future prospects. *Discover Applied Sciences, 7*(3), 191.

Alcorta, A., Porta, A., Tárrega, A., Alvarez, M. D., & Vaquero, M. P. (2021). Foods for plant-based diets: Challenges and innovations. *Food, 10*(2), 293.

Aunon-Lopez, A., Strauss, M., Hinterreiter-Kern, E., Klein, A., Varga, E., & Pignitter, M. (2025). Influence of processing of seitan, tempeh, and firm regular tofu on protein and lipid oxidation and Maillard reaction products formation. *Food Chemistry, 467*, 142273.

Aydar, A. Y. (2023). An overview of plant-based food alternatives (PBFAs): Classification, textural and sensory characteristics. In *Plant-Based Foods: Ingredients, Technology and Health Aspects* (pp. 1–17).

Bakhsh, A., Lee, E. Y., Ncho, C. M., Kim, C. J., Son, Y. M., Hwang, Y. H., & Joo, S. T. (2022). Quality characteristics of meat analogs through the incorporation of textured vegetable protein: A systematic review. *Food, 11*(9), 1242.

Bassin, S. R., De Carvalho, J. F., & Gulati, M. (2024). A review of plant-based diets for obesity management. *Endocrine Practice, 30*(8), 779–789.

Carreno, I., & Dolle, T. (2018). Tofu steaks? Developments on the naming and marketing of plant-based foods in the aftermath of the TofuTown judgement. *European Journal of Risk Regulation, 9*(3), 575–584.

Chen, C. C., Hsieh, J. F., & Kuo, M. I. (2023). Insight into the processing, gelation and functional components of tofu: A review. *Processes, 11*(1), 202.

Choi, H. W., Lee, Y. Y., Ryoo, C., Yoon, H. I., Hahn, J., & Choi, Y. J. (2022). Influence of a post-processing heat treatment method on the textural properties of textured vegetable protein. *Journal of Food Science, 87*(12), 5340–5348.

Cornet, S. H., Snel, S. J., Schreuders, F. K., van der Sman, R. G., Beyrer, M., & van der Goot, A. J. (2022). Thermo-mechanical processing of plant proteins using shear cell and high-moisture extrusion cooking. *Critical Reviews in Food Science and Nutrition, 62*(12), 3264–3280.

Craig, W. J., & Fresán, U. (2021). International analysis of the nutritional content and a review of health benefits of non-dairy plant-based beverages. *Nutrients, 13*(3), 842.

Das, S., Kathiriya, M. R., Gawai, K., & Hati, S. (2023). Fermented plant-based foods (eg, tofu, sauerkraut, sourdough). In *Engineering plant-based food systems* (pp. 233–251). Academic Press.

DeClercq, V., Nearing, J. T., & Sweeney, E. (2022). Plant-based diets and cancer risk: What is the evidence? *Current Nutrition Reports, 11*(2), 354–369.

Di Sotto, A., Vitalone, A., & Di Giacomo, S. (2020). Plant-derived nutraceuticals and immune system modulation: An evidence-based overview. *Vaccine, 8*(3), 468.

Dupuis, J. H., Cheung, L. K., Newman, L., Dee, D. R., & Yada, R. Y. (2023). Precision cellular agriculture: The future role of recombinantly expressed protein as food. *Comprehensive Reviews in Food Science and Food Safety, 22*(2), 882–912.

Finnigan, T. J. A. (2011). Mycoprotein: Origins, production and properties. In *Handbook of food proteins* (pp. 335–352).

Ghadiri, M., Cheshmazar, E., Shateri, Z., Gerami, S., Nouri, M., & Gargari, B. P. (2023). Healthy plant-based diet index as a determinant of bone mineral density in osteoporotic postmenopausal women: A case-control study. *Frontiers in Nutrition, 9*, 1083685.

Górska, K., Pejcz, E., & Harasym, J. (2025). Tempeh and fermentation—Innovative substrates in a classical microbial process. *Applied Sciences, 15*(16), 8888.

Guan, X., Zhong, X., Lu, Y., Du, X., Jia, R., Li, H., & Zhang, M. (2021). Changes of soybean protein during tofu processing. *Foods, 10*(7), 1594.

Guerrero, M., Stone, A. K., Singh, R., Lui, Y. C., Koksel, F., & Nickerson, M. T. (2025). Effect of extrusion conditions on the characteristics of texturized vegetable protein from a Faba bean protein mix and its application in vegan and hybrid burgers. *Food, 14*(4), 547.

Hashempour-Baltork, F., Khosravi-Darani, K., Hosseini, H., Farshi, P., & Reihani, S. F. S. (2020). Mycoproteins as safe meat substitutes. *Journal of Cleaner Production, 253*, 119958.

Hemler, E. C., & Hu, F. B. (2019). Plant-based diets for cardiovascular disease prevention: All plant foods are not created equal. *Current Atherosclerosis Reports, 21*(5), 18.

Hong, S., Shen, Y., & Li, Y. (2022). Physicochemical and functional properties of texturized vegetable proteins and cooked patty textures: Comprehensive characterization and correlation analysis. *Food, 11*(17), 2619.

Hoxha, L., & Taherzadeh, M. J. (2025). Alternative proteins from filamentous fungi: Drivers of transformative change in future food systems. *Fermentation, 12*(1), 7.

Hu, F. B. (2003). Plant-based foods and prevention of cardiovascular disease: An overview. *The American Journal of Clinical Nutrition, 78*(3), 544S–551S.

Huling, R. (2023). Applications of fungi for alternative protein. In *Fungi and fungal products in human welfare and biotechnology* (pp. 365–395). Springer Nature Singapore.

Ivanova, S., Delattre, C., Karcheva-Bahchevanska, D., Benbasat, N., Nalbantova, V., & Ivanov, K. (2021). Plant-based diet as a strategy for weight control. *Food, 10*(12), 3052.

Jiang, L., Zhang, G., Li, Y., Shi, G., & Li, M. (2021). Potential application of plant-based functional foods in the development of immune boosters. *Frontiers in Pharmacology, 12*, 637782.

Johannesen, C. O., Dale, H. F., Jensen, C., & Lied, G. A. (2020). Effects of plant-based diets on outcomes related to glucose metabolism: A systematic review. *Diabetes, Metabolic Syndrome and Obesity, 13*, 2811–2822.

Kelly, L., Kelly, H., & Park, J. (2024). *The healthy bones plant-based nutrition plan and cookbook: How to prepare and combine plant foods to maintain optimal bone density and prevent osteoporosis*. Chelsea Green Publishing.

Ketnawa, S., Reginio, F. C., Jr., Thuengtung, S., & Ogawa, Y. (2022). Changes in bioactive compounds and antioxidant activity of plant-based foods by gastrointestinal digestion: A review. *Critical Reviews in Food Science and Nutrition, 62*(17), 4684–4705.

Kharnaior, P., & Tamang, J. P. (2022). Metagenomic-metabolomic mining of kinema, a naturally fermented soybean food of the Eastern Himalayas. *Frontiers in Microbiology, 13*, 868383.

Langyan, S., Yadava, P., Khan, F. N., Dar, Z. A., Singh, R., & Kumar, A. (2022). Sustaining protein nutrition through plant-based foods. *Frontiers in Nutrition, 8*, 772573.

Leonard, W., Zhang, P., Ying, D., & Fang, Z. (2020). Application of extrusion technology in plant food processing byproducts: An overview. *Comprehensive Reviews in Food Science and Food Safety, 19*(1), 218–246.

Linder, T. (2024). What next for mycoprotein? *Current Opinion in Food Science, 58*, 101199.

Lonchamp, J., Stewart, K., Munialo, C. D., Evans, L., Akintoye, M., Gordon, S., et al. (2022). Mycoprotein as novel functional ingredient: Mapping of functionality, composition and structure throughout the Quorn fermentation process. *Food Chemistry, 396*, 133736.

London, C. (2010). Functional foods that boost the immune system. In *Functional food product development* (pp. 293–321).

Maningat, C. C., Jeradechachai, T., & Buttshaw, M. R. (2022). Textured wheat and pea proteins for meat alternative applications. *Cereal Chemistry, 99*(1), 37–66.

Mariotti, F. (2025). Nutritional and health benefits and risks of plant-based substitute foods. *Proceedings of the Nutrition Society, 84*(1), 110–123.

McClements, D. J., & Grossmann, L. (2021a). The science of plant-based foods: Constructing next-generation meat, fish, milk, and egg analogs. *Comprehensive Reviews in Food Science and Food Safety, 20*(4), 4049–4100.

McClements, D. J., & Grossmann, L. (2021b). A brief review of the science behind the design of healthy and sustainable plant-based foods. *npj Science of Food, 5*(1), 17.

McClements, D. J., & Grossmann, L. (2022). Next-generation plant-based foods. *Next-Generation Plant-Based Foods, 57*(177), 198–207.

Mokhtari, E., Mirzaei, S., Asadi, A., Akhlaghi, M., & Saneei, P. (2022). Association between plant-based diets and metabolic health status in adolescents with overweight and obesity. *Scientific Reports, 12*(1), 13772.

Mulla, M. Z., Subramanian, P., & Dar, B. N. (2022). Functionalization of legume proteins using high pressure processing: Effect on technofunctional properties and digestibility of legume proteins. *LWT, 158*, 113106.

Muneer, N., Ahmad, S., Sibt-e-Abbas, M., Shehzad, M. A., Usman, M., Yaqoob, S., et al. (2025). Evaluation of nutritional qualities of Seitan meat analogue produce from wheat, barley and peas composite flour. *European Food Research and Technology, 251*(10), 3327–3337.

Munekata, P. E., Domínguez, R., Budaraju, S., Roselló-Soto, E., Barba, F. J., Mallikarjunan, K., et al. (2020). Effect of innovative food processing technologies on the physicochemical and nutritional properties and quality of non-dairy plant-based beverages. *Food, 9*(3), 288.

Munialo, C. D., & Andrei, M. (2023). General health benefits and sensory perception of plant-based foods. In *Engineering plant-based food systems* (pp. 13–26). Academic Press.

Nolden, A. A., & Forde, C. G. (2023). The nutritional quality of plant-based foods. *Sustainability, 15*(4), 3324.

Nout, M. J. R., Bakshi, D., & Sarkar, P. K. (1998). Microbiological safety of kinema, a fermented soya bean food. *Food Control, 9*(6), 357–362.

Ogutu, F. O., Okiko, G., Wanjala, G., Luvitaa, S., Obong'o, B. O., Vriesekoop, F., & Munialo, C. D. (2024). Unlocking the potential of plant-based foods in sub-Saharan Africa: A review of the opportunities and challenges. *International Journal of Food Science and Technology, 59*(8), 5326–5342.

Öncü Öner, T. (2023). Health effects of plant-based foods and their components. In *Plant-based foods: Ingredients, technology and health aspects* (pp. 137–178). Springer International Publishing.

Park, K. Y., & Wi, S. J. (2016). Potential of plants to produce recombinant protein products. *Journal Of Plant Biology, 59*(6), 559–568.

Prakash, S., Bhandari, B., & Gaiani, C. (Eds.). (2022). *Engineering plant-based food systems*. Academic Press.

Qin, P., Wang, T., & Luo, Y. (2022). A review on plant-based proteins from soybean: Health benefits and soy product development. *Journal of Agriculture and Food Research, 7*, 100265.

Rahman, M. A., Bulbul, M. R. H., & Kabir, Y. (2020). Plant-based products in cancer prevention and treatment. In *Functional foods in cancer prevention and therapy* (pp. 237–259). Academic Press.

Riaz, M. N. (2011). Texturized vegetable proteins. In *Handbook of food proteins* (pp. 395–418). Woodhead Publishing.

Ribas-Agustí, A., Martín-Belloso, O., Soliva-Fortuny, R., & Elez-Martínez, P. (2018). Food processing strategies to enhance phenolic compounds bioaccessibility and bioavailability in plant-based foods. *Critical Reviews in Food Science and Nutrition, 58*(15), 2531–2548.

Rousseau, S., Kyomugasho, C., Celus, M., Hendrickx, M. E., & Grauwet, T. (2020). Barriers impairing mineral bioaccessibility and bioavailability in plant-based foods and the perspectives for food processing. *Critical Reviews in Food Science and Nutrition, 60*(5), 826–843.

Saeed, F., Afzaal, M., Khalid, A., Shah, Y. A., Ateeq, H., Islam, F., et al. (2023). Role of mycoprotein as a non-meat protein in food security and sustainability: A review. *International Journal of Food Properties, 26*(1), 683–695.

Salehin, S., Rasmussen, P., Mai, S., Mushtaq, M., Agarwal, M., Hasan, S. M., et al. (2023). Plant based diet and its effect on cardiovascular disease. *International Journal of Environmental Research and Public Health, 20*(4), 3337.

Samtiya, M., Aluko, R. E., Dhewa, T., & Moreno-Rojas, J. M. (2021). Potential health benefits of plant food-derived bioactive components: An overview. *Food, 10*(4), 839.

Shahid, M., Shah, P., Mach, K., Rodgers-Hunt, B., Finnigan, T., Frost, G., et al. (2024). The environmental impact of mycoprotein-based meat alternatives compared to plant-based meat alternatives: A systematic review. *Future Foods, 10*, 100410.

Sim, S. Y. J., Srv, A., Chiang, J. H., & Henry, C. J. (2021). Plant proteins for future foods: A roadmap. *Food, 10*(8), 1967.

Souza Filho, P. F., Nair, R. B., Andersson, D., Lennartsson, P. R., & Taherzadeh, M. J. (2018). Vegan-mycoprotein concentrate from pea-processing industry byproduct using edible filamentous fungi. *Fungal Biology and Biotechnology, 5*(1), 5.

Sridhar, K., Bouhallab, S., Croguennec, T., Renard, D., & Lechevalier, V. (2023). Recent trends in design of healthier plant-based alternatives: Nutritional profile, gastrointestinal digestion, and consumer perception. *Critical Reviews in Food Science and Nutrition, 63*(30), 10483–10498.

Subali, D., Christos, R. E., Givianty, V. T., Ranti, A. V., Kartawidjajaputra, F., Antono, L., et al. (2023). Soy-based tempeh rich in paraprobiotics properties as functional sports food: More than a protein source. *Nutrients, 15*(11), 2599.

Švarc, P. L., Jensen, M. B., Langwagen, M., Poulsen, A., Trolle, E., & Jakobsen, J. (2022). Nutrient content in plant-based protein products intended for food composition databases. *Journal of Food Composition and Analysis, 106*, 104332.

Tamang, J. P. (2012). Plant-based fermented foods and beverages of Asia. In *Handbook of plant-based fermented food and beverage technology* (pp. 49–90).

Tamang, J. P. (2015). Naturally fermented ethnic soybean foods of India. *Journal of Ethnic Foods, 2*(1), 8–17.

Tamang, J. P. (2024). Unveiling kinema: Blending tradition and science in the Himalayan fermented soya delicacy. *Journal of Ethnic Foods, 11*(1), 29.

Tamang, J. P., & Nikkuni, S. (1996). Selection of starter cultures for the production of kinema, a fermented soybean food of the Himalaya. *World Journal of Microbiology and Biotechnology, 12*(6), 629–635.

Thwe, S. Y., Au, D. J. E., Tan, K. L., Neo, Y. P., & Chew, L. Y. (2025). Development of plant-based burger patties using vegan Staples like tofu and tempeh: An exploration study among the young adults. *Journal of Culinary Science & Technology, 23*(4), 904–918.

Usman, M., & Xu, M. (2024). Plant-based proteins: Plant source, extraction, food applications, and challenges. In *Flavor-Associated Applications in Health and Wellness Food Products* (pp. 253–294).

Viroli, G., Kalmpourtzidou, A., & Cena, H. (2023). Exploring benefits and barriers of plant-based diets: Health, environmental impact, food accessibility and acceptability. *Nutrients, 15*(22), 4723.

Wang, M., Li, D., Zang, Z., Sun, X., Tan, H., Si, X., et al. (2022). 3D food printing: Applications of plant-based materials in extrusion-based food printing. *Critical Reviews in Food Science and Nutrition, 62*(26), 7184–7198.

Watzl, B. (2008). Anti-inflammatory effects of plant-based foods and of their constituents. *International Journal for Vitamin and Nutrition Research, 78*(6), 293–298.

Webb, D., Dogan, H., Li, Y., & Alavi, S. (2023). Physico-chemical properties and texturization of pea, wheat and soy proteins using extrusion and their application in plant-based meat. *Food, 12*(8), 1586.

Whittaker, J. A., Johnson, R. I., Finnigan, T. J., Avery, S. V., & Dyer, P. S. (2020). The biotechnology of quorn mycoprotein: Past, present and future challenges. In *Grand challenges in fungal biotechnology* (pp. 59–79). Springer International Publishing.

Wilson, J. W., Thompson, T. W., Wei, Y., Chaparro, J. M., Stull, V. J., Nair, M. N., & Weir, T. L. (2024). Comparison of the in vitro iron bioavailability of tempeh made with Tenebrio molitor to beef and plant-based meat alternatives. *Nutrients, 16*(16), 2756.

Yan, X., Zeng, Z., McClements, D. J., Gong, X., Yu, P., Xia, J., & Gong, D. (2023). A review of the structure, function, and application of plant-based protein–phenolic conjugates and complexes. *Comprehensive Reviews in Food Science and Food Safety, 22*(2), 1312–1336.

Yano, H., & Fu, W. (2022). Effective use of plant proteins for the development of "new" foods. *Foods, 11(9)*, 1185.

Yemets, A. I., Tanasienko, I. V., Krasylenko, Y. A., & Blume, Y. B. (2014). Plant-based biopharming of recombinant human lactoferrin. *Cell Biology International, 38*(9), 989–1002.

Yusuf, E. H. (2023). Comparison of life cycle assessments and nutritional contents of soy protein and wheat protein (seitan) based vegan bacon products for human and environmental health. *Journal of the Science of Food and Agriculture, 103*(7), 3315–3321.

Zhang, J., Liu, L., Liu, H., Yoon, A., Rizvi, S. S., & Wang, Q. (2019). Changes in conformation and quality of vegetable protein during texturization process by extrusion. *Critical Reviews in Food Science and Nutrition, 59*(20), 3267–3280.

Zhao, D., Huang, L., Li, H., Ren, Y., Cao, J., Zhang, T., & Liu, X. (2022). Ingredients and process affect the structural quality of recombinant plant-based meat alternatives and their components. *Food, 11*(15), 2202.

Zheng, L., Regenstein, J. M., Teng, F., & Li, Y. (2020). Tofu products: A review of their raw materials, processing conditions, and packaging. *Comprehensive Reviews in Food Science and Food Safety, 19*(6), 3683–3714.

Chapter 3
Cell-Based Foods

Abbreviations

Akt	Protein kinase B
EGF	Epidermal growth factor
ERK	Extracellular signal-regulated kinase
FBS	Fetal bovine serum
FGF	Fibroblast growth factor
GRAS	Generally Recognized as Safe
HACCP	Hazard Analysis and Critical Control Point
IGF	Insulin-like growth factor
MAPK	Mitogen-activated protein kinase
mTOR	Mechanistic (mammalian) target of rapamycin
pH	Potential of Hydrogen
PI3K	Phosphoinositide 3-kinase
STRs	Stirred tank bioreactors
TGF-β	Transforming growth factor-beta
WHO	World Health Organization
3D	Three-dimensional

3.1 Introduction

Cell-based foods, also known as cell-cultivated or cultured foods, represent a novel and rapidly advancing approach to food production based on cellular agriculture (WHO, 2023). In this system, animal cells are isolated and grown in controlled in vitro environments using nutrient-rich culture media, allowing the production of meat, seafood, and other animal-derived foods without conventional animal farming or slaughter (Datar & Betti, 2010). By replicating the biological processes

Y. Kumar, S. Dey, *Emerging Technology in Food Processing*,
https://doi.org/10.1007/978-3-032-24505-2_3

responsible for tissue formation, cell-based foods aim to closely mimic the sensory, nutritional, and functional properties of traditional animal products while offering greater control over composition and safety (Broad, 2019).

The growing interest in cell-based foods is driven by increasing concerns regarding the environmental footprint of conventional livestock production, animal welfare, and global food security (WHO, 2023). Traditional animal agriculture is a major contributor to greenhouse gas emissions, land degradation, water consumption, and biodiversity loss (Broad, 2019). In contrast, cell-based food production has the potential to significantly reduce land and water use and minimize environmental impacts, although energy efficiency and scalability remain active areas of research. Furthermore, the controlled production environment reduces risks associated with zoonotic diseases, antibiotic resistance, and microbial contamination (Rubio et al., 2020).

Recent scientific advancements have focused on the development of serum-free and animal component-free culture media, scalable bioreactor systems, three-dimensional tissue structuring, and cost-effective bioprocessing strategies (Barzee et al., 2022). Alongside technological progress, regulatory frameworks for cell-based foods are evolving, with countries such as Singapore and the United States establishing safety assessment and approval pathways. As these innovations mature, cell-based foods are increasingly recognized as a promising component of sustainable and resilient future food systems (WHO, 2023).

3.2 Type of Cell-Based Foods

Cell-based foods constitute a rapidly evolving category within cellular agriculture, encompassing a diverse range of food products produced through the in vitro cultivation of cells rather than traditional agricultural practices (Rubio et al., 2020). By leveraging advances in cell biology, tissue engineering, and bioprocessing, cell-based food systems aim to reproduce the structural, nutritional, and sensory attributes of conventional animal-derived foods while addressing sustainability, ethical, and food security challenges (Rubio et al., 2020). Based on the source of cells, intended food product, and degree of structural complexity, cell-based foods can be broadly classified into several major categories, including cell-based meat, seafood, dairy, eggs, and emerging hybrid or multicellular products. Each category presents distinct technological requirements, benefits, and challenges (Barzee et al., 2022). Table 3.1 provides a comprehensive overview of the major categories of cell-based foods and their defining characteristics.

Table 3.1 Classification of cell-based foods based on source, structure, cellular composition, and processing approach

Classification basis	Category	Description	Representative examples
Based on biological source	Cell-based meat	Foods produced from cultured animal muscle and fat cells obtained from livestock species	Cultured beef, chicken, pork
	Cell-based seafood	Products derived from cultured cells of aquatic organisms	Cultured fish fillets, shrimp, shellfish
	Cell-based dairy	Milk components synthesized using animal mammary cells or precision cell cultures	Cultured milk proteins, fats, cheese analogues
	Cell-based eggs	Egg proteins produced from cultured avian cells without whole egg formation	Cultured egg white proteins
Based on product structure	Unstructured products	Products composed of dispersed or aggregated cells without defined tissue architecture	Minced meat, nuggets, patties
	Semi-structured products	Products with partial alignment and organization of cells	Sausages, hybrid meat products
	Structured/whole cut products	Highly organized tissues mimicking natural muscle structure	Steaks, cutlets, fillets
Based on cellular composition	Muscle based	Products primarily composed of myocytes or muscle fibers	Lean cultured meat
	Fat enhanced	Products enriched with cultured adipocytes to improve flavor and mouthfeel	Marbled cultured meat
	Co-cultured systems	Products developed using multiple cell types including muscle, fat, and connective cells	Whole cut cultured meat
Based on processing approach	Scaffold based	Cells grown on edible scaffolds that provide structural support and guide tissue formation	Structured cultured meat using collagen or alginate scaffolds
	Scaffold free	Cells self-aggregate without external support, suitable for unstructured products	Cell pellets, aggregates
	Hybrid products	Combination of cultured cells with plant-based matrices or ingredients	Cell-based plant blended foods

3.2.1 Cell-Based Meat

Cell-based meat, often referred to as cultivated or cultured meat, is the most extensively researched and commercially visible category of cell-based foods (Ong et al., 2020). It is produced by isolating animal muscle stem cells (such as satellite cells, myoblasts, or induced pluripotent stem cells) and expanding them in a controlled environment using nutrient-rich culture media. These cells proliferate and differentiate into muscle fibers, fat cells, and connective tissue, which together form meat-like structures (Post et al., 2020).

Cell-based meat products can be further categorized based on their structural complexity. First-generation products primarily consist of unstructured or minimally structured tissues, such as minced meat, nuggets, burgers, and sausages (Zhang et al., 2020). These products are comparatively easier to manufacture, as they do not require highly organized tissue architecture. In contrast, whole cut products such as steaks or chops require advanced scaffolding materials, vascularization strategies, and three-dimensional tissue engineering to replicate the fibrous texture and mouthfeel of conventional meat (Kadim et al., 2015).

From a sustainability perspective, cell-based meat has the potential to significantly reduce land use, water consumption, and greenhouse gas emissions compared to conventional livestock production (Zhang et al., 2020). Additionally, production under sterile and controlled conditions minimizes the risks of zoonotic diseases, antibiotic residues, and microbial contamination (Azhar et al., 2023). However, challenges remain related to cost reduction, scale-up, energy demand, and consumer acceptance.

3.2.2 Cell-Based Seafood

Cell-based seafood represents another important category of cell-based foods, addressing growing concerns over overfishing, marine ecosystem degradation, and contamination of seafood with heavy metals and microplastics (Ovissipour et al., 2024). Similar to cell-based meat, this approach involves the cultivation of fish or shellfish muscle cells, fat cells, and connective tissue in vitro to produce seafood products such as fish fillets, shrimp, crab, and tuna (Rubio et al., 2020).

Cell-based seafood production offers unique advantages over terrestrial cell-based meat. Fish muscle cells often exhibit faster proliferation rates and lower culture temperature requirements, which may reduce energy inputs during production (Ovissipour et al., 2024). Moreover, cell-based seafood can be engineered to contain optimized fatty acid profiles, such as enhanced omega-3 polyunsaturated fatty acids, improving the nutritional value of the final product (Datar & Betti, 2010).

Currently, most cell-based seafood products under development are unstructured or semi-structured, including fish cakes, minced fish, and seafood analogues (Ovissipour et al., 2024). The development of structured products like fillets remains technically complex due to challenges in recreating muscle fiber alignment and connective tissue composition. Nevertheless, with rising consumer demand for sustainable seafood and increasing pressure on wild fish stocks, cell-based seafood is expected to play a crucial role in future protein systems (Ovissipour et al., 2024).

3.2.3 Cell-Based Dairy Products

Cell-based dairy products differ fundamentally from cell-based meat and seafood, as they do not involve the cultivation of animal muscle cells (Mendly-Zambo et al., 2021). Instead, they rely on precision fermentation and cell culture techniques to produce milk components such as casein, whey proteins, and milk fats. Mammalian cells or genetically engineered microorganisms are used to synthesize dairy identical proteins, which are then combined with plant-based fats, sugars, and minerals to recreate milk, cheese, yogurt, and ice cream (Stephens et al., 2018).

Cell-based dairy offers several advantages over conventional dairy farming, including reduced methane emissions, lower land and water requirements, and elimination of animal welfare concerns. Additionally, these products are free from lactose, antibiotics, and hormonal residues, making them suitable for lactose-intolerant consumers and those seeking cleaner label foods (Hadi & Brightwell, 2021).

Despite these benefits, replicating the complex microstructure, flavor development, and functional properties of traditional dairy products remains challenging. Processes such as cheese ripening and flavor formation depend on intricate biochemical interactions that are difficult to fully reproduce in vitro (Mendly-Zambo et al., 2021). Nevertheless, continued advances in fermentation technology and food formulation are rapidly narrowing the gap between conventional and cell-based dairy products (Geciova et al., 2002).

3.2.4 Cell-Based Eggs

Cell-based eggs represent an emerging category aimed at producing egg components without conventional poultry farming. Using avian cell cultures or microbial fermentation systems, egg proteins such as ovalbumin, ovotransferrin, and ovomucoid can be synthesized and assembled into egg-like products (Dong & Zhang, 2021).

These products are particularly attractive for food processing applications, as egg proteins play critical functional roles in emulsification, foaming, binding, and texture development (Humbird, 2021). Cell-based egg ingredients can be tailored for specific functionalities, offering greater consistency and performance compared to conventional eggs, which may vary due to feed, season, or bird health (Dong & Zhang, 2021).

In addition to sustainability benefits, cell-based eggs reduce the risk of foodborne illnesses such as *Salmonella* contamination and eliminate ethical concerns associated with battery cage systems. However, scaling up production and achieving cost parity with conventional eggs remain significant challenges (Dong & Zhang, 2021).

3.2.5 Cell-Based Fats and Ingredients

Beyond whole food products, cell-based technologies are increasingly applied to the production of specific food ingredients, particularly fats (Schüler et al., 2024). Cultivated animal fat plays a crucial role in enhancing the flavor, juiciness, and mouthfeel of cell-based meat and hybrid food products. Adipocytes grown in vitro can be blended with plant-based proteins or cultivated muscle cells to improve sensory quality (Zhang et al., 2020).

Cell-based fats can also be engineered to possess customized lipid profiles, such as reduced saturated fat content or enhanced levels of beneficial fatty acids. This flexibility offers opportunities to design healthier food products without compromising sensory appeal. In addition, cell-based ingredients such as collagen, gelatin, and bioactive peptides are being explored for applications in food, nutraceuticals, and functional foods (Schüler et al., 2024).

3.2.6 Hybrid Cell-Based Foods

Hybrid cell-based foods combine cultivated cells with plant-based or microbial ingredients to create products that balance functionality, sustainability, and cost-effectiveness. In these systems, cell-based components provide key sensory attributes such as flavor and texture, while plant proteins contribute bulk and structure (Bryant & Barnett, 2018).

Hybrid approaches are particularly attractive in the early stages of commercialization, as they reduce reliance on expensive cell culture inputs while maintaining consumer appeal (Bhat et al., 2015). Examples include plant-based burgers enhanced with cultivated fat or protein fractions. Such products also serve as a transitional step toward fully cell-based foods, helping to familiarize consumers with cellular agriculture technologies (Allan et al., 2019).

3.2.7 Emerging and Future Cell-Based Food Categories

As cellular agriculture technologies continue to evolve, new categories of cell-based foods are expected to emerge (Ben-Arye & Levenberg, 2019). These include cell-based leather and gelatin for food contact materials, personalized nutrition products tailored to individual dietary needs, and functional foods enriched with bioactive compounds. Advances in stem cell biology, synthetic biology, and bioreactor design are likely to expand the diversity and complexity of cell-based food products (Ben-Arye & Levenberg, 2019).

Regulatory frameworks are also evolving to accommodate these innovations. Countries such as Singapore and the United States have established approval pathways for cell-based foods, setting precedents for safety assessment, labelling, and commercialization. As regulatory clarity improves, broader market adoption of diverse cell-based food categories is anticipated (Allan et al., 2019). Table 3.2 summarizes the current regulatory approaches, key authorities, and approval requirements governing cell-based foods in major jurisdictions.

3.3 Processing of Cell-Based Foods

Cell-based foods, also referred to as cultured, cultivated, or lab-grown foods, represent a transformative innovation in the food processing and biotechnology sectors (Ben-Arye & Levenberg, 2019). These foods are produced by cultivating animal or plant cells in controlled environments rather than relying on conventional agriculture or animal husbandry (Allan et al., 2019). The processing of cell-based foods integrates principles of cell biology, bioprocess engineering, food science, and materials science to convert living cells into safe, nutritious, and consumer-acceptable food products. As global food systems face mounting challenges related to population growth, climate change, resource scarcity, and ethical concerns, cell-based food processing has emerged as a promising alternative to conventional food production systems (Allan et al., 2019).

Unlike traditional food processing, which primarily focuses on post-harvest or post-slaughter handling of raw materials, cell-based food processing begins at the cellular level (Barzee et al., 2022). It involves cell sourcing, cell proliferation, differentiation, harvesting, and downstream processing to create structured or unstructured food products such as cultured meat, seafood, dairy analogues, and functional ingredients. Each stage requires precise control of environmental parameters to ensure cell viability, functionality, safety, and quality (Allan et al., 2019). The overall workflow of these stages is illustrated in Fig. 3.1 providing a visual overview of this interdisciplinary and technologically sophisticated process.

3.3.1 Cell Sourcing and Selection

The processing of cell-based foods begins with the selection and isolation of appropriate cell types. For animal-based products, stem cells, satellite cells, or progenitor cells are typically harvested through minimally invasive biopsies. These cells possess the capacity for self-renewal and differentiation into specific tissues such as muscle, fat, or connective tissue (European Food Safety Authority (EFSA) et al. 2024). In the case of plant-based cell culture, meristematic or undifferentiated plant cells are commonly used due to their high proliferative capacity. Cell selection is a critical step because the growth characteristics, metabolic efficiency, and differentiation potential of the cells directly influence processing efficiency and final product quality (WHO, 2023). Cells must be robust, genetically stable, and

Table 3.2 Comparative analysis of cell-based foods, conventional animal-based foods, and plant-based alternatives

Parameter	Cell-based foods	Conventional animal-based foods	Plant-based foods
Biological source	Cultured animal cells (muscle, fat, connective tissues)	Whole animals raised through livestock farming	Plant proteins, oils, and carbohydrates
Production system	Controlled in vitro cultivation using bioreactors	Animal rearing, slaughtering, and processing	Crop cultivation, extraction, and food processing
Land use	Significantly reduced land requirement	High land use for grazing and feed crops	Moderate land use depending on crop type
Water consumption	Lower water demand compared to livestock systems	High water usage for animal feed and husbandry	Generally lower than animal-based foods
Greenhouse gas emissions	Potentially lower emissions with renewable energy integration	High emissions (methane, nitrous oxide)	Low to moderate emissions
Antibiotic use	No routine antibiotics required	Widely used for disease prevention and growth promotion	Not applicable
Pathogen risk	Reduced risk of zoonotic pathogens	Risk of contamination from animals (salmonella, E. Coli)	Minimal zoonotic risk
Nutritional customization	Nutrient profile can be tailored (fat type, micronutrients)	Fixed nutritional composition	Can be fortified but lacks some animal-specific nutrients
Protein quality	Complete proteins identical to animal meat	Complete proteins with high bioavailability	Often incomplete; requires protein blending
Fat composition	Adjustable saturated and unsaturated fat levels	Typically high in saturated fat	Primarily unsaturated fats
Texture and sensory attributes	Designed to closely mimic conventional meat	Natural meat texture and flavor	Requires extensive formulation to mimic meat
Ethical considerations	Minimal animal slaughter	Involves animal slaughter	Animal free
Scalability challenges	High production cost and technological complexity	Established and scalable	Highly scalable and cost-effective
Food safety control	High process control and traceability	Variable due to biological systems	High control during processing
Consumer acceptance	Emerging; influenced by cost and perception	Widely accepted	Increasing acceptance among flexitarians and vegans
Regulatory status	Emerging regulatory frameworks	Well-established regulations	Fully regulated and widely marketed

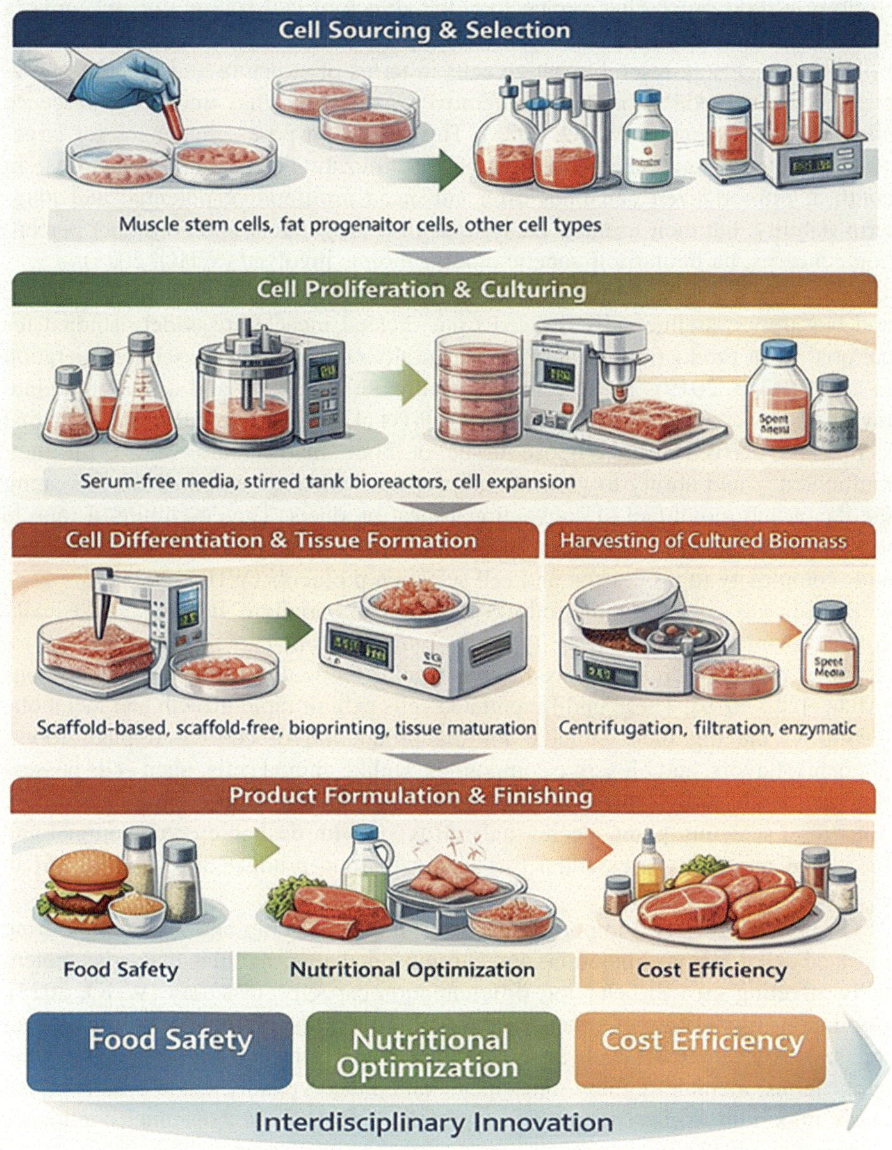

Fig. 3.1 Stages in the processing of cell-based foods, from cell selection and proliferation to differentiation, tissue formation, and harvesting

capable of sustained proliferation under in vitro conditions. Following isolation, cells are characterized and banked to create master and working cell banks, ensuring consistency and traceability throughout the production process.

From a food processing perspective, the choice of cell source directly impacts scalability, cost, safety, and consumer acceptance. Primary cells derived directly from tissues closely resemble native cells in terms of structure and function; however, they often exhibit limited proliferative capacity and may undergo senescence after repeated passages (WHO, 2023). This limitation poses challenges for large-scale production and necessitates careful optimization of culture conditions. In contrast, immortalized cell lines offer enhanced proliferative potential and long-term stability, but their use in food systems raises regulatory and consumer perception concerns, particularly if genetic modification is involved (WHO, 2023).

Animal cell sourcing strategies prioritize minimal animal harm, reproducibility, and biosafety. Satellite cells isolated from skeletal muscle are widely studied for cultured meat production because they naturally contribute to muscle regeneration in vivo (Broad, 2019). These cells can proliferate efficiently and differentiate into myotubes under appropriate conditions (Santo et al., 2020). Similarly, mesenchymal stem cells derived from adipose tissue or bone marrow are valued for their multipotency and ability to generate adipocytes, which are essential for recreating the flavor and mouthfeel of conventional meat products. The co-culture of muscle and fat precursor cells is increasingly explored to improve sensory quality, but it also adds complexity to processing and cell selection protocols (WHO, 2023).

Plant-based cell sourcing follows a different paradigm but remains equally critical. Plant cell cultures are typically initiated from explants such as leaves, stems, or roots, which are induced to form callus tissue under controlled conditions (Allan et al., 2019). These undifferentiated cells exhibit rapid growth and metabolic flexibility, making them suitable for the production of cell-based plant foods, pigments, flavors, and bioactive compounds. Unlike animal cells, plant cells possess rigid cell walls, which influence downstream processing, texture, and digestibility. Therefore, selecting plant species and cell types with desirable compositional and functional properties is essential for successful food product development (Allan et al., 2019).

Genetic stability is a important criterion in cell sourcing and selection. During repeated cell divisions, mutations and chromosomal abnormalities may arise, potentially affecting growth behavior, differentiation capacity, or safety (WHO, 2023). Continuous monitoring of genetic integrity through karyotyping, genomic sequencing, and molecular markers is an integral part of the cell selection process. For food applications, maintaining non-tumorigenic and phenotypically stable cells is particularly important to meet regulatory expectations and ensure consumer confidence (Allan et al., 2019).

Another important consideration in cell sourcing is metabolic efficiency. Cells selected for food processing should efficiently convert nutrients into biomass with minimal by-product formation (Allan et al., 2019). High metabolic efficiency reduces media consumption, waste generation, and overall production costs. Advances in cell screening and metabolic profiling allow researchers to identify cell lines with superior nutrient utilization and growth kinetics, thereby improving the economic feasibility of cell-based food processing (Ong et al., 2020).

Cell sourcing is also closely linked to ethical and sustainability considerations. The use of non-lethal biopsy techniques aligns with animal welfare principles and reduces dependence on livestock farming (Allan et al., 2019). Additionally, sourcing cells from rare or endangered species for specialty food products raises ethical and conservation-related questions that must be addressed through responsible processing practices and regulatory oversight. Transparent documentation of cell origin and handling procedures is therefore essential (Allan et al., 2019).

Once suitable cells are identified, standardized isolation protocols are applied to ensure reproducibility and hygiene. Enzymatic digestion, mechanical dissociation, and selective adhesion techniques are commonly used to separate target cells from surrounding tissues (Ong et al., 2020). These operations must be performed under aseptic conditions to prevent microbial contamination at the earliest stage of processing. From an industrial standpoint, simplifying and automating cell isolation procedures is a major objective to support large-scale production. Following isolation, extensive characterization is conducted to confirm cell identity, purity, and functional potential (Barzee et al., 2022). This includes morphological assessment, growth curve analysis, marker expression profiling, and differentiation assays. Such characterization not only ensures process consistency but also supports quality assurance and regulatory compliance. Cells that meet predefined performance criteria are then cryopreserved to establish master cell banks and working cell banks (Barzee et al., 2022).

Cell banking is a cornerstone of cell-based food processing. Master cell banks serve as the original, well-characterized cell source, while working cell banks are derived from the master bank for routine production (Bennie et al., 2025). This hierarchical system minimizes variability, reduces the risk of contamination, and enables long-term production planning. Cryopreservation protocols must be carefully optimized to maintain cell viability and functionality upon thawing, as damage at this stage can compromise downstream processing efficiency. Cell sourcing and selection form the biological foundation of cell-based food processing. Decisions made at this stage influence not only the technical feasibility of cultivation and scale-up but also the nutritional quality, safety, and consumer acceptance of the final product. As the field advances, improved cell discovery, screening, and banking strategies are expected to play a decisive role in enabling the commercial viability of cell-based foods (Bennie et al., 2025).

3.3.2 Cell Proliferation and Culture Media

Once selected, cells are expanded through proliferation in controlled culture systems. This stage relies heavily on the formulation of culture media, which supplies essential nutrients such as amino acids, carbohydrates, lipids, vitamins, minerals, and growth factors (O'Neill et al., 2021). Traditionally, fetal bovine serum (FBS) was used as a culture supplement; however, ethical, economic, and safety concerns have driven the development of serum-free and chemically defined media for food applications. From a processing perspective, culture media optimization is essential for cost reduction and scalability (Barzee et al., 2022). Media composition affects cell growth rates, metabolic pathways, and biomass yield. Food-grade media components are increasingly prioritized to ensure regulatory compliance and consumer acceptance. Advances in metabolic engineering and systems biology are also being applied to tailor media formulations that enhance cell efficiency while minimizing waste generation (Bennie et al., 2025).

3.3.2.1 Principles of Cell Proliferation in Culture Systems

Cell proliferation refers to the process by which cells undergo repeated cycles of growth and division, leading to an exponential increase in cell number (O'Neill et al., 2021). In vitro proliferation is regulated by a combination of intrinsic cellular factors and extrinsic environmental cues, including nutrient availability, growth factor signaling, oxygen tension, temperature, pH, and mechanical stimuli. Unlike traditional food fermentation systems, which typically rely on robust microbial species, animal and plant cells used in cell-based food production are comparatively sensitive and require precisely balanced growth environments (Barzee et al., 2022).

The proliferative capacity of cells is strongly influenced by their lineage and differentiation state. Stem cells and progenitor cells generally exhibit higher proliferation rates and longer expansion potential compared to fully differentiated cells (O'Neill et al., 2021). However, uncontrolled proliferation can lead to genetic instability, senescence, or undesired phenotypic drift, necessitating stringent monitoring of cell health, morphology, and metabolic activity throughout the culture process. Consequently, culture media formulation and environmental control are central to maintaining a balance between rapid cell expansion and cellular integrity (O'Neill et al., 2021).

3.3.2.2 Role and Composition of Culture Media

Culture media serve as the biochemical foundation for cell proliferation by providing essential nutrients and regulatory molecules required for cellular metabolism, biosynthesis, and division. At a fundamental level, culture media must supply macronutrients, micronutrients, and bioactive factors in concentrations that support sustained growth without inducing metabolic stress (O'Neill et al., 2021). Macronutrients include amino acids, carbohydrates, and lipids, all of which contribute to energy production and biomass synthesis. Amino acids function not only as building blocks for protein synthesis but also as signaling molecules and precursors for nucleotides and other metabolites. Essential amino acids must be supplied exogenously, whereas non-essential amino acids may be synthesized by cells but are often supplemented to reduce metabolic burden and enhance proliferation efficiency (Bennie et al., 2025). Glucose remains the most widely used carbohydrate source, serving as the primary energy substrate through glycolysis and oxidative phosphorylation. However, excessive glucose concentrations can lead to lactate accumulation, resulting in acidification of the culture environment and inhibition of cell growth (Bennie et al., 2025).

Lipids are another crucial component of culture media, as they contribute to membrane biosynthesis, energy storage, and signaling pathways. Fatty acids, cholesterol, and phospholipids are typically provided either directly or through lipid-rich supplements (Schüler et al., 2024). Given the importance of lipid composition in determining the nutritional profile of cell-based foods, increasing attention is being paid to tailoring lipid supplementation strategies to achieve desirable fatty acid profiles in the final product. Micronutrients such as vitamins and minerals play indispensable roles as enzyme cofactors and regulators of cellular homeostasis (Schüler et al., 2024). Vitamins from the B complex group are particularly important for energy metabolism and nucleic acid synthesis, while minerals such as iron, zinc, magnesium, and selenium are required for redox balance, enzymatic activity, and structural stability of proteins. The bioavailability and chemical form of these micronutrients must be carefully optimized to avoid toxicity or precipitation in the culture medium (O'Neill et al., 2021).

3.3.2.3 Growth Factors and Hormonal Regulation

Beyond basic nutrients, growth factors and hormones are essential for regulating cell cycle progression, survival, and differentiation. These bioactive molecules interact with specific cell surface receptors to activate intracellular signaling pathways such as the PI3K/Akt, MAPK/ERK, and mTOR pathways, which collectively control cell proliferation and metabolism. Commonly used growth factors include fibroblast growth factor (FGF), insulin-like growth factor (IGF), epidermal growth factor (EGF), and transforming growth factor-beta (TGF-β), among others (Di & Jia, 2024).

In traditional cell culture systems, many of these factors were supplied indirectly through complex biological supplements such as serum. However, reliance on undefined supplements introduces variability and complicates process standardization. For cell-based food applications, precise control over growth factor concentrations is essential not only for reproducibility but also for regulatory transparency and consumer trust (Di & Jia, 2024).

3.3.2.4 Transition from Serum-Based to Serum-Free Media

Historically, fetal bovine serum (FBS) was widely used as a universal supplement in cell culture due to its rich composition of growth factors, hormones, attachment proteins, and nutrients (Amirvaresi & Ovissipour, 2024). While effective in promoting cell proliferation, FBS presents significant drawbacks for food-related applications. Ethical concerns surrounding animal welfare, high batch-to-batch variability, potential contamination with pathogens or adventitious agents, and escalating costs have rendered FBS unsuitable for large-scale, food-grade production systems (Amirvaresi & Ovissipour, 2024).

In response, the field has progressively shifted toward serum-free and chemically defined media, wherein all components are known, quantified, and sourced from non-animal or food-grade origins (Amirvaresi & Ovissipour, 2024). Serum-free media formulations rely on recombinant growth factors, plant-derived hydrolysates, algal extracts, or microbial fermentation products to replace the functional roles of serum. Chemically defined media go a step further by eliminating complex hydrolysates altogether, using only purified compounds with known molecular compositions (Amirvaresi & Ovissipour, 2024).

The adoption of serum-free media offers several advantages, including improved reproducibility, enhanced biosafety, and greater alignment with regulatory requirements for novel foods. Moreover, chemically defined systems facilitate more accurate metabolic modelling and process optimization, as nutrient consumption and waste generation can be quantitatively monitored and controlled (Ovissipour et al., 2024).

3.3.2.5 Cost and Scalability Considerations

From a processing and industrial perspective, culture media represent one of the largest cost contributors in cell-based food production. Growth factors, in particular, account for a substantial portion of media expenses, often exceeding the cost of all other components combined. Consequently, media optimization is a central focus of research aimed at improving the economic feasibility of cultured food products (Ovissipour et al., 2024).

Strategies to reduce media costs include lowering growth factor concentrations through synergistic combinations, developing more stable or potent analogues, and engineering cells to exhibit reduced dependence on exogenous signals (Barzee et al., 2022). Additionally, the use of food-grade bulk nutrients, such as plant-derived sugars and amino acids produced through microbial fermentation, can significantly reduce raw material costs (Bennie et al., 2025).

Scalability also imposes constraints on media formulation. Components that perform well in small-scale laboratory cultures may exhibit solubility issues, instability, or inconsistent performance in large-volume bioreactors. Therefore, media designed for industrial-scale proliferation must be compatible with high-density cultures, efficient mixing, and long-term storage without degradation (Bennie et al., 2025).

3.3.2.6 Metabolic Engineering and Systems Biology Approaches

Recent advances in metabolic engineering and systems biology have enabled a more rational approach to culture media design. By integrating omics data, including transcriptomics, proteomics, and metabolomics, researchers can gain detailed insights into cellular nutrient requirements, metabolic bottlenecks, and waste accumulation patterns (Allan et al., 2019). This information can be used to tailor media compositions that maximize cell growth while minimizing by-product formation such as lactate and ammonia. Metabolic flux analysis and computational modelling further allow prediction of how changes in nutrient availability influence cellular behavior. These tools support the development of dynamic feeding strategies, wherein nutrient concentrations are adjusted over time to match the evolving metabolic demands of proliferating cells. Such approaches not only improve biomass yield but also enhance process robustness and consistency (Allan et al., 2019).

3.3.2.7 Waste Management and Environmental Sustainability

An often overlooked aspect of culture media optimization is waste generation and environmental impact. Inefficient nutrient utilization leads to the accumulation of metabolic by-products that can inhibit cell growth and necessitate frequent media replacement (Sadhukhan et al., 2020). This, in turn, increases water usage and waste disposal requirements, raising sustainability concerns. Optimized media formulations aim to improve nutrient uptake efficiency and reduce excess supplementation. Recycling strategies, such as partial media exchange or nutrient recovery systems, are also being explored to minimize waste. From a life cycle assessment perspective, reducing media consumption and waste generation is essential for improving the overall environmental footprint of cell-based food production (Sadhukhan et al., 2020).

3.3.2.8 Regulatory and Consumer Acceptance Considerations

For cell-based foods to achieve widespread adoption, culture media components must meet regulatory standards for food safety and labelling. Regulatory agencies increasingly scrutinize the origin, purity, and processing history of media ingredients, particularly growth factors and additives. The use of food-grade or Generally Recognized as Safe (GRAS) components is therefore prioritized during media development (WHO, 2023).

Consumer perception also plays a critical role. Transparency regarding media composition and the elimination of animal-derived inputs are often viewed positively, reinforcing the ethical and sustainability claims associated with cultured foods. As a result, culture media design is not solely a technical challenge but also a socio-economic and regulatory consideration (WHO, 2023).

3.3.3 *Bioreactor-Based Cultivation*

Bioreactors serve as the core processing units in cell-based food production. These systems provide a sterile, controlled environment for large-scale cell cultivation. Common bioreactor types include stirred tank bioreactors (STRs), perfusion bioreactors, hollow fiber systems, and fixed bed bioreactors (Allan et al., 2019). The choice of bioreactor depends on the target product, cell type, and desired tissue structure. The main processing parameters such as temperature, pH, dissolved oxygen, shear stress, and nutrient availability must be carefully regulated to maintain optimal cell growth and functionality. Excessive shear forces can damage cells, while inadequate oxygen transfer can limit proliferation. Process monitoring and automation play a crucial role in maintaining consistency and reproducibility at industrial scales (Allan et al., 2019).

3.3.3.1 Types of Bioreactors for Cell-Based Food Applications

Several bioreactor configurations have been adapted or developed specifically for cell-based food production, each offering distinct advantages and limitations depending on the target cell type and desired product characteristics. Stirred tank bioreactors (STRs) are among the most widely used systems due to their operational flexibility, scalability, and extensive industrial precedent in biopharmaceutical manufacturing (Allan et al., 2019). STRs employ mechanical agitation to ensure homogenous mixing, efficient mass transfer, and uniform distribution of nutrients and gases. However, the impeller-induced shear forces generated during mixing can be detrimental to shear-sensitive cells, such as animal muscle progenitors, necessitating careful impeller design and optimized agitation speeds (Zhang et al., 2020).

Perfusion bioreactors represent an alternative approach that supports high cell densities by continuously supplying fresh culture medium while simultaneously removing metabolic waste (Allan et al., 2019). This dynamic exchange enhances nutrient availability and minimizes the accumulation of inhibitory by-products such as lactate and ammonia. Perfusion systems are particularly advantageous for prolonged cultivation and for processes requiring stable metabolic conditions. Nevertheless, their operational complexity and increased media consumption pose challenges for cost-effective large-scale implementation (Zhang et al., 2020).

Hollow fiber bioreactors provide a three-dimensional culture environment that closely mimics in vivo tissue architecture (Mendly-Zambo et al., 2021). In these systems, cells grow on or around semi-permeable fibers through which nutrients and oxygen are delivered, while waste products diffuse out. This configuration enables high surface area to volume ratios and reduced shear exposure, making it suitable for delicate cell types and structured tissue formation. However, limitations related to scalability, oxygen diffusion gradients, and process monitoring currently restrict widespread industrial adoption (Mendly-Zambo et al., 2021).

Fixed-bed bioreactors, in which cells are immobilized on solid carriers or porous matrices, offer another strategy for minimizing shear stress and enhancing cell retention. These systems are particularly relevant for anchorage-dependent cells and for applications requiring partial tissue structuring. While fixed-bed bioreactors can achieve high cell densities, challenges related to mass transfer limitations and uneven nutrient distribution must be carefully addressed through reactor design and flow optimization (Mendly-Zambo et al., 2021).

3.3.3.2 Control of Hydrodynamic and Mass Transfer Conditions

One of the most critical engineering challenges in bioreactor-based cultivation is balancing efficient mass transfer with minimal mechanical stress (Allan et al., 2019). Adequate oxygen transfer is essential for sustaining aerobic metabolism and high proliferation rates, yet oxygen has limited solubility in aqueous culture media. To overcome this constraint, bioreactors employ strategies such as increased agitation, gas sparging, or the use of oxygen-enriched gas mixtures (Allan et al., 2019). However, these approaches can inadvertently increase shear forces, potentially compromising cell viability and function. Shear stress arises from fluid motion, bubble formation, and impeller rotation, and its impact on cells depends on both magnitude and exposure duration (Allan et al., 2019). Excessive shear can disrupt cell membranes, alter cytoskeletal organization, and trigger stress-related signaling pathways. Consequently, bioreactor design increasingly incorporates low shear mixing technologies, including marine impellers, wave-induced agitation, and air-lift systems, to create more cell-friendly hydrodynamic environments (Sadhukhan et al., 2020).

Nutrient availability and waste removal are equally important considerations. Gradients in nutrient concentration or pH can develop in poorly mixed systems, leading to heterogeneous cell populations and inconsistent product quality. Advanced bioreactor configurations and optimized feeding strategies aim to minimize such gradients, ensuring uniform growth conditions throughout the culture volume (Allan et al., 2019).

3.3.3.3 Process Monitoring, Automation, and Scale Up

To achieve consistent and reproducible outcomes at industrial scales, bioreactor-based cultivation relies heavily on process monitoring and automation. Sensors for pH, dissolved oxygen, temperature, and biomass concentration provide real-time data that enable precise control of operating conditions (Allan et al., 2019). Integration of these sensors with automated control systems allows for rapid adjustments in gas flow, agitation rate, and nutrient feeding in response to dynamic cellular demands. Emerging technologies such as soft sensors, digital twins, and model predictive control are further enhancing the robustness of bioreactor operations (Allan et al., 2019). These tools leverage mathematical models and historical process data to anticipate deviations and optimize performance proactively. In the context of cell-based food production, such approaches are particularly valuable for reducing batch-to-batch variability and ensuring compliance with quality and safety standards (Amirvaresi & Ovissipour, 2024).

Scale up from laboratory or pilot scale bioreactors to industrial volumes presents additional challenges, as hydrodynamic conditions and mass transfer characteristics do not scale linearly with reactor size (Amirvaresi & Ovissipour, 2024). Successful scale-up strategies, therefore, focus on maintaining critical dimensionless parameters, such as oxygen transfer coefficients and power input per unit volume, rather than simply increasing reactor size. This engineering-driven approach is essential for preserving cell behavior and productivity across different scales (Amirvaresi & Ovissipour, 2024).

3.3.3.4 Relevance to Product Quality and Sustainability

Ultimately, bioreactor-based cultivation has a direct impact on the quality, safety, and sustainability of cell-based food products. The cultivation environment influences not only cell proliferation but also metabolic composition, structural properties, and functional attributes of the harvested biomass (Bennie et al., 2025). Efficient bioreactor design and operation can reduce energy consumption, minimize media usage, and lower overall production costs, thereby improving the environmental and economic viability of cultured food systems (Allan et al., 2019).

Bioreactors serve as the technological backbone of cell-based food production, integrating biological requirements with engineering principles to enable scalable and controlled cell cultivation. Continued innovation in bioreactor design, monitoring, and automation will be instrumental in advancing the commercial feasibility and consumer acceptance of cell-based foods (Allan et al., 2019).

3.3.4 Cell Differentiation and Tissue Formation

For structured cell-based foods such as cultured meat or seafood, cell differentiation is a vital processing step. Differentiation involves guiding proliferated cells to develop into specific tissue types, including muscle fibers and adipose tissue (Allan et al., 2019). This is typically achieved by modifying culture conditions, such as reducing growth factors, altering nutrient composition, or applying mechanical and biochemical stimuli. Scaffold-based and scaffold-free approaches are used to support tissue formation. Edible scaffolds made from biomaterials such as alginate, collagen, chitosan, or plant-derived polymers provide structural support and promote cell attachment and alignment. From a processing standpoint, scaffolds must be food grade, biodegradable, and capable of withstanding downstream processing operations (Allan et al., 2019).

3.3.4.1 Principles of Cell Differentiation in Cell-Based Foods

Cell differentiation is the biological process through which proliferated cells undergo phenotypic specialization, acquiring the structural, biochemical, and functional properties of specific tissue types (Barzee et al., 2022). In the context of cultured meat and seafood, the primary target tissues include skeletal muscle fibers, adipose tissue, and, in some cases, connective tissue. Each tissue type contributes distinct attributes to the final product, such as protein content, lipid profile, juiciness, and mouthfeel. Differentiation is regulated by tightly coordinated changes in gene expression, cytoskeletal organization, and metabolic activity (Post et al., 2020). These changes are influenced by both intrinsic cellular programming and extrinsic environmental cues. In vitro, differentiation is typically initiated by withdrawing proliferative signals and introducing conditions that mimic aspects of the in vivo tissue microenvironment. Careful temporal control is required, as premature or incomplete differentiation can compromise tissue structure and functionality (Post et al., 2020).

3.3.4.2 Modulation of Culture Conditions to Induce Differentiation

One of the most common strategies to induce differentiation involves modifying culture media composition. During the proliferation phase, culture media are rich in growth factors and nutrients that promote rapid cell division (Post et al., 2020). In contrast, differentiation media are formulated with reduced growth factor concentrations and altered nutrient profiles to signal cells to exit the cell cycle and initiate tissue-specific gene expression. For example, lowering levels of fibroblast growth factor or insulin-like growth factor can trigger myogenic differentiation in muscle progenitor cells (Schüler et al., 2024).

Nutrient availability also plays a significant role in differentiation. Adjustments in glucose, amino acid ratios, and lipid supplementation can influence metabolic pathways associated with tissue maturation (Schüler et al., 2024). In adipogenic differentiation, for instance, specific fatty acids and lipid precursors are supplied to promote lipid droplet formation and the development of mature adipocytes. These metabolic interventions not only support tissue formation but also allow partial control over the nutritional composition of the final product (Schüler et al., 2024).

3.3.4.3 Mechanical and Biochemical Stimuli

Beyond biochemical cues, mechanical and physical stimuli are increasingly recognized as powerful regulators of cell differentiation and tissue organization. In vivo, muscle cells are exposed to mechanical loading, stretching, and contraction, which are essential for proper fiber alignment and maturation (Datar & Betti, 2010). To replicate these conditions in vitro, bioreactor systems and scaffolds may incorporate cyclic stretching, shear stress, or electrical stimulation. Mechanical stimulation has been shown to enhance myotube formation, increase protein synthesis, and improve the structural integrity of engineered muscle tissues (Datar & Betti, 2010). Similarly, controlled shear forces can influence cell alignment and extracellular matrix deposition. Electrical stimulation, although more complex to implement, can further promote muscle fiber maturation and functional properties, bringing engineered tissues closer to their natural counterparts. Biochemical stimuli, including hormones and signaling molecules, also play an important role (Allan et al., 2019). Compounds such as dexamethasone, insulin, and thyroid hormones are commonly used in research-scale differentiation protocols. However, for food applications, there is growing emphasis on replacing pharmaceutical-grade additives with food-grade or naturally derived alternatives to ensure regulatory compliance and consumer acceptance (Allan et al., 2019).

3.3.4.4 Scaffold-Based Tissue Engineering Approaches

Scaffold-based approaches are widely employed to support cell attachment, differentiation, and three-dimensional tissue formation. Scaffolds provide a physical framework that mimics the extracellular matrix, guiding cell organization and facilitating nutrient and oxygen diffusion. In cell-based food systems, scaffolds must meet additional criteria beyond those used in biomedical tissue engineering, including edibility, food safety, biodegradability, and sensory neutrality (Vargas-Ovalle et al., 2025).

Edible scaffolds derived from biomaterials such as alginate, collagen, chitosan, gelatin, cellulose, and plant-derived polymers are commonly explored. These materials offer varying degrees of mechanical strength, porosity, and biochemical compatibility. For example, collagen and gelatin promote strong cell adhesion and alignment due to their natural bioactivity, while alginate and cellulose-based scaffolds provide tunable mechanical properties and are suitable for large-scale processing (Vargas-Ovalle et al., 2025).

Scaffold architecture, including pore size, fiber orientation, and stiffness, significantly influences tissue development. Aligned scaffolds can guide muscle fiber orientation, improving textural attributes, while porous structures enhance mass transfer and cell infiltration. Advanced fabrication techniques such as extrusion-based bioprinting, electrospinning, and freeze drying are increasingly used to produce scaffolds with precisely controlled microstructures (Vargas-Ovalle et al., 2025).

3.3.4.5 Scaffold Free and Self-Assembly Strategies

In addition to scaffold-based methods, scaffold-free approaches are gaining attention as a means of simplifying processing and reducing material costs (Kim et al., 2024). These strategies rely on the intrinsic ability of cells to self-assemble into three-dimensional aggregates or tissues through cell-to-cell interactions and endogenous extracellular matrix production (Vargas-Ovalle et al., 2025). Techniques such as spheroid formation, cell sheet engineering, and aggregate fusion allow the generation of tissue-like structures without exogenous support materials. Scaffold-free systems offer advantages in terms of compositional purity and regulatory simplicity, as they eliminate the need for additional biomaterials (Kim et al., 2024). However, challenges related to mechanical stability, thickness limitations, and nutrient diffusion must be carefully managed. Hybrid approaches that combine minimal scaffolding with self-assembly principles are being explored to balance structural support with biological simplicity (Kim et al., 2024).

3.3.4.6 Integration of Multiple Cell Types

To more accurately replicate the complexity of conventional meat or seafood, co-culture systems incorporating multiple cell types are increasingly utilized (Ovissipour et al., 2024). Muscle cells are often combined with adipocytes to improve flavor and juiciness, while fibroblasts may be included to support extracellular matrix formation. Achieving the correct spatial distribution and ratio of these cell types is essential for replicating the sensory and nutritional properties of traditional products (Ovissipour et al., 2024). Co-culture differentiation requires carefully synchronized protocols, as different cell types may respond differently to environmental cues. Sequential differentiation strategies, in which one cell population is differentiated prior to the introduction of another, are commonly employed to achieve controlled tissue architecture (Kim et al., 2024).

3.3.4.7 Processing and Scale-Up Considerations

From a food processing perspective, differentiation and tissue formation must be compatible with scalable, reproducible manufacturing workflows. Differentiation timelines, media consumption, and scaffold materials directly impact production costs and throughput (Allan et al., 2019). Additionally, engineered tissues must withstand downstream processing operations such as harvesting, shaping, cooking, and packaging without compromising structural integrity or safety. Ensuring uniform differentiation across large tissue volumes remains a key challenge, particularly for thick or highly structured products. Innovations in perfusion systems, dynamic culture platforms, and real-time monitoring are helping to address these limitations by improving mass transfer and environmental control during tissue formation (Allan et al., 2019).

3.3.5 *Harvesting of Cultured Biomass*

Harvesting marks the transition from upstream to downstream processing. Cells or tissues are separated from the culture medium using methods such as centrifugation, filtration, or enzymatic detachment (Allan et al., 2019). This step must be designed to preserve cell integrity while ensuring efficient recovery of biomass. In food processing, harvesting also involves the removal or recycling of spent media to reduce production costs and environmental impact. The harvested biomass may consist of single cells, cell aggregates, or partially differentiated tissues, depending on the intended product format. Hygienic design and aseptic handling are essential to prevent microbial contamination and ensure food safety (Humbird, 2021).

3.3.5.1 Objectives and Challenges of Harvesting

The primary objective of harvesting is to separate viable cellular biomass from the culture medium while maintaining cellular integrity, structural characteristics, and food safety. Unlike microbial biomass harvesting, animal and plant cells used in cell-based foods are relatively fragile and highly sensitive to mechanical and chemical stress (Humbird, 2021). Excessive shear forces, abrupt changes in osmolarity, or harsh chemical treatments can compromise cell membranes, disrupt tissue structure, and negatively affect texture and nutritional value (Allan et al., 2019).

Another major challenge lies in achieving high recovery efficiency without excessive media loss. Culture media represent a significant cost component in cell-based food production; therefore, harvesting strategies must be designed to allow partial or complete recovery and recycling of spent media where possible. Balancing efficient biomass recovery with minimal process complexity and contamination risk remains a central engineering consideration (Allan et al., 2019).

3.3.5.2 Harvesting Techniques for Cell-Based Foods

Several harvesting methods are employed depending on cell type, culture format, and product requirements (Bhat et al., 2015). Centrifugation is one of the most commonly used techniques for separating suspended cells or small aggregates from liquid culture media. By applying controlled centrifugal forces, cells can be efficiently concentrated without prolonged exposure to mechanical stress (Bhat et al., 2015). However, high-speed centrifugation may damage shear-sensitive cells, necessitating careful optimization of operating parameters such as rotational speed and residence time (Bhat et al., 2015).

Filtration-based methods, including membrane filtration and tangential flow filtration, offer a gentler alternative for cell separation (Allan et al., 2019). These systems allow continuous or semi-continuous harvesting while maintaining sterility and scalability. Filtration techniques are particularly well suited for perfusion cultures and high-density cell systems, as they enable simultaneous biomass retention and media exchange. Challenges associated with filtration include membrane fouling and pressure-induced stress, which must be mitigated through appropriate membrane selection and process control (Allan et al., 2019).

For adherent cells or scaffold-based tissues, harvesting often requires enzymatic or mechanical detachment. Enzymes such as trypsin or collagenase are commonly used in research settings to dissociate cells from surfaces (Allan et al., 2019). However, for food applications, there is increasing emphasis on identifying food-grade or enzyme-free detachment methods to ensure regulatory compliance and consumer acceptance. Mechanical scraping, temperature-induced detachment, or biodegradable scaffold dissolution are among the alternatives being explored (Allan et al., 2019).

3.3.5.3 Harvesting of Structured and Partially Differentiated Tissues

When harvesting partially differentiated tissues or structured biomass, additional considerations arise. These tissues may possess defined architecture, mechanical strength, and spatial organization that must be preserved during recovery (Bryant & Barnett, 2018). Gentle handling, low shear transfer systems, and minimal processing steps are therefore essential to maintain tissue integrity. In some cases, structured tissues are harvested together with edible scaffolds, eliminating the need for scaffold removal and simplifying downstream processing (Vargas-Ovalle et al., 2025). In other cases, scaffolds are designed to degrade or dissolve under specific conditions, allowing selective recovery of the cellular component. The chosen approach depends on the intended product format, such as minced products, hybrid foods, or whole cut analogues (Allan et al., 2019).

3.3.5.4 Media Removal, Recycling, and Sustainability

From a sustainability and cost reduction perspective, media removal and recycling are integral components of the harvesting stage. Spent culture media contain residual nutrients, metabolites, and bioactive components that may be partially recoverable through filtration, adsorption, or purification techniques (Sadhukhan et al., 2020). Recycling strategies aim to reduce raw material consumption, wastewater generation, and overall environmental impact. However, media reuse must be carefully managed to avoid the accumulation of inhibitory by-products or contaminants that could impair subsequent cultivation cycles. Quality monitoring, sterilization, and supplementation are therefore essential components of any media recycling strategy. Advances in process analytics and real-time monitoring are enabling more effective implementation of closed-loop systems in cell-based food manufacturing (Sadhukhan et al., 2020).

3.3.5.5 Hygienic Design and Food Safety Considerations

Given that harvested biomass is intended for human consumption, hygienic design and aseptic handling are paramount throughout the harvesting process. Equipment must be constructed from food-grade materials, designed to minimize dead zones, and compatible with clean in place and sterilize in place procedures (Sadhukhan et al., 2020). Closed system harvesting is preferred to reduce the risk of microbial contamination and to maintain consistent product quality. Compliance with food safety regulations requires rigorous validation of harvesting protocols, including microbial testing, traceability, and documentation. As cell-based foods move toward commercialization, alignment with hazard analysis and critical control point (HACCP) principles will be essential to ensure safe and reliable production (O'Neill et al., 2021).

3.3.6 Downstream Processing and Product Formation

Downstream processing converts harvested cell biomass into consumable food products. This stage is analogous to conventional food processing operations such as mixing, structuring, texturizing, and thermal treatment (Kumar & Soumya, 2025). For unstructured products like minced cultured meat, the biomass may be blended with fats, binders, and flavoring agents to enhance sensory attributes. For structured products, advanced processing techniques such as extrusion, 3D bioprinting, and molding are employed to replicate the texture and appearance of conventional foods (Schüler et al., 2024). Thermal treatments, including mild heating or pasteurization, may be applied to ensure microbial safety while preserving nutritional and functional properties. The challenge lies in balancing safety requirements with the preservation of cell-derived proteins and lipids, which may be sensitive to excessive heat (Schüler et al., 2024).

3.3.7 Quality and Safety Considerations

Ensuring the safety and quality of cell-based foods is a central aspect of processing. Potential hazards include microbial contamination, residual growth factors, and chemical contaminants from processing equipment (Santo et al., 2020). Hazard Analysis and Critical Control Point (HACCP) principles are increasingly adapted for cell-based food systems to identify and control risks throughout the production chain (Schüler et al., 2024). Quality attributes such as texture, flavor, color, and nutritional composition are influenced by processing conditions at every stage. Lipid oxidation, protein denaturation, and moisture loss can occur during downstream processing if not properly controlled. Therefore, integrated quality management systems are essential to maintain product consistency and consumer acceptability (Schüler et al., 2024).

3.3.8 Nutritional Optimization

Processing of cell-based foods also offers unique opportunities for nutritional enhancement. By controlling cell metabolism and culture conditions, it is possible to tailor the composition of proteins, lipids, and micronutrients (Schüler et al., 2024). For example, fatty acid profiles can be modified to increase the proportion of beneficial omega-3 fatty acids, while sodium content can be reduced compared to conventional processed foods. From a food processing perspective, nutritional optimization must be aligned with regulatory standards and labelling requirements. Post-harvest fortification and blending with plant-based ingredients are additional strategies used to enhance nutritional value while improving texture and flavor (Schüler et al., 2024).

3.3.9 Packaging and Storage

Packaging plays a crucial role in maintaining the quality and safety of cell-based foods during storage and distribution. Due to their high moisture and protein content, these products are susceptible to spoilage and oxidation (Post et al., 2020). Modified atmosphere packaging, vacuum packaging, and active packaging technologies are increasingly explored to extend shelf life. Cold chain management is typically required to preserve product quality, particularly for fresh or minimally processed cell-based foods. Research is ongoing to develop preservation techniques such as high-pressure processing and pulsed electric fields that can enhance shelf life without compromising sensory or nutritional quality (Bhat et al., 2015).

3.3.10 Regulatory and Ethical Implications in Processing

Processing of cell-based foods must comply with evolving regulatory frameworks that address food safety, labelling, and production standards. Regulatory agencies require detailed documentation of processing steps, raw materials, and quality control measures to ensure consumer safety (WHO, 2023). Ethical considerations also influence processing decisions, particularly regarding the use of animal-derived inputs and genetic modification. Transparent processing practices and clear communication are essential to build consumer trust and facilitate market acceptance (WHO, 2023).

3.4 Health Benefits

Cell-based foods, also referred to as cultured or cultivated foods, represent a transformative innovation in the global food system, offering not only sustainability and ethical advantages but also significant potential health benefits (Santo et al., 2020). By producing food products directly from cultured cells under controlled conditions, cell-based food technologies enable unprecedented precision in nutrient composition, safety, and functional quality (Santo et al., 2020). Unlike conventional animal-derived foods, whose nutritional profiles are influenced by genetics, feed, environment, and slaughter practices, cell-based foods can be engineered and optimized to meet specific health and dietary requirements. As a result, cell-based foods are increasingly recognized as a promising strategy for improving public health outcomes while addressing nutrition-related challenges associated with traditional food systems (Rubio et al., 2020).

3.4.1 Enhanced Nutritional Control and Customization

One of the most compelling health benefits of cell-based foods lies in their ability to precisely control nutritional composition. Conventional animal products often exhibit variability in protein quality, fat content, micronutrient levels, and bioactive compounds (Rubio et al., 2020). In contrast, cell-based foods are produced in tightly regulated environments where nutrient inputs, metabolic pathways, and cellular outputs can be modulated to achieve consistent and targeted nutritional profiles. Protein derived from cultured animal cells is biologically complete, containing all essential amino acids required for human health. Moreover, amino acid composition can be optimized through media formulation and metabolic regulation to enhance protein digestibility and bioavailability. This is particularly relevant for vulnerable populations such as children, the elderly, and individuals with increased protein needs due to illness or recovery (Rubio et al., 2020).

In addition to protein, the lipid profile of cell-based foods can be tailored to promote cardiovascular health. Traditional red meat products often contain high levels of saturated fatty acids and cholesterol, which are associated with increased risk of cardiovascular disease (Cheli & Baldi, 2011). Cell-based foods, however, allow for modulation of lipid synthesis pathways, enabling the production of foods with reduced saturated fat, lower cholesterol content, and enhanced levels of beneficial polyunsaturated fatty acids, including omega-3 and omega-6 fatty acids. Such customization positions cell-based foods as a healthier alternative to conventional meat without compromising sensory quality (Cheli & Baldi, 2011).

3.4.2 *Reduced Exposure to Harmful Compounds*

Another major health advantage of cell-based foods is the reduction or elimination of harmful contaminants commonly associated with conventional animal agriculture (Mendly-Zambo et al., 2021). Traditional meat production systems are susceptible to contamination with pathogens such as *Salmonella*, *Listeria monocytogenes*, and *Escherichia coli*, which pose significant food safety risks (Rubio et al., 2020). In addition, residues of antibiotics, hormones, and environmental pollutants can accumulate in animal tissues and enter the human food chain. Cell-based food production occurs in sterile, closed bioreactor systems that significantly reduce the risk of microbial contamination. Because these systems do not require routine antibiotic use to prevent disease, the risk of antibiotic residues in the final product is minimized (Rubio et al., 2020). This is particularly important in the context of growing global concern over antimicrobial resistance, which has been linked to the widespread use of antibiotics in livestock production. Furthermore, cell-based foods are not exposed to environmental contaminants such as heavy metals, dioxins, or pesticide residues that may be present in animal feed or soil. By eliminating these exposure pathways, cell-based foods offer a cleaner and more predictable safety profile, contributing to improved long-term health outcomes (Rubio et al., 2020).

3.4.3 *Lower Risk of Zoonotic Diseases*

Zoonotic diseases that are transmitted from animals to humans have been responsible for numerous public health crises, including outbreaks of avian influenza, swine flu, and other emerging infectious diseases (Abebe et al., 2020). Conventional animal farming, particularly in high-density systems, creates conditions that facilitate the transmission and mutation of pathogens. Cell-based food production bypasses the need for live animals and slaughter processes, thereby decoupling food production from zoonotic disease reservoirs. The absence of animal hosts and reduced human-animal interaction substantially lowers the risk of pathogen spillover events. From a public health perspective, this represents a profound advantage, particularly in an increasingly interconnected global food system (Abebe et al., 2020).

3.4.4 *Improved Digestibility and Bioavailability*

The digestibility and bioavailability of nutrients are critical determinants of a food's health impact. Cell-based foods can be engineered to optimize tissue composition, protein structure, and fat distribution, thereby enhancing nutrient absorption and utilization by the human body (Ogawa et al., 2018). For example, connective tissue content which can reduce digestibility in conventional meat can be minimized or precisely controlled in cultured meat products. Additionally, cell-based foods offer opportunities to incorporate functional components such as bioactive peptides, antioxidants, and micronutrients in bioavailable forms. These enhancements can support metabolic health, immune function, and overall well-being. Such functional optimization is particularly relevant for populations with specific nutritional deficiencies or increased dietary needs (Ogawa et al., 2018).

3.4.5 Potential Role in Managing Chronic Diseases

Diet-related chronic diseases, including cardiovascular disease, type 2 diabetes, obesity, and certain cancers, represent major global health challenges (Ogawa et al., 2018). Cell-based foods may contribute to the prevention and management of these conditions by enabling healthier dietary patterns without requiring drastic changes in consumer behavior (Ogawa et al., 2018). By reducing saturated fat and cholesterol while maintaining high-quality protein content, cell-based meat alternatives can support heart health and improved lipid profiles. Lower energy density and improved fat composition may also aid in weight management and metabolic control. Furthermore, the absence of certain carcinogenic compounds associated with high-temperature cooking of conventional meat, such as heterocyclic amines and polycyclic aromatic hydrocarbons, may reduce long-term cancer risk (Abebe et al., 2020).

3.4.6 Addressing Food Allergies and Dietary Sensitivities

Food allergies and intolerances pose significant challenges for affected individuals and healthcare systems. Cell-based food technologies offer the potential to reduce allergenic risks by eliminating or modifying specific allergenic proteins (Ogawa et al., 2018). Through careful selection of cell lines and controlled protein expression, it may be possible to produce foods that are better tolerated by individuals with sensitivities to conventional animal products. Additionally, cell-based foods can be designed to meet the needs of specific dietary groups, including low-cholesterol diets, high-protein therapeutic diets, and specialized nutrition for clinical or geriatric populations. This level of personalization represents a paradigm shift in how foods can support individualized health outcomes (Ogawa et al., 2018).

3.4.7 *Nutritional Security and Global Health Implications*

From a broader public health perspective, cell-based foods have the potential to enhance nutritional security, particularly in regions facing food scarcity or limited access to high-quality protein sources (WHO, 2023). Controlled production systems can operate independently of climatic conditions, arable land availability, and water constraints, enabling more reliable food supply chains. Improved access to nutrient-dense foods can help combat malnutrition, micronutrient deficiencies, and protein energy malnutrition, which remain pressing global health issues (WHO, 2023). By delivering consistent nutritional quality at scale, cell-based foods may contribute to reducing health disparities and supporting global nutrition goals (WHO, 2023). The diverse health benefits of cell-based foods, including precise nutrient composition, reduced pathogen risk, elimination of antibiotic residues, and decreased exposure to environmental contaminants, are summarized in Fig. 3.2. This figure illustrates how cell-based food technologies not only provide safe and nutritious dietary options but also reinforce public health objectives through consistent, controlled, and sustainable production practices.

3.5 Conclusion

Cell-based food technology represents a paradigm shift in the way food is produced, processed, and perceived, integrating advances in cell biology, food engineering, and nutritional science to address some of the most pressing challenges of contemporary food systems. Throughout this chapter, the foundational components of cell-based food production, including cell sourcing and selection, cell proliferation and culture media optimization, bioreactor-based cultivation, and associated health benefits, have been systematically examined to provide a comprehensive understanding of this emerging field.

The initial stages of cell-based food production, encompassing cell isolation and selection, underscore the importance of choosing robust, ethically sourced, and highly proliferative cell lines capable of sustaining large-scale cultivation. Advances in stem cell biology and tissue engineering have enabled the identification of cell types that balance rapid growth with phenotypic stability, laying a strong biological foundation for scalable food production. These developments highlight the critical role of interdisciplinary research in bridging biological feasibility with industrial applicability. Equally pivotal is the optimization of cell proliferation through precisely formulated culture media. The transition from serum-based systems to serum-free and chemically defined media marks a significant advancement toward ethical acceptability, process reproducibility, and regulatory transparency. By supplying cells with carefully balanced macronutrients, micronutrients, and bioactive factors, culture media not only support efficient biomass expansion but also influence metabolic pathways, nutritional composition, and overall product quality. Continued

Fig. 3.2 Main health benefits of cell-based foods highlighting nutritional safety, pathogen reduction, and reduced contaminant exposure

innovation in media design, supported by metabolic engineering and systems biology, is essential for reducing production costs and improving sustainability, thereby enhancing the commercial viability of cell-based foods.

Bioreactor-based cultivation emerges as the technological backbone of cell-based food manufacturing, translating biological processes into controlled, scalable production systems. The diversity of bioreactor configurations, including stirred tank, perfusion, hollow fiber, and fixed bed systems, reflects the need to accommodate different cell types and product formats. Precise control of environmental parameters such as temperature, pH, oxygen availability, and shear stress is crucial for maintaining cell viability and functional integrity. Furthermore, advancements in process monitoring, automation, and scale-up strategies contribute significantly to consistency, reproducibility, and safety at industrial scales, reinforcing the feasibility of large-scale cell-based food production. Beyond processing and engineering considerations, the potential health benefits of cell-based foods constitute a compelling rationale for their development and adoption. The ability to tailor nutritional profiles, reduce harmful components such as saturated fats and cholesterol, and eliminate exposure to antibiotics, pathogens, and environmental contaminants positions cell-based foods as a promising alternative to conventional animal-derived products. Moreover, the decoupling of food production from live animal farming reduces the risk of zoonotic diseases and aligns with public health objectives aimed at improving food safety and disease prevention. These attributes suggest that cell-based foods may play a meaningful role in addressing diet-related chronic diseases, enhancing nutritional security, and supporting global health outcomes.

Despite these promising advantages, the widespread adoption of cell-based foods is contingent upon overcoming several remaining challenges. Economic scalability, regulatory harmonization, consumer acceptance, and long-term health validation remain critical areas requiring continued research and policy support. Transparent communication regarding production processes, ingredient sourcing, and nutritional benefits will be essential for building consumer trust. Additionally, life cycle assessments and sustainability evaluations must continue to inform technological improvements to ensure that cell-based foods deliver tangible environmental and societal benefits alongside health advantages.

In conclusion, cell-based foods represent a transformative innovation with the potential to reshape future food systems by combining scientific precision with ethical and health-oriented objectives. While still in an early stage of commercialization, ongoing advancements in cell culture technology, bioreactor engineering, and nutritional optimization are steadily addressing existing limitations. As research, regulation, and industry efforts converge, cell-based foods are poised to emerge as a viable and health-conscious component of sustainable global diets. Their successful integration into mainstream food systems will depend on continued interdisciplinary collaboration, evidence-based policymaking, and proactive engagement with consumers and stakeholders alike.

References

Abebe, E., Gugsa, G., & Ahmed, M. (2020). Review on major food-borne zoonotic bacterial pathogens. *Journal of Tropical Medicine, 2020*(1), 4674235.

Allan, S. J., De Bank, P. A., & Ellis, M. J. (2019). Bioprocess design considerations for cultured meat production with a focus on the expansion bioreactor. *Frontiers in Sustainable Food Systems, 3*, 44.

Amirvaresi, A., & Ovissipour, R. (2024). Assessment of plant-and microbial-derived protein hydrolysates as sustainable for fetal bovine serum in seafood cell culture media. *Future Foods, 10*, 100443.

Azhar, A., Zeyaullah, M., Bhunia, S., Kacham, S., Patil, G., Muzammil, K., et al. (2023). Cell-based meat: The molecular aspect. *Frontiers in Food Science and Technology, 3*, 1126455.

Barzee, T. J., El Mashad, H. M., Cao, L., Chio, A., Pan, Z., & Zhang, R. (2022). Cell-cultivated food production and processing: A review. *Food Bioengineering, 1*(1), 4–25.

Ben-Arye, T., & Levenberg, S. (2019). Tissue engineering for clean meat production. *Frontiers in Sustainable Food Systems, 3*, 46.

Bennie, R. Z., Ogilvie, O. J., Loo, L. S. W., Zhou, H., Ng, S. K., Jin, A., et al. (2025). A risk-based approach can guide safe cell line development and cell banking for scaled-up cultivated meat production. *Nature Food, 6*(1), 25–30.

Bhat, Z. F., Kumar, S., & Fayaz, H. (2015). In vitro meat production: Challenges and benefits over conventional meat production. *Journal of Integrative Agriculture, 14*(2), 241–248.

Broad, G. M. (2019). Plant-based and cell-based animal product alternatives: An assessment and agenda for food tech justice. *Geoforum, 107*, 223–226.

Bryant, C., & Barnett, J. (2018). Consumer acceptance of cultured meat: A systematic review. *Meat Science, 143*, 8–17.

Cheli, F., & Baldi, A. (2011). Nutrition-based health: Cell-based bioassays for food antioxidant activity evaluation. *Journal of Food Science, 76*(9), R197–R205.

Datar, I., & Betti, M. (2010). Possibilities for an in vitro meat production system. *Innovative Food Science & Emerging Technologies, 11*(1), 13–22.

Di, C., & Jia, W. (2024). Food-derived bioactive peptides as momentous food components: Can functional peptides passed through the PI3K/Akt/mTOR pathway and NF-κB pathway to repair and protect the skeletal muscle injury? *Critical Reviews in Food Science and Nutrition, 64*(25), 9210–9227.

Dong, X., & Zhang, Y. Q. (2021). An insight on egg white: From most common functional food to biomaterial application. *Journal of Biomedical Materials Research Part B: Applied Biomaterials, 109*(7), 1045–1058.

European Food Safety Authority, Afonso, A. L., Gelbmann, W., Germini, A., Fernández, E. N., Parrino, L., et al. (2024). EFSA scientific colloquium 27: Cell culture-derived foods and food ingredients. *EFSA Supporting Publications, 21*(3), 8664E.

Geciova, J., Bury, D., & Jelen, P. (2002). Methods for disruption of microbial cells for potential use in the dairy industry—A review. *International Dairy Journal, 12*(6), 541–553.

Hadi, J., & Brightwell, G. (2021). Safety of alternative proteins: Technological, environmental and regulatory aspects of cultured meat, plant-based meat, insect protein and single-cell protein. *Foods, 10(6)*, 1226.

Humbird, D. (2021). Scale up economics for cultured meat. *Biotechnology and Bioengineering, 118*(8), 3239–3250.

Kadim, I. T., Mahgoub, O., Baqir, S., Faye, B., & Purchas, R. (2015). Cultured meat from muscle stem cells: A review of challenges and prospects. *Journal of Integrative Agriculture, 14*(2), 222–233.

Kim, W. J., Kim, Y., Ovissipour, R., & Nitin, N. (2024). Plant-based biomaterials as scaffolds for cellular agriculture. *Future Foods, 10*, 100468.

Kumar, K. A., & Soumya, S. (2025). Introduction to downstream processes. In *Bioprocess engineering and technology* (pp. 1–14). CRC Press.

Mendly-Zambo, Z., Powell, L. J., & Newman, L. L. (2021). Dairy 3.0: Cellular agriculture and the future of milk. *Food, Culture & Society, 24*(5), 675–693.

Ogawa, Y., Donlao, N., Thuengtung, S., Tian, J., Cai, Y., Reginio, F. C., Jr., et al. (2018). Impact of food structure and cell matrix on digestibility of plant-based food. *Current Opinion in Food Science, 19*, 36–41.

O'Neill, E. N., Cosenza, Z. A., Baar, K., & Block, D. E. (2021). Considerations for the development of cost-effective cell culture media for cultivated meat production. *Comprehensive Reviews in Food Science and Food Safety, 20*(1), 686–709.

Ong, S., Choudhury, D., & Naing, M. W. (2020). Cell-based meat: Current ambiguities with nomenclature. *Trends in Food Science & Technology, 102*, 223–231.

Ovissipour, R., Yang, X., Saldana, Y. T., Kaplan, D. L., Nitin, N., Shirazi, A., et al. (2024). Cell-based fish production case study for developing a food safety plan. *Heliyon, 10*(13).

Post, M. J., Levenberg, S., Kaplan, D. L., Genovese, N., Fu, J., Bryant, C. J., et al. (2020). Scientific, sustainability and regulatory challenges of cultured meat. *Nature Food, 1*(7), 403–415.

Rubio, N. R., Xiang, N., & Kaplan, D. L. (2020). Plant-based and cell-based approaches to meat production. *Nature Communications, 11*(1), 6276.

Sadhukhan, J., Dugmore, T. I., Matharu, A., Martinez-Hernandez, E., Aburto, J., Rahman, P. K., & Lynch, J. (2020). Perspectives on "game changer" global challenges for sustainable 21st century: Plant-based diet, unavoidable food waste biorefining, and circular economy. *Sustainability, 12*(5), 1976.

Santo, R. E., Kim, B. F., Goldman, S. E., Dutkiewicz, J., Biehl, E., Bloem, M. W., et al. (2020). Considering plant-based meat substitutes and cell-based meats: A public health and food systems perspective. *Frontiers in Sustainable Food Systems, 4*, 569383.

Schüler, K., Marques, D. M., Gusmao, A., Jabouille, M., Leite, M., Cabral, J. M., et al. (2024). 3D printing of plant-based fat inks towards manufacturing complex cellular agriculture products with fatty structures. *Food Hydrocolloids, 157*, 110369.

Stephens, N., Di Silvio, L., Dunsford, I., Ellis, M., Glencross, A., & Sexton, A. (2018). Bringing cultured meat to market: Technical, socio-political, and regulatory challenges in cellular agriculture. *Trends in Food Science & Technology, 78*, 155–166.

Vargas-Ovalle, M. I., Demitri, C., & Madaghiele, M. (2025). Plant-based scaffolds for tissue engineering: A review. *Polymers, 17*(19), 2705.

World Health Organization. (2023). *Food safety aspects of cell-based food*. World Health Organization.

Zhang, G., Zhao, X., Li, X., Du, G., Zhou, J., & Chen, J. (2020). Challenges and possibilities for bio-manufacturing cultured meat. *Trends in Food Science & Technology, 97*, 443–450.

Chapter 4
Space Foods

Abbreviations

3D	Three-Dimensional
AA	Acetic Acid
APACHE	Advanced Plant Habitat and Controlled Environment (simulation environment used by NASA)
ARGOS	Active Response Gravity Offload System
ASTP	Apollo–Soyuz Test Project
BES	Bioelectrochemical System
BLSS	Bioregenerative Life Support Systems
CEA	Controlled Environment Agriculture
CHP	Crew Health and Performance
CNES	Centre National d'Études Spatiales (French Space Agency)
CO_2	Carbon Dioxide
CSRM	Crew State and Risk Model
DCS	Decompression Sickness
DLR	German Aerospace Center (Deutsches Zentrum für Luft—und Raumfahrt)
EAB	Electroactive Bacteria
EET	Extracellular Electron Transfer
ESM	Equivalent System Mass
EVA	Extravehicular Activity
EVOH	Ethylene Vinyl Alcohol
FFF	Fused Filament Fabrication
HACCP	Hazard Analysis and Critical Control Point
HDPE	High-Density Polyethylene
IMEA	Injury Modes and Effects Analysis
IMF	Intermediate Moisture Foods
ISRO	Indian Space Research Organization

Y. Kumar, S. Dey, *Emerging Technology in Food Processing*,
https://doi.org/10.1007/978-3-032-24505-2_4

ISS	International Space Station
LED	Light Emitting Diode
LEO	Low Earth Orbit
LLDPE	Linear Low-Density Polyethylene
LSS	Life Support Systems
MATS	Microwave-Assisted Thermal Sterilization
MELiSSA	Micro Ecological Life Support System Alternative
MES	Microbial Electrosynthesis
NASA	National Aeronautics and Space Administration
PACES	Physical and Cognitive Exploration Simulations
PATS	Pressure-Assisted Thermal Sterilization
SDGs	Sustainable Development Goals
SMEAT	Skylab Medical Experiments Altitude Test
US	United States
USD	United States Dollar

4.1 Introduction

Human spaceflight has always been constrained not only by propulsion, navigation, and life-support technologies, but also by a more fundamental requirement: food. From the earliest suborbital flights of the late 1950s to contemporary plans for sustained human presence on the Moon and Mars, the development of space foods has mirrored the broader evolution of space exploration itself. What began as a question of whether humans could safely eat and digest food in microgravity has expanded into a multidisciplinary challenge encompassing nutrition science, food engineering, packaging technology, microbiology, psychology, and sustainability. Space food systems are no longer viewed merely as sources of calories; they are now recognized as integral components of crew health, mission success, and long-term human survival beyond Earth (Oluwafemi et al., 2018).

The earliest space missions, such as Project Mercury, were characterized by extreme uncertainty. With mission durations measured in minutes to hours, food scientists faced unprecedented questions: Would swallowing function normally in microgravity? Could crumbs or liquids pose hazards to spacecraft systems or crew safety? Would digestion and nutrient absorption be impaired? These concerns drove the development of highly simplified food forms, such as small cubes, semi-liquids, pastes, and powders, processed through freezing, drying, and freeze-drying. Packaging solutions emphasized containment, shelf stability, and ease of consumption, relying heavily on aluminum, plastics, and tube-based delivery systems. Food was functional, minimal, and experimental, reflecting the cautious approach of an era when human survival in space was not yet assured (Venturini et al., 2025).

As missions extended in duration during the Gemini and Apollo programs, food systems evolved in both complexity and ambition. Multi-day missions necessitated greater variety, improved palatability, and longer shelf life. Advances in vacuum drying, freeze drying, and thermostabilization enabled the inclusion of recognizable meals such as soups, meats, vegetables, and desserts. Packaging technologies likewise progressed, incorporating multilayer flexible laminates, rehydratable pouches with one-way valves, and overwrap systems designed to withstand the rigors of launch, storage, and microgravity use. By the time of the Apollo lunar missions, space food had become a carefully engineered balance of nutrition, safety, and crew acceptance capable of supporting astronauts not only in Earth orbit, but also on the lunar surface (Perchonok & Bourland, 2002).

The Skylab program marked a significant turning point by introducing longer mission durations and a space-station-like environment. For the first time, astronauts could eat meals resembling those on Earth, including steak, bread, yogurt, and even ice cream. Processing technologies such as irradiation, pressure-assisted thermal sterilization (PATS), and microwave-assisted thermal sterilization (MATS) extended shelf life to several years, while aluminum cans and advanced polymeric pouches provided robust packaging solutions. These developments highlighted an emerging recognition: food quality, variety, and mealtime experience had measurable effects on crew morale, psychological well-being, and overall mission performance (Cooper et al., 2011).

International collaboration during the Apollo–Soyuz Test Project (ASTP) and subsequent Space Shuttle and International Space Station (ISS) programs further diversified space food systems. With multinational crews came the need to accommodate cultural preferences, dietary restrictions, and shared meal traditions. At the same time, logistical realities demanded increasingly efficient packaging, reduced mass and volume, and compatibility with complex spacecraft systems. The ISS era consolidated decades of innovation, combining thermostabilized, irradiated, dehydrated, and ready-to-eat foods with sophisticated rehydration and beverage delivery systems. Space food had become a mature technological domain, yet one still fundamentally reliant on Earth-based resupply (Abercromby et al., 2022).

Today, as humanity prepares for a new phase of exploration, returning to the Moon, journeying to Mars, and contemplating deep-space missions, the limitations of traditional space food systems are becoming increasingly apparent. Long-duration missions introduce challenges that cannot be solved solely through improved packaging or longer shelf life. Nutrient degradation over time, exposure to cosmic radiation, altered human physiology in microgravity or partial gravity, and the psychological toll of isolation and confinement all demand a more adaptive and regenerative approach to food. Future missions may last years rather than months, with limited or no opportunity for resupply, forcing a paradigm shift from stored food to integrated food production systems (Perchonok & Bourland, 2002).

Emerging technologies point toward this transformation. Hydroponic and controlled-environment agriculture systems have already demonstrated the feasibility of growing fresh produce in microgravity, providing not only nutrients but also psychological benefits and contributions to life-support loops. Cellular agriculture and microbial food production offer the possibility of generating proteins, fats, vitamins, and flavor compounds using compact bioreactors and minimal inputs.

Three-dimensional (3D) food printing introduces unprecedented flexibility, enabling personalized nutrition, efficient use of raw materials, and enhanced sensory variety. At a systems level, bio-regenerative life-support concepts aim to close the loop entirely, recycling waste into nutrients, producing oxygen, purifying water, and sustaining human life indefinitely (Zabel et al., 2016; Häder et al., 2018; Oluwafemi et al., 2018).

Beyond their technical functions, space foods carry profound social and cultural significance. Meals structure daily routines, reinforce cultural identity, and foster social cohesion factors that become critically important during long missions in confined environments. The ability to share familiar foods, celebrate cultural traditions, or participate in food cultivation and preparation can significantly influence crew morale and mental health. As space exploration becomes increasingly international and inclusive, food systems must reflect and support this diversity (Douglas et al., 2021).

The implications of advanced space food systems extend far beyond spaceflight. Technologies developed to feed astronauts, such as water-efficient hydroponics, high-barrier packaging, microbial protein production, and waste-recycling ecosystems, have direct applications on Earth. In a world facing climate change, population growth, and resource scarcity, these innovations offer pathways toward more sustainable, resilient, and equitable food systems. Space thus serves as both a proving ground and an accelerator for transformative food technologies (Oluwafemi et al., 2018).

This chapter explores the evolution of space foods from the earliest missions to future long-duration expeditions, examining the interplay between food processing, packaging, safety, nutrition, and human factors. By tracing historical developments and emerging innovations, it highlights how space food systems have progressed from simple survival tools to complex, bio-regenerative ecosystems. Ultimately, the study of space foods is not merely about feeding astronauts; it is about understanding how humanity can sustain itself in the most extreme environments imaginable, both beyond Earth and on our home planet.

4.2 Type of Space Foods Processing and Packaging in Exiting Missions

The evolution of space food processing and packaging reflects the increasing duration, complexity, and objectives of human space missions. During the Mercury mission (1958–1963), mission durations ranged from 15 min to 34 h, requiring simple, compact, and easily consumable foods. Foods were developed as cubes, powders, semi-liquids, and pastes using freezing, drying, and freeze-drying techniques. Packaging materials such as aluminum, kraft paper, plexiglass, and plastic laminates ensured containment and shelf stability, while polystyrene tube dispensers facilitated the consumption of semi-liquid foods in microgravity (Rahul, 2023; Pandith et al., 2023).

The Gemini missions (1961–1966) extended flight durations to 2–14 days, necessitating improved shelf-stable foods and a greater variety. Vacuum drying, freeze drying, and conventional drying were employed to produce meals such as soups, eggs, meats, vegetables, and desserts. These foods were packaged in clear four-ply flexible laminate films and could be rehydrated directly within the package using spring-loaded valves connected to water dispensers, enhancing convenience and safety (Grover et al., 2022).

During the Apollo missions (1961–1969), food systems supported lunar landings and longer missions. Dehydrated foods, intermediate moisture foods (IMF), and thermostabilized products were introduced, providing extended shelf life at ambient conditions. Packaging included flexible laminates for dehydrated foods and heat-processable laminates, aluminum tubes, and cans for thermostabilized items, with non-flammable overwrap materials for added safety (Perchonok & Bourland, 2002).

The Skylab mission (1973) marked a major advancement, offering over 70 food items processed through irradiation, pressure-assisted thermal sterilization, and microwave-assisted thermal sterilization, achieving shelf lives of up to 5 years. Foods were mainly packaged in aluminum cans, while beverages utilized collapsible polymeric pouches with reconstitution valves (Perchonok & Bourland, 2002).

Subsequent programs, including the Apollo–Soyuz Test Project and the Space Shuttle and ISS eras, built upon these technologies, incorporating flexible retort pouches, multilayer polymer laminates, and vacuum-sealed packaging to ensure food safety, extended shelf life, and improved crew acceptance during long-duration missions (Perchonok & Bourland, 2002). Table 4.1 represents the highlights of all the existing space missions from the start to the advancement of space food processing and packaging.

4.2.1 Mercury Mission (1958–1963): Space Food Processing and Packaging

Project Mercury marked the first attempt by the United States to send humans into space, and it presented unprecedented challenges for food scientists, physiologists, and engineers. At the time, very little was known about how the human body would function in a microgravity environment, particularly with respect to eating, swallowing, digestion, and nutrient absorption. Early research papers from NASA's Life Sciences Program emphasized concerns that food crumbs could float freely and damage sensitive spacecraft instruments, liquids might not flow normally, and swallowing could be impaired due to the absence of gravity-driven movement of food through the esophagus. As a result, the primary objective of food development during the Mercury program was safety and functionality rather than variety or palatability (Evans, 2023).

Table 4.1 Evolution of space food processing and packaging across human space missions

Mission/ program	Mission duration and era	Major food types	Food processing technologies	Packaging materials and systems	Key Contributions to space food systems
Mercury Mission (1958–1963)	15 min–34 h (First human spaceflight)	Cubes, powders, semi-liquids, pastes	Freezing, drying, freeze-drying	Aluminum foil, kraft paper laminates, plexiglass dispensers, clear three-ply plastic laminates, polystyrene pontubes	Demonstrated feasibility of eating and digestion in microgravity; established basic food safety and containment principles
Gemini Mission (1961–1966)	2–14 days (Extended orbital missions)	Soups, meals, fried eggs, oats, pudding, chicken, vegetables, applesauce	Vacuum drying, freeze-drying, drying	Clear four-ply flexible laminates (polyethylene–polyester–fluorohalocarbon), rehydratable pouches with one-way valves, polyolefin–aluminum foil–polyester overwraps	Introduced in-package rehydration, expanded menu variety, improved shelf stability and ease of consumption
Apollo Mission (1961–1969)	Up to ~14 days (Lunar missions)	Dehydrated foods, intermediate moisture foods (IMF), thermostabilized foods, bread, spreads, juices	Drying, freeze-drying, IMF stabilization, thermal sterilization	Clear flexible laminates, heat-processable laminated pouches, collapsible aluminum tubes, aluminum cans, non-flammable fluorohalocarbon overwraps	Enabled long-duration and lunar surface missions; balanced mass constraints with improved palatability and nutrition
Skylab Mission (1973)	Up to 84 days (First U.-S. space station)	Steak, ice cream, yogurt, soups, cheese, espresso, cakes, bread, dried vegetables	Irradiation, pressure-assisted thermal sterilization (PATS), microwave-assisted thermal sterilization (MATS)	Aluminum cans (solid foods), transparent collapsible polymeric beverage pouches with reconstitution valves	Major shift toward crew comfort and variety; achieved 2–5-year shelf life; improved psychological well-being
Apollo–Soyuz Test Project (1975)	~9 days (First international mission)	Ham, corned beef, turkey slices, char-broiled beef steak	Dehydration, intermediate moisture stabilization, thermal processing	Flexible four-ply laminate pouches, aluminum cans, modified collapsible polymeric beverage containers	Demonstrated interoperability of food systems for multinational crews; reinforced standardization

Space Shuttle Era/ISS (1998–2011)	1–6 months (Long-duration orbital missions)	Natural foods, ready-to-eat meals, thermostabilized and irradiated foods, condiments, beverages	Dehydration, thermostabilization, irradiation, retort processing	Injection-molded HDPE containers with Saran coating, flexible retort pouches, aluminum and bimetallic cans, five-layer co-extruded films (nylon/EVOH/PE/LLDPE), foil overwraps	Matured space food systems; supported long-term habitation; optimized safety, shelf life, waste management, and crew autonomy

Lane and Schoeller (1999); Perchonok and Bourland (2002); Clément and Slenzka (2006); Ley et al. (2009); Macdonald and Badescu (2014); Shayanfar and Pillai (2015); Grover et al. (2022); Evans (2023); Rollock (2023); Schmidt and Legato (2025)

Mission durations ranged from short suborbital flights of approximately 15 min to orbital missions lasting up to 34 h. Given this limited timeframe, food was not considered essential for survival but was included to evaluate physiological feasibility and astronaut tolerance. Research conducted at the US Air Force School of Aerospace Medicine and NASA laboratories led to the development of highly controlled food forms designed to minimize risk. These foods were produced as bite-sized cubes, powders, semi-liquids, and pastes, which could be easily consumed without generating crumbs or free-floating particles (Shayanfar & Pillai, 2015).

Freeze-drying emerged as one of the most significant processing technologies used during the Mercury missions. Studies described freeze-drying as a dehydration technique based on sublimation, where frozen water in food is removed under reduced pressure during primary drying, followed by secondary drying to eliminate bound moisture. This process preserved nutritional quality, reduced weight, and significantly extended shelf life, making it ideal for space applications. Powdered foods produced through freeze-drying could be reconstituted with water or consumed directly in semi-liquid form. Conventional drying methods were also employed to remove moisture from foods intended to be formed into compact cubes or pastes, further improving stability and reducing microbial growth (Rahul, 2023; Pandith et al., 2023).

Packaging research was equally critical during the Mercury program. Since refrigeration was not available and food needed to remain stable under launch vibrations, temperature fluctuations, and radiation exposure, multiple materials were evaluated. Aluminum foil was widely used due to its excellent barrier properties against oxygen, moisture, and light. Kraft paper laminated with aluminum provided additional structural support while maintaining lightweight characteristics. Plastics, particularly clear three-ply laminated films, were introduced to allow visual inspection of food while maintaining containment. Plexiglass was primarily used in the construction of food dispensers and feeding systems rather than as direct food-contact packaging (Evans, 2023).

One of the most distinctive innovations of the Mercury food system was the development of "pontubes," or polystyrene extension tubes. Research articles describe these tubes as toothpaste-like containers designed for semi-liquid and paste foods, such as meat purées or fruit blends. The astronaut would squeeze the tube, dispensing food directly into the mouth, eliminating the risk of floating particles. This method also allowed precise portion control and reduced handling complexity in microgravity. Bite-sized cubes were coated with gelatin to prevent crumbling, a design choice supported by experimental studies conducted during parabolic flight tests (Shayanfar & Pillai, 2015).

Despite their technical success, Mercury foods were often criticized by astronauts for poor taste and texture. However, the scientific value of these early experiments was substantial. Data collected from Mercury missions confirmed that humans could chew, swallow, and digest food normally in space, dispelling early concerns and providing a foundation for more advanced food systems in subsequent programs. Research from this era laid the groundwork for later developments in space food processing, packaging, and safety standards, making the Mercury mission a critical milestone in the history of human space nutrition.

4.2.2 Gemini Mission (1961–1966): Advances in Space Food Processing and Packaging

The Gemini mission represented a critical transition phase in the US human spaceflight program, bridging the gap between the short-duration Mercury missions and the extended lunar missions of the Apollo era. With mission durations increasing significantly from a few hours to as long as 14 days, food systems could no longer be treated as experimental or optional. Instead, they became essential components of life support, requiring improvements in nutritional adequacy, shelf stability, safety, and ease of use in microgravity. Research conducted during this period focused on developing reliable food processing and packaging technologies capable of sustaining astronauts for longer periods while minimizing mass, volume, and waste (Clément & Slenzka, 2006).

One of the most significant advancements during the Gemini program was the expansion of food variety. Based on astronaut feedback and physiological studies, scientists developed a wider range of food products, including soups, complete meals, fried eggs, oats, puddings, chicken, vegetables, and fruit-based products such as applesauce. These foods were designed to meet caloric and nutritional requirements while improving palatability and reducing menu fatigue. Research articles from NASA's Space Food Systems Laboratory emphasized the importance of sensory acceptance, as prolonged consumption of unappealing foods could negatively affect intake and crew performance (Schmidt & Legato, 2025).

To achieve the required shelf life and stability, multiple food processing techniques were employed. Vacuum drying became a important method during the Gemini missions. In this process, foods were dried in the absence of air under reduced pressure, which lowered the boiling point of water and allowed moisture removal at relatively low temperatures. This helped preserve flavor, texture, and nutritional quality while significantly reducing microbial growth. Immediately after drying, foods were sealed in vacuum packages to prevent oxygen exposure and moisture reabsorption. Freeze-drying was also widely used, particularly for complete meals and fruit products. As in the Mercury program, freeze-drying relied on sublimation to remove water, but refinements in process control resulted in improved rehydration properties and better sensory quality (Clément & Slenzka, 2006; Schmidt & Legato, 2025).

A major innovation of the Gemini food system was the ability to rehydrate foods directly within the packaging. Research literature highlights the development of specialized food packages equipped with one-way, spring-loaded valves. These valves could be connected to an onboard water dispenser, allowing astronauts to inject a controlled amount of hot or cold water into the package without opening it. This design minimized contamination risk and prevented free-floating liquids. Once rehydrated, food was consumed through a polyethylene feeding tube attached to the opposite end of the package, enabling safe and efficient consumption in microgravity (Evans, 2023).

Packaging technology advanced substantially during the Gemini era. The primary packaging material was a clear four-ply flexible laminate, consisting of polyethylene layers on the inside and outside, with polyester and fluorohalocarbon layers in between. This multilayer structure provided excellent barriers against oxygen, moisture, and aroma loss while maintaining flexibility and low weight. The transparency of the packaging allowed astronauts to visually inspect food before consumption, which research showed improved acceptance and confidence in food quality. Additionally, overwrap packaging made of polyolefin, aluminum foil, and polyester film was used to provide extra mechanical protection, thermal stability, and extended shelf life (Evans, 2023).

The Gemini program also addressed operational challenges such as waste management and packaging disposal. Flexible pouches generated less waste volume compared to rigid containers and could be compacted after use, an important consideration for confined spacecraft environments. Findings from Gemini missions demonstrated that astronauts could successfully prepare, rehydrate, and consume meals independently, validating the practicality of more complex food systems (Rahul, 2023).

The Gemini mission marked a major step forward in space food technology. The processing and packaging innovations developed during this period laid the foundation for the Apollo program, proving that diverse, shelf-stable, and nutritionally adequate foods could support humans during extended spaceflight while maintaining safety and operational efficiency.

4.2.3 Apollo Mission (1961–1969): Space Food Processing and Packaging for Lunar Exploration

The Apollo program represents one of the most significant milestones in human history, achieving the first successful landing of humans on the Moon and their safe return to Earth. Unlike earlier missions, Apollo missions involved extended durations, complex mission profiles, lunar surface activities, and greater physical and psychological demands on astronauts. Consequently, the space food system had to evolve beyond basic sustenance to provide nutritionally balanced, safe, shelf-stable, and acceptable foods capable of supporting crew health and performance (CHP) during Earth orbit, translunar flight, lunar surface operations, and return missions. Research during this period built extensively on the Mercury and Gemini food programs while introducing new processing and packaging innovations (Rollock, 2023).

A wide range of food types was developed for Apollo missions to address dietary variety, nutritional adequacy, and crew acceptability. These included dehydrated foods such as oatmeal; intermediate moisture foods (IMF) such as dried peaches, pears, apricots, and beef jerky; and thermostabilized foods such as fruit puddings and tuna fish puddings. In addition, astronauts were provided with bread slices, sandwich spreads, cheddar cheese spreads, frankfurters, and fruit juices, reflecting a move toward more familiar and comforting food items. Research articles from NASA's Food Systems Division emphasize that increased menu diversity was critical for maintaining food intake and preventing appetite loss during longer missions (Ley et al., 2009).

Dehydrated foods were produced by removing most of the water content through controlled drying or freeze-drying processes, significantly reducing weight and inhibiting microbial growth. These foods required rehydration before consumption and were valued for their long shelf life and efficient storage. Intermediate moisture foods represented a major innovation of the Apollo era. IMF products retained a moderate level of moisture but were stabilized through heat processing, humectants, and careful control of water activity. This combination inhibited microbial growth while allowing foods to be consumed directly without rehydration. Studies showed that IMF products offered improved texture and palatability compared to fully dehydrated foods, making them particularly popular among astronauts (Lane & Schoeller, 1999).

Thermostabilized foods were another important component of the Apollo food system. These foods were heat processed using techniques similar to commercial canning, achieving commercial sterility and enabling storage at ambient temperatures for extended periods. Research indicated that thermostabilized foods maintained nutritional quality and safety for the duration of Apollo missions, provided packaging integrity was maintained. However, these foods were generally heavier than dehydrated alternatives, requiring careful selection to balance mass constraints and crew preference (Rahul, 2023).

Packaging technology played a crucial role in ensuring food safety, stability, and usability. Dehydrated and intermediate moisture foods were packaged in clear flexible laminate pouches, which provided effective barriers against oxygen and moisture while allowing visual inspection of contents. Thermostabilized foods were packaged in heat-processable laminated materials, typically composed of polyolefin, aluminum foil, and polyester layers. These multilayer structures offered excellent resistance to heat, pressure, and mechanical stress during thermal processing and spaceflight. In some cases, collapsible aluminum tubes and rigid aluminum cans were used, particularly for semi-liquid foods and spreads (Perchonok & Bourland, 2002).

An additional safety innovation during the Apollo missions was the use of non-flammable fluorohalocarbon film as a meal overwrap material. This addressed spacecraft fire safety concerns while providing secondary protection against contamination and mechanical damage. Packaging waste management was also considered, with flexible packaging preferred for its low mass and reduced storage volume after use (Rahul, 2023).

The Apollo food system represented a mature and highly reliable approach to space nutrition. Through advances in food processing, stabilization, and packaging, Apollo missions demonstrated that astronauts could be sustained safely and effectively during complex, long-duration missions beyond Earth orbit. The technologies and knowledge developed during the Apollo era laid the foundation for future space stations and long-term exploration missions, making it a cornerstone in the history of space food science.

4.2.4 Skylab Mission (1973): Advancements in Space Food Processing and Packaging

The Skylab mission marked a major milestone in the evolution of human spaceflight by introducing the United States' first space station and enabling astronauts to live and work in space for extended durations of up to 84 days. Unlike earlier missions, Skylab provided a relatively spacious, station-like environment that allowed greater flexibility in daily routines, food preparation, and eating practices. As a result, the food system developed for Skylab represented a significant departure from earlier, highly constrained space food programs, placing greater emphasis on menu variety, nutritional adequacy, crew comfort, and long-term shelf stability. Research conducted during and after Skylab demonstrated that food quality and mealtime experience played a critical role in maintaining crew health, morale, and productivity during long-duration missions (Macdonald & Badescu, 2014).

One of the most notable features of the Skylab food system was the unprecedented variety of available food items. Astronauts had access to approximately 72 different food products, including steak, vanilla ice cream, yogurt, soups, cheddar cheese, espresso, cakes, breads, dried leafy vegetables, and various meat-based products. This wide selection was intentionally designed to reduce menu fatigue and ensure adequate caloric intake over missions lasting several months. Research articles from NASA's Skylab Medical Experiments Altitude Test (SMEAT) program highlighted that increased food variety significantly improved consumption patterns and overall crew satisfaction compared to earlier missions (Perchonok & Bourland, 2002).

To achieve the required shelf life and safety, Skylab foods employed advanced processing technologies. Irradiation was one of the primary methods used, involving non-ionizing radiation to sterilize food by inactivating microorganisms and parasites without significantly raising the temperature. Studies showed that irradiation could extend the shelf life of foods to approximately 2–3 years while preserving nutritional quality and sensory attributes. In addition, pressure-assisted thermal sterilization (PATS) and microwave-assisted thermal sterilization (MATS) were explored and implemented for certain food products. PATS combined high pressure with moderate heat to achieve microbial inactivation more efficiently than conventional thermal processing, while MATS utilized microwave energy alongside heat to rapidly and uniformly sterilize packaged foods. These techniques enabled shelf lives of up to 5 years, an important requirement for long-duration missions and future space stations (Grover et al., 2022).

Packaging innovations were equally significant during the Skylab mission. Most solid and semi-solid foods were packaged in aluminum cans, which provided excellent mechanical strength, gas and moisture barrier properties, and compatibility with advanced thermal processing techniques. The use of rigid containers also allowed foods to be heated and consumed more similarly to terrestrial meals, enhancing the overall eating experience. Research indicated that this "normalcy" in meal presentation had positive psychological effects on crew members (Evans, 2023).

Beverages, however, required a different packaging approach. Drinks were stored in transparent, collapsible polymeric pouches designed to expand during reconstitution. These pouches were equipped with integrated reconstitution valves that allowed astronauts to add water directly into the package in a controlled manner. The transparency of the packaging enabled visual confirmation of proper mixing, while the collapsible design minimized waste volume after use and prevented free-floating liquids in microgravity (Clément & Slenzka, 2006).

The Skylab food system demonstrated that long-duration space missions could be supported by highly varied, safe, and palatable foods using advanced processing and packaging technologies. Lessons learned from Skylab directly influenced the design of food systems for the Space Shuttle and International Space Station, establishing new standards for shelf life, food quality, and crew-centered design. The Skylab mission represents a critical advancement in space food science, shifting the focus from mere survival toward sustained health, comfort, and operational efficiency in space.

4.2.5 Apollo–Soyuz Test Project (1975): Space Food Systems in the First International Mission

The Apollo–Soyuz Test Project (ASTP) of 1975 was a landmark achievement in the history of human spaceflight, representing the first joint space mission conducted by the United States and the Soviet Union. Beyond its political and technological significance, ASTP also introduced new challenges in the design of space food systems, as it required compatibility between two different spacecraft, life-support systems, and cultural approaches to food. The mission highlighted the importance of standardized, reliable, and culturally acceptable food processing and packaging technologies capable of supporting multinational crews in a shared orbital environment (Perchonok & Bourland, 2002).

The duration and mission profile of ASTP were comparable to late Apollo missions, allowing NASA to build upon the well-established food technologies developed during the Apollo program. Consequently, the majority of food products used in the Apollo–Soyuz mission were derived from proven Apollo-era formulations. Research literature indicates that this approach minimized technical risk while ensuring food safety and crew familiarity. The menu included a range of meat-based items such as ham slices, corned beef, turkey slices, and char-broiled beef steak, reflecting an emphasis on high-protein foods to support astronaut energy requirements and muscle maintenance during spaceflight (Rahul, 2023).

From a processing standpoint, ASTP foods were primarily classified as dehydrated foods and intermediate moisture foods (IMF). Dehydrated foods underwent extensive moisture removal through controlled drying or freeze-drying processes, significantly reducing mass and inhibiting microbial growth. These foods required rehydration before consumption and offered long shelf life with minimal refrigeration requirements. Intermediate moisture foods retained a moderate level of water but were stabilized through a combination of heat processing and water activity control, allowing them to be consumed directly without rehydration. Research studies conducted by NASA's food science teams showed that IMF products provided improved texture, flavor, and overall acceptability compared to fully dehydrated alternatives, making them particularly suitable for short- to medium-duration missions such as ASTP (Macdonald & Badescu, 2014).

Packaging technologies used during the Apollo–Soyuz Test Project closely mirrored those employed in the Apollo and Gemini programs. Dehydrated and IMF foods were packaged in flexible, four-ply laminate films designed to provide high barriers against oxygen, moisture, and light. These multilayer laminates typically consisted of polyethylene layers combined with polyester and other barrier materials, ensuring product stability throughout the mission. The flexible nature of the packaging allowed for reduced launch mass and efficient storage, as well as compact disposal after use, an important consideration in the confined spacecraft environment (Evans, 2023).

Wet foods, including certain meat products and prepared dishes, were packaged either in the same flexible laminate pouches or in aluminum cans. Aluminum cans provided excellent mechanical strength and compatibility with thermal processing methods, ensuring commercial sterility and extended shelf life. Their rigid structure also protected food integrity during launch vibrations and docking maneuvers. Beverages were packaged in a modified form of clear, collapsible polymeric containers, similar to those used in earlier missions but adapted to meet the specific requirements of the ASTP mission. These packages allowed controlled rehydration and consumption while preventing the release of free-floating liquids in microgravity (Clément & Slenzka, 2006).

The Apollo–Soyuz Test Project demonstrated that standardized space food processing and packaging systems could successfully support international crews without compromising safety or performance. Lessons learned from ASTP reinforced the importance of interoperability, reliability, and cultural sensitivity in space food design. The mission served as a foundation for future international collaborations, including the Space Shuttle Mir program and the International Space Station, where multinational crews rely on shared and interoperable food systems for long-duration missions in space.

4.2.6 *International Space Station (1998–2011): Evolution of Space Food Processing and Packaging Systems*

The International Space Station (ISS) represents the most complex and long-lasting human presence in space, requiring food systems capable of supporting astronauts for missions ranging from several months to over a year. The development of ISS food systems is closely linked to the "Space Shuttle Era," which began with the launch of the first Space Shuttle on 12 April 1981. During this period, NASA focused on refining food processing, preservation, and packaging technologies to meet the operational, nutritional, and psychological needs of astronauts while ensuring compatibility with spacecraft constraints. The ISS food system is therefore the culmination of decades of innovation that began with Mercury and matured through Gemini, Apollo, Skylab, and the Apollo–Soyuz Test Project (Perchonok & Bourland, 2002).

The initial Space Shuttle missions, known as the Orbital Flight Tests, relied heavily on food types and packaging technologies developed for Apollo and Skylab missions. These included dehydrated, intermediate moisture, and thermostabilized foods packaged in flexible laminates and aluminum containers. As mission frequency and duration increased, feedback from astronauts and operational experience drove significant improvements in food preparation, packaging convenience, and waste management (Grover et al., 2022).

A major advancement occurred during the fifth Space Shuttle mission, when rehydrated spoon-bowl packaging and beverage containers were redesigned. Injection-molded, high-density polyethylene (HDPE) containers coated with Saran were introduced, along with thermoformed flexible lids made from Saran-coated polyethylene film. These packages incorporated septum-based rehydration ports that allowed the automatic injection of preset volumes of hot or cold water. This innovation improved consistency in food preparation and reduced crew workload. Foods and beverages were consumed through polyethylene straws or feeding tubes inserted through the same septum used for rehydration, while plastic clamps ensured leak-free storage when the straw was not in use. Research studies highlighted that these systems significantly enhanced safety and ease of use in microgravity by preventing free-floating liquids and food particles (Evans, 2023).

During the Space Shuttle Era, the food system expanded to include a wide range of packaging formats. These included plastic pouches for natural and ready-to-eat foods, flexible retort pouches for thermostabilized items, aluminum and bimetallic cans for heat-processed foods, and commercially sized condiment packets containing mustard, ketchup, mayonnaise, and hot sauce. Liquid seasonings such as salt and pepper were supplied in polyethylene dropper bottles to avoid dispersion in microgravity. Meals were organized using single-meal or multi-meal overwraps made from polyethylene bags, simplifying storage, inventory management, and waste containment (Rahul, 2023).

The Space Shuttle Era concluded on 21 July 2011, marking a transition to what is often referred to as the "Space Station Era." During this phase, food systems were optimized specifically for long-duration ISS missions. Thermostabilized and irradiated foods were increasingly packaged in commercial-style flexible pouches, reflecting advances in high-barrier polymer materials and thermal processing compatibility. Beverage packaging retained the basic design used during the Shuttle Era but incorporated improved foil and plastic laminations to enhance barrier properties and extend shelf life (Shayanfar & Pillai, 2015).

More advanced packaging structures were also introduced for certain food products. These included five-layer co-extruded films composed of nylon, ethylene vinyl alcohol (EVOH), polyethylene tie layers, and linear low-density polyethylene (LLDPE). These multilayer films provided superior resistance to oxygen and moisture transmission and were often overwrapped with aluminum foil laminates and vacuum sealed to maximize shelf life and food safety (Evans, 2023).

The ISS food system represents a mature and highly adaptable approach to space nutrition. Through the integration of advanced processing methods, innovative packaging designs, and crew-centered functionality, the ISS program has demonstrated the feasibility of sustaining human life in space for extended periods. The technologies and operational lessons learned from the ISS continue to inform the design of food systems for future lunar, Martian, and deep-space missions, underscoring the central role of food science in long-term human space exploration.

4.3 Space Foods in a Long-Term Future Missions

As humanity prepares to embark on extended missions beyond low-Earth orbit toward the Moon, Mars, and potentially deep-space destinations, the question of how to sustain astronaut health and performance becomes increasingly complex. Space food systems, once focused primarily on caloric sufficiency and packaging efficiency, must now evolve toward holistic, adaptive, and sustainable nutrition strategies capable of supporting crews for months or years without reliance on resupply. Long-duration missions introduce unique physiological, psychological, logistical, and technological challenges. Radiation exposure, microgravity, isolation, resource scarcity, closed-loop life support systems (LSS), and variable environmental conditions all influence the composition, production, and delivery of food (Oluwafemi et al., 2018). Table 4.2 represents the various space food technologies and their roles in long-duration space missions.

4.3.1 *Hydroponics and Plant Growth Systems in Space*

Long-duration space missions and extraterrestrial settlements require reliable, efficient, and sustainable food production systems. Traditional soil-based agriculture is impractical in space due to mass constraints, contamination risks, and the absence of naturally fertile soil on planetary bodies such as the Moon or Mars. As a result,

Table 4.2 The space food technologies and their roles in long duration space missions

Food system/technology	Core principle	Advantages for space missions	Major challenges	Mission relevance	References
Hydroponics and plant growth systems	Soil-less cultivation using nutrient solutions under controlled environments	High water efficiency (up to 90% reduction), fresh food supply, oxygen production, CO_2 removal, psychological benefits	Microgravity fluid behavior, power demand for lighting, system complexity	Mars surface missions, long-term habitats, deep-space settlements	Oluwafemi et al. (2018); Esfandabadia and Bannovab (2019)
Cellular agriculture (tissue and fermentation-based)	Production of animal-equivalent foods from cells or microorganisms	High protein yield, ethical sustainability, compact systems, integration with BLSS	Regulatory uncertainty, bioreactor complexity, limited flight validation	Mars surface missions, deep-space missions	Stephens and Ellis (2020); Allan et al. (2019); Miller (2019)
Microbial food production (EAB and MES)	Conversion of electricity and CO_2 into edible biomass or compounds	High energy efficiency, low mass, independence from light, CO_2 recycling	Limited palatability, early technology readiness, acceptance	Mars missions, deep-space and contingency systems	Bell et al. (2022); Zabel et al. (2016); Kaschubek (2021)
3D-printed foods	Additive manufacturing of foods from base nutrient "inks"	Customizable nutrition, reduced menu fatigue, efficient use of stored ingredients	Taste acceptance, technical reliability, regulatory gaps	ISS, lunar missions, Mars transit	Douglas et al. (2021); Anandharamakrishnan et al. (2022)
Bioregenerative life support systems (BLSS)	Closed-loop biological recycling of air, water, food, and waste	Reduced resupply, integrated life support, ecological resilience	System instability risk, biological complexity, long-term reliability	Mars surface missions, deep-space settlements	Häder et al. (2018); Massa et al. (2017)
Shelf-stable prepackaged foods	Thermally processed or freeze-dried meals	Proven safety, long shelf life, minimal crew time	Nutrient degradation, menu fatigue, limited freshness	Lunar missions, Mars transit	Douglas et al. (2021); Perchonok and Bourland (2002)

hydroponics, the cultivation of plants using nutrient-rich water solutions without soil, has emerged as a leading candidate for space-based agriculture and bioregenerative life support systems (BLSS) (Esfandabadia & Bannovab, 2019).

Hydroponic systems allow plants to grow either directly in a liquid nutrient solution or in inert substrates such as perlite, mineral wool, expanded clay, or coconut coir. These media provide physical support while nutrients, oxygen, and water are delivered precisely to the root zone. In the controlled environment of a spacecraft or space habitat, critical parameters such as light intensity, photoperiod, temperature, humidity, carbon dioxide concentration, and nutrient composition can be tightly regulated. This high level of control leads to faster plant growth, higher yields, and efficient use of water and nutrients compared to conventional farming methods (Oluwafemi et al., 2018).

One of the most significant advantages of hydroponics in space missions is its compatibility with bioregenerative life support systems (BLSS). Plants play a dual role by producing food and maintaining atmospheric balance. Through photosynthesis, plants absorb carbon dioxide exhaled by astronauts and release oxygen, contributing directly to air revitalization. Additionally, plant transpiration aids in humidity control, while harvested biomass provides fresh food rich in vitamins and antioxidants that are difficult to preserve in stored rations (Oluwafemi et al., 2018).

Microgravity presents unique challenges to plant growth, including altered root orientation, fluid behavior, and gas exchange. Research conducted on the International Space Station (ISS), such as NASA's Veggie and Advanced Plant Habitat experiments, has demonstrated that plants can successfully grow in hydroponic and aeroponic systems under microgravity conditions. These studies have shown that with appropriate lighting (often LED-based), nutrient delivery, and root-zone aeration, crops such as lettuce, radishes, wheat, and dwarf tomatoes can complete their growth cycles in space (Esfandabadia & Bannovab, 2019).

From a resource-efficiency perspective, hydroponics is highly advantageous. Closed-loop systems recycle water, minimizing losses a critical factor where resupply from Earth is costly or impossible. Compared to soil agriculture, hydroponic systems can reduce water consumption by up to 90%, a benefit particularly relevant for arid extraterrestrial environments. Furthermore, nutrient solutions can be formulated using recycled waste streams, especially when hydroponics is integrated with systems such as aquaponics or waste-processing bioreactors (Oluwafemi et al., 2018).

For planetary missions, including proposed Mars habitats, hydroponic plant growth modules can be pre-deployed and operated autonomously before crew arrival. Early plant establishment shortens the time required to achieve food self-sufficiency and enhances mission safety. Over time, these systems may evolve into fully closed ecological systems that integrate plant growth, waste recycling, and atmospheric control (Oluwafemi et al., 2018).

The hydroponics represents a cornerstone technology for sustainable human presence in space (Fig. 4.1). Supported by decades of terrestrial use and growing evidence from spaceflight research, hydroponic plant growth offers a practical,

Fig. 4.1 Hydroponics and plant growth systems in space

scalable, and resilient solution to food production and life support for future space exploration and colonization (Oluwafemi et al., 2018).

4.3.2 Cellular Agriculture in Space

Cellular agriculture represents a transformative approach to food and material production that is particularly well suited to the extreme constraints of space exploration. Rather than relying on whole-animal farming, cellular agriculture produces animal-derived products such as meat, milk, fats, and leather directly from cells or microorganisms using controlled biotechnological processes. For long-duration space missions, lunar bases, or Martian settlements, these technologies offer compelling advantages in terms of resource efficiency, system controllability, and ethical and environmental sustainability (Stephens & Ellis, 2020).

Cellular agriculture can be broadly divided into two technological paradigms: tissue-engineering-based cellular agriculture and fermentation-based cellular agriculture. Tissue-engineering approaches focus on the cultivation and differentiation of animal cells, such as muscle or fat cells, in bioreactors to form structured or semi-structured foods like cultured meat. Fermentation-based approaches use engineered microorganisms, often yeast or bacteria, to produce specific compounds, including proteins, fats, or growth factors, that are functionally equivalent to animal-derived products (Stephens & Ellis, 2020).

In the context of space missions, tissue-engineered cultured meat offers the possibility of producing high-quality protein without the need to raise and manage livestock, which is infeasible in reduced gravity and closed habitats. Advances in bioreactor design, scaffold development, and serum-free media, many pioneered in terrestrial research programs, are directly applicable to space. Compact, modular bioreactors can be integrated into life-support systems, allowing precise control of nutrient delivery, waste removal, and environmental conditions. Understanding cellular metabolic stoichiometry, including nutrient uptake and waste generation, is critical for efficient scale-up and closed-loop operation in space habitats where resources are tightly limited (Allan et al., 2019).

Fermentation-based cellular agriculture may be even more immediately applicable to space due to its robustness and scalability. Engineered yeast and bacterial systems can convert simple feedstocks into complex nutrients such as lipids, vitamins, and growth factors. These systems can operate continuously, tolerate a wide range of conditions, and require relatively simple infrastructure. For example, microbial production of edible fats or milk proteins could complement hydroponic crop production, significantly expanding dietary variety and nutritional completeness for astronauts (Miller, 2019).

An important advantage of cellular agriculture in space is its compatibility with bioregenerative life support systems. Carbon dioxide exhaled by crew members can be utilized by photosynthetic organisms or converted into microbial feedstocks, while waste biomass and by-products can be recycled into nutrients. Cellular agriculture systems can therefore function as both food production units and integral components of closed-loop ecological systems, reducing dependence on resupply from Earth. Operational autonomy is another critical consideration. As with proposed autonomous plant-growth modules, cellular agriculture units could be deployed and activated before crew arrival. Early establishment of microbial cultures or cell lines would shorten the time required to achieve steady-state food production. Automation, real-time monitoring, and minimal human intervention are essential, particularly during transit or early mission phases (Miotto et al., 2017).

Finally, cellular agriculture in space raises important regulatory and ethical questions. While terrestrial regulation of cultured foods is still evolving, space missions may require mission-specific safety and quality standards. Lessons from early regulatory frameworks on Earth will inform how these products are validated for astronaut consumption (Stephens & Ellis, 2020).

The cellular agriculture offers a powerful complement to hydroponics and aquaponics for space exploration (Fig. 4.2). By enabling efficient, scalable, and controllable production of animal-equivalent foods and biomaterials, cellular agriculture has the potential to play a central role in sustaining human life beyond Earth.

4.3.3 Microbial Food Production in Space

The cost of supplying food for human spaceflight remains prohibitively high, particularly for long-duration missions beyond low Earth orbit. Current space food systems rely heavily on prepackaged meals and limited fresh produce transported from Earth, approaches that are neither scalable nor sustainable for missions to the Moon, Mars, or deep space. These limitations, combined with the risks posed by global catastrophic events on Earth, have intensified interest in alternative food production systems that are resilient, compact, and resource-efficient. Among emerging solutions, microbial food production using electroactive bacteria (EAB) has gained attention as a promising pathway for sustainable nutrition in space (Bell et al., 2022).

Agriculture in space is constrained by multiple factors, including microgravity, ionizing radiation, restricted habitable volume, and the presence of airborne contaminants. Plant-based systems require extensive shielding, large surface areas, and artificial lighting, which consume significant electrical energy while offering relatively low conversion efficiency from light to edible biomass. Since fuel mass dominates the cost of space missions, reducing the mass and energy requirements of food production systems is critical. Microbial food systems offer a compelling alternative because they bypass photosynthesis and rely instead on electrochemical energy inputs that can be directly supplied by onboard power systems (Zabel et al., 2016).

Electroactive bacteria are microorganisms capable of transferring electrons to and from solid-state electron donors or acceptors through extracellular electron transfer (EET) mechanisms. This metabolic flexibility allows EAB to grow in diverse and

Fig. 4.2 Cellular agriculture in space exploration

extreme environments, making them well suited for space applications. Many EAB are autotrophic or lithoautotrophic, meaning they can fix carbon dioxide without requiring organic substrates or oxygen—both valuable resources in closed space habitats. Their widespread presence in terrestrial ecosystems, including wastewater treatment systems and even the human gastrointestinal tract, further underscores their robustness and adaptability.

A major platform enabling microbial food production in space is the bioelectrochemical system (BES). BES technologies harness microbial redox reactions to convert electricity and simple molecules, such as CO_2 and water, into biomass or food-relevant compounds. Of particular interest is microbial electrosynthesis (MES), which reverses the conventional microbial fuel cell process by supplying electrical energy to drive microbial production of organic compounds. Compared to photosynthetic systems, MES offers higher energy efficiency due to direct electron transfer and favorable electrode surface area-to-cell ratios (Waters et al., 2002).

One promising MES product is acetic acid (AA), a central metabolic intermediate that can serve both as a caloric food source and as vinegar when diluted. Acetic acid has additional value as a preservative, cleaning agent, and precursor for pharmaceuticals and vitamins, enhancing its multifunctional utility in space habitats. The production of AA via MES also contributes to CO_2 recycling, integrating seamlessly with life support systems and reducing atmospheric waste accumulation (Kaschubek, 2021).

Prior terrestrial studies have explored MES as a food source during global catastrophic risk scenarios, where sunlight-dependent agriculture may fail. While these studies operate at much larger scales, their findings are highly relevant to small, isolated crews in space. Compared to alternative microbial foods such as hydrogen-oxidizing bacteria or microalgae like *Spirulina platensis*, MES-based systems demonstrate competitive caloric efficiencies and potentially lower equivalent system mass (ESM), a NASA-developed metric for evaluating space technologies (Bell et al., 2022).

The microbial food production using electroactive bacteria and bioelectrochemical systems offers a novel and efficient approach to sustaining human life in space. By converting electricity and CO_2 into edible products with minimal mass and energy inputs, these systems align strongly with NASA's strategic objectives for resilient, long-term human exploration beyond Earth.

4.3.4 3D-Printed Foods for Space Missions

Space foods are uniquely engineered to meet the demanding requirements of human spaceflight. Beyond being nutritious, safe, and palatable, foods for astronauts must be lightweight, shelf-stable, and easy to prepare and consume in microgravity environments (Douglas et al., 2021; Cooper et al., 2011). Microgravity significantly alters human sensory perception, digestion, and nutrient absorption, often leading to reduced food intake and nutritional deficiencies during long-duration missions (Viejo et al., 2024). Moreover, astronauts' food preferences tend to change over the course of a mission, contributing to menu fatigue and appetite loss (Douglas et al., 2021). These challenges highlight the need for innovative food manufacturing approaches that enhance variety, customization, and sustainability while minimizing resource use.

Traditionally, space foods are thermally processed or dehydrated most commonly via freeze-drying to ensure long shelf life and safety. While effective, such methods often result in monotonous food options that can negatively impact crew morale and health. Consequently, both conventional and emerging food processing technologies are being explored to diversify space diets. Among these, three-dimensional (3D) food printing has emerged as a promising solution (Fig. 4.3).

3D printing, an additive manufacturing technique, has already demonstrated transformative potential in aerospace and space missions. Space agencies such as NASA and ISRO have successfully employed 3D printing to fabricate tools, spare parts, and even rocket components, significantly reducing payload mass, resupply costs, and mission risks (Kechagias & Chaidas, 2023). In microgravity environments like the International Space Station (ISS), fused filament fabrication (FFF) based 3D printers have proven capable of producing functional components on demand (National Research Council, 2014). These successes provide a strong foundation for extending 3D printing applications to food manufacturing in space.

In food systems, material extrusion is currently the most widely used 3D printing technique. It involves depositing semi-solid or fluid food "inks" layer by layer through a nozzle, followed by solidification. This approach is particularly suitable for space missions, as a limited number of base food inks can be transformed into a wide variety of shapes, textures, flavors, and nutritional profiles (Anandharamakrishnan et al., 2022). Such flexibility allows for personalized nutrition tailored to individual astronauts' physiological needs and sensory preferences, which is critical for long-duration missions to the Moon, Mars, and beyond.

Despite its promise, several challenges hinder the widespread adoption of 3D food printing. These include technological stability, limited consumer awareness, lack of standardized regulations, and the need for trained professionals (Anandharamakrishnan et al., 2022). Taste remains the most critical acceptance factor, yet systematic sensory evaluations of 3D-printed foods are still limited. Addressing these gaps through interdisciplinary research, education, and investment is essential for commercial and space-based deployment.

Nevertheless, interest in 3D food printing continues to grow. The global market was valued at approximately USD 475 million in 2020 and is projected to reach USD 1.3 billion by 2027 (Anandharamakrishnan et al., 2022). Companies such as FoodInk, BeeHex, and ByFlow have already demonstrated practical applications, including NASA-supported pizza printing systems. Collectively, these advances position 3D food printing as a important enabling technology for future space exploration, offering sustainable, customizable, and psychologically satisfying food solutions.

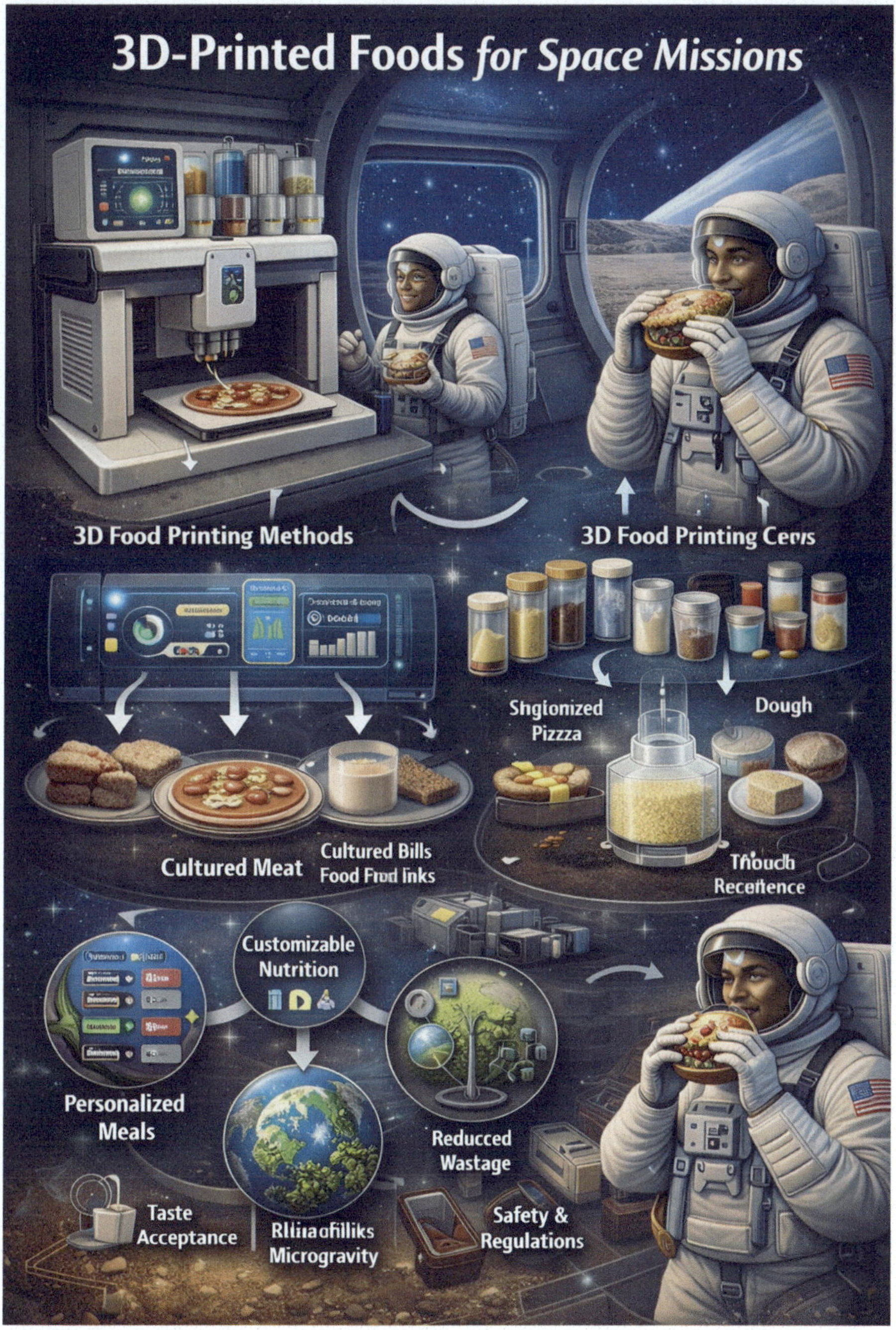

Fig. 4.3 3D printed foods for space missions

4.3.5 Bioregenerative Life Support Systems in Space Missions

Long-duration human space missions, such as prospective crewed missions to Mars, pose significant challenges related to sustaining life in environments where resupply from Earth is impractical or impossible. Astronauts undertaking months- or years-long journeys require continuous access to food, oxygen, and clean water, while spacecraft mass and volume constraints severely limit the amount of consumables that can be transported. Although water recycling technologies are already operational on the International Space Station (ISS), food and oxygen production remain critical bottlenecks for extended missions. To address these limitations, bioregenerative life support systems (BLSS) have been proposed as a sustainable and integrated solution (Häder et al., 2018).

Bioregenerative life support systems are based on the concept of closed-loop ecological systems in which biological components recycle essential nutrients and gases. Unlike physicochemical life support systems that rely primarily on mechanical and chemical processes, BLSS utilize living organisms to regenerate air, water, and food. Central to these systems are photosynthetic organisms, which play a key role in gas exchange. When exposed to light, photosynthetic microorganisms such as flagellates, as well as higher plants, produce oxygen through photosynthesis while simultaneously consuming carbon dioxide exhaled by astronauts and other organisms within the system (Häder et al., 2018). This natural coupling of oxygen production and carbon dioxide removal is fundamental to maintaining a breathable atmosphere in closed-space habitats.

In addition to gas exchange, nutrient recycling is a critical function of bioregenerative systems. Human metabolic waste products, including urea and ammonia, are toxic if allowed to accumulate. Specialized bacteria within BLSS can detoxify these compounds by converting them into forms that can be reused as nutrients by plants or other organisms. This microbial processing closes the nitrogen loop, reducing waste and minimizing the need for external inputs. Such microbial subsystems are essential for maintaining chemical stability and preventing contamination in long-term missions (Häder et al., 2018).

To further enhance system efficiency and food production capacity, bioregenerative life support concepts often incorporate multiple trophic levels. Beyond primary producers such as algae and plants, the inclusion of primary consumers has been explored. For example, ciliates or other small aquatic organisms can feed on microorganisms and plant matter, converting biomass into higher-quality protein sources. These primary consumers could, in turn, support secondary consumers such as fish, providing astronauts with a nutritionally valuable and psychologically beneficial food source. The integration of aquatic food chains within BLSS represents a step toward creating self-sustaining, miniature ecosystems in space (Häder et al., 2018).

Various ecological model systems with differing levels of biological complexity have been investigated in ground-based experiments and tested during short-term space flights. These studies aim to understand organism interactions under microgravity, radiation, and confined conditions, as well as to identify stable configurations capable of long-term operation. While fully closed bioregenerative systems have not yet been deployed for extended missions, accumulated research demonstrates their feasibility and potential advantages over conventional life support approaches.

The bioregenerative life support systems offer a promising pathway for sustaining human life during long-duration space missions. By integrating photosynthetic organisms, microbial recyclers, and higher trophic levels into closed nutrient and gas loops, BLSS can significantly reduce dependence on Earth-based resupply. Continued research and testing will be essential to optimize system stability, productivity, and resilience, paving the way for future human exploration of Mars and beyond.

4.3.6 Mission-Specific Food System Design

4.3.6.1 Lunar Surface Missions (2–6 Months)

Initial lunar surface missions under programs such as NASA's Artemis are expected to involve small crews of four to six astronauts with mission durations of less than 3 months. The primary objectives of these missions are exploration, system validation, and scientific research, leaving little crew time for food preparation or system maintenance. Consequently, lunar surface food systems will rely almost entirely on shelf-stable, prepackaged foods similar to those currently used on the International Space Station (ISS). These foods must meet strict requirements for multi-year shelf life, minimal mass and volume, ease of preparation, and compatibility with constrained water recycling systems.

Given the short mission duration and limited infrastructure, in situ food production is not considered viable for early lunar missions. The mass, power, and crew-time requirements associated with agricultural systems would outweigh any nutritional or psychological benefits. Instead, research efforts focus on reducing the mass of food systems through strategies such as increased nutrient density, reduced water content, and optimized packaging. Increasing fat content is one approach to improving caloric density, though this introduces trade-offs related to oxidative stability and long-term health impacts, particularly concerning saturated versus unsaturated fats (Douglas et al., 2021).

Nutrient-dense meal replacement bars have been explored as a mass-reduction strategy; however, results from ground-based analog studies indicate potential risks. A 30-day simulation replacing one daily meal with such bars resulted in reduced caloric intake, weight loss, and declines in mood and cognitive performance, largely due to monotony and lack of variety (Cooper et al., 2011). These findings suggest that while such approaches may be feasible for very short missions, they are not optimal without careful integration and variety management.

Despite the lack of reliance on bioregenerative food systems, early lunar missions present unique opportunities for biological research. The lunar surface and cis-lunar space offer access to partial gravity and deep-space radiation environments unavailable in low Earth orbit. Small-scale experiments involving plants, algae, fungi, or microbes may be conducted to assess their response to these conditions, potentially at platforms such as the Gateway. Although these experiments will be highly competitive with other scientific priorities, they represent a critical step in validating future food production technologies for long-duration exploration missions.

4.3.6.2 Mars Transit Missions (6–9 Months Each Way)

Mars transit missions represent one of the most demanding environments for human food systems due to their extended duration, isolation, and lack of resupply capability. Transit times ranging from 6 to 9 months each way necessitate food systems with exceptional stability, nutritional adequacy, and acceptability over multiple years. Crews are expected to be small, typically four to six astronauts, with limited available volume, power, and crew time for food-related activities.

As with lunar missions, Mars transit food systems will rely predominantly on shelf-stable, ready-to-eat meals. However, the extended duration magnifies challenges related to nutrient degradation, menu fatigue, and psychological well-being. Vitamins such as C and D, as well as flavor quality and texture, are known to degrade over time, potentially impacting crew health and morale (Perchonok & Bourland, 2002). Therefore, transit food systems must be carefully designed to maintain nutritional integrity while offering sufficient variety.

Supplemental bioregenerative food production during transit is theoretically attractive but remains highly constrained. Microgravity, limited growing volume, and competition for power and crew time restrict production to small-scale systems, if any. Lessons learned from plant growth experiments on the ISS and lunar missions may enable targeted production of fresh foods such as leafy greens to supplement prepackaged diets, primarily for psychological and micronutrient benefits rather than caloric contribution (Zabel et al., 2016).

Food preparation infrastructure during transit is expected to remain minimal, reinforcing the requirement that food systems resemble those used in everyday terrestrial kitchens, simple, reliable, and low-effort. The absence of food specialists means all systems must be intuitive and robust. Additionally, food for Mars transit may need to be preselected years in advance, limiting personalization based on final crew composition.

Ultimately, Mars transit missions will serve as a proving ground for the durability and acceptability of long-duration food systems. While largely dependent on Earth-supplied foods, these missions will benefit from incremental integration of validated bioregenerative components, setting the stage for greater food autonomy once crews reach the Martian surface.

4.3.6.3 Mars Surface Missions (18–30 Months)

Initial Mars surface missions will mark a transition from purely exploratory missions toward sustained human presence. Although early surface crews are expected to be small, their surface stays may last several months to over a year, increasing the importance of food system resilience, variety, and partial autonomy. Due to communication delays and the impossibility of resupply, food systems must be prepositioned and highly reliable.

Shelf-stable foods will continue to provide the majority of calories during early surface missions. However, unlike transit missions, the Martian surface offers partial gravity, increased available volume, and the potential for surface-based infrastructure, enabling the gradual introduction of bioregenerative food systems. These systems are expected to supplement, rather than replace, prepackaged foods, providing fresh produce that enhances diet quality, flavor, and crew morale (Massa et al., 2017).

Bioregenerative food production on Mars will likely begin with crops that can be consumed with minimal processing, such as leafy greens, herbs, and salad vegetables. These crops are well suited to controlled environment agriculture (CEA) systems and have been shown to be acceptable and safe in ground analogs and ISS experiments. Research conducted on the Moon will inform system design, particularly regarding the effects of partial gravity and deep-space radiation on plant growth and microbial communities.

Novel food production methods, including algae cultivation and fermentation-based systems, may also be evaluated during this phase. However, many such products are not inherently palatable and may require additional processing, automation, or specialist oversight to become acceptable for routine consumption (Tang et al., 2021). As a result, extensive ground testing will be required to assess feasibility, resource requirements, and integration with life support systems.

Mars surface missions will also highlight the interdependence between food production and environmental control systems. Plant-based systems can contribute to atmospheric revitalization, water recycling, and waste management, but failures can rapidly impact habitat health. Therefore, early systems must prioritize reliability, repairability, and clear failure modes.

The Mars surface missions represent a critical evolutionary step toward food autonomy. While Earth-supplied foods will remain essential, validated bioregenerative systems will begin to play a meaningful supporting role in sustaining human life on Mars.

4.3.6.4 Deep Space and Interstellar Concepts

Concepts for deep space settlements and interstellar missions represent the furthest extension of human exploration and impose fundamentally different requirements on food systems. Mission durations may span decades or generations, making reliance on Earth-supplied food infeasible. In these scenarios, food production systems must be fully autonomous, regenerative, and deeply integrated into closed-loop life support architectures.

Unlike exploration missions, such settlements would include food production specialists among the initial population. This enables diversification of crops, processing methods, and culinary practices. Food systems would function not only as sources of nutrition but also as critical components of atmospheric revitalization, water purification, and waste recycling. As such, they would form the backbone of bioregenerative life support systems (BLSS).

Controlled environment agriculture systems, such as the modular cultivation concepts developed by CNES and its partners, demonstrate the feasibility of high-yield, space-efficient food production. Recent results indicate that optimized, crop-specific growth modules can provide more than 50% of the caloric needs of four astronauts within an 80 m^2 surface area, with yields comparable to or exceeding advanced terrestrial systems. The modularity of these systems allows scalability and redundancy, both essential for long-duration missions.

However, significant research gaps remain. Many existing trade studies rely on outdated assumptions regarding crop yields, resource consumption, and system integration. Updated models are needed to accurately assess mass, power, labor, and reliability requirements under space conditions (Drysdale et al., 2000). Additionally, long-term effects of radiation, altered gravity, and closed ecological feedback loops on plants, microbes, and humans remain insufficiently understood.

Deep space and interstellar food systems will also need to address cultural, psychological, and social dimensions of food. Variety, culinary creativity, and shared food experiences will be essential for maintaining mental health over extended isolation. While interstellar food systems remain largely conceptual, research conducted today on the Moon, Mars, and Earth analogs lays the foundation for sustainable human life beyond the solar system (Fig. 4.4).

4.4 Applications of Advanced Space Food Systems (Table 4.3)

4.4.1 Improved Crew Health and Performance

Human space exploration missions expose crews to multiple interacting hazards, including radiation, reduced gravity, isolation, confinement, and long-term habitation within artificially controlled life-support environments. These stressors present significant risks to crew health and performance (CHP), necessitating targeted research and capability development to enable safe and effective exploration missions (Abercromby et al., 2022). Among these capabilities, advanced food systems play a vital role in mitigating physiological and cognitive risks associated with long-duration missions.

Fig. 4.4 Bioregenerative life support systems

Table 4.3 Applications of advanced space food systems in human space exploration

Application area	Role of advanced space food systems	Benefits	Representative technologies/ approaches	References
Improved crew health and performance	Nutrition as a countermeasure integrated into crew health and performance (CHP) frameworks	Supports musculoskeletal health, immune function, cognitive performance, fatigue mitigation	Mission-flexible nutrition systems, PACES simulations, integrated CHP modeling	Abercromby et al. (2022)
Psychological well-being and crew morale	Food systems reduce stress, injury-related anxiety, and operational uncertainty	Improved morale, confidence in EVA performance, reduced cognitive load	Suit-integrated monitoring, CSRM, EVA simulations, injury mitigation strategies	Abercromby et al. (2022); Williams et al. (2009); Conkin (2010)
Sustainable life support systems (BLSS)	Integration of food production with air, water, and waste recycling	Reduced resupply, closed-loop sustainability, nutritional diversity	Higher plants, microalgae, microbial recyclers, ecosystem-based BLSS	De Micco et al. (2023); Häder et al. (2018)
Reduction of launch mass and mission costs	Regenerative food systems reduce consumable mass and resupply frequency	Lower launch mass, increased autonomy, reduced mission risk	Bioregenerative food production, nutrient recycling, partial resupply strategies	Jones (2018)
Enhanced food safety and microbial control	Advanced monitoring and control of microbial risks in closed habitats	Protection of crew health, system integrity, long-term mission safety	HACCP adaptations, biosensors, nucleic-acid detection, in-flight microbial monitoring	Lee et al. (2021)
Synthetic biology for on-demand nutrient production	In situ biosynthesis of vitamins and micronutrients	Mitigates nutrient degradation, reduces resupply dependence	Engineered microbes, BioNutrients program, microbial bioreactors	Llorente et al. (2022); Ball (2022); Onofri et al. (2025)
Integration of 3D food printing systems	On-demand fabrication of customized foods from shelf-stable inputs	Increased variety, personalization, reduced packaging mass	Material extrusion 3D printing, printable food inks	Zhang et al. (2022); Enfield et al. (2023); Santhoshkumar et al. (2024)
Planetary habitat development (Moon and Mars)	Transition from Earth-dependent logistics to autonomous food ecosystems	Partial food self-sufficiency, atmospheric revitalization, psychological benefits	Controlled environment agriculture, modular cultivation chambers	Duri et al. (2022); Pluchon et al. (2024); Oluwafemi et al. (2018)

(continued)

Table 4.3 (continued)

Application area	Role of advanced space food systems	Benefits	Representative technologies/ approaches	References
Earth applications and spin-off technologies	Transfer of space food technologies to terrestrial sustainability	Improved food security, water efficiency, urban agriculture	Vertical farming, CEA, closed-loop water and nutrient recycling	Zeidler and Schubert (2014); Maiwald et al. (2021); Venturini et al. (2025)
Educational and scientific applications	Space food systems as experimental and training platforms	Advances in nutrition science, human factors, systems engineering	ISS experiments, closed-chamber studies, long-duration mission analogs	Oluwafemi et al. (2018); Allan et al. (2019); Duri et al. (2022)

NASA employs a structured, capability-driven approach to CHP development, beginning with the identification of human system risks and capability gaps, followed by roadmap-driven technology maturation and validation in ground-based and flight-relevant environments (Abercromby et al., 2022). Food and nutrition systems are developed as "suit-agnostic" and mission-flexible capabilities, ensuring adaptability across vehicles, habitats, and operational scenarios. These systems function not only to meet caloric needs but also as countermeasures supporting musculoskeletal integrity, metabolic efficiency, immune response, and cognitive performance under reduced-gravity and high-workload conditions.

The integration of food systems into standardized simulation frameworks, such as Physical and Cognitive Exploration Simulations (PACES), enables quantitative assessment of nutrition impacts on workload tolerance, fatigue, and decision-making during representative exploration tasks. Hybrid-reality environments and unpressurized suit simulators further support evaluation of food timing, preparation constraints, and nutritional adequacy during cognitively and physically demanding operations (Abercromby et al., 2022). Data generated across these environments feed into crew state and risk models, strengthening evidence-based dietary requirements.

Extended habitation studies conducted in hypobaric chambers provide additional insight into food system performance under altered atmospheric conditions, supporting validation of nutrition strategies for exploration class missions. Collectively, advanced food systems developed in concert with CHP simulation, modeling, and testing serve as a foundational element for sustaining crew health, resilience, and operational effectiveness beyond low Earth orbit.

4.4.2 Psychological Well-Being and Crew Morale in Suited Space Exploration

Physiological well-being and crew morale are critical determinants of mission success during long-duration human space exploration. Suited mission phases, particularly extravehicular activities (EVAs) on planetary surfaces, impose substantial physical and cognitive demands on astronauts. Historical evidence from the Apollo program and ongoing International Space Station (ISS) operations indicates that musculoskeletal injuries associated with spacesuit use are not only possible but likely unless improved injury prediction, monitoring, and mitigation capabilities are developed and implemented (Abercromby et al., 2022; Williams et al., 2009). Such injuries can degrade individual performance, compromise mission objectives, and negatively affect long-term crew health and morale.

Future exploration missions require spacesuits that accommodate a wide range of anthropometries, ensuring that both male and female astronauts of varying body sizes can perform mission-critical tasks without discomfort or elevated injury risk. This requirement is further complicated by anthropometric changes that occur during spaceflight, such as fluid shifts, muscle atrophy, and bone density loss, which must be identified and accommodated in suit design and operations (Abercromby et al., 2022). Failure to address these issues can lead to pain, fatigue, reduced task efficiency, and psychological stress, all of which directly affect crew morale.

To systematically address these risks, NASA established an Injury Modes and Effects Analysis (IMEA) in 2021–2022. The IMEA identifies potential suited injury modalities and assigns likelihood and consequence estimates, enabling prioritization of mitigation strategies and tracking of capability gap closure (Abercromby et al., 2022). In parallel, a suite of integrated injury models, including finite element, ergonomic, musculoskeletal, and anthropometric models, is being developed. These models adapt established human injury frameworks, such as those from the Global Human Body Models Consortium, to high-mobility spacesuit systems. Validation of these models requires parabolic flight testing, the only terrestrial analog that realistically simulates human weight within a suit (Abercromby et al., 2022).

Beyond physical injury, physiological stressors such as elevated inspired carbon dioxide (CO_2), thermal strain, metabolic load, and decompression stress significantly influence cognitive performance and psychological well-being. Crew capability and constraint monitoring technologies are therefore essential for EVA planning, real-time decision support, and system design. Studies conducted in simulation environments such as APACHE and ARGOS evaluate astronaut performance under emergency scenarios, altered CO_2 levels, simulated lunar gravity, and post-mission sensorimotor readaptation. These investigations reduce uncertainty in performance prediction and inform acceptable operational limits for future spacesuits (Abercromby et al., 2022).

The aggregation of physiological and cognitive data across standardized Physical and Cognitive Exploration Simulations (PACES) enables the development of the

Crew State and Risk Model (CSRM). This model provides individualized predictions of fatigue, metabolic cost, thermal strain, CO_2 dose, and decompression sickness (DCS) risk, integrating both experimental and archived flight data (Abercromby et al., 2022). Reliable prediction and mitigation of DCS are especially important for crew confidence and morale, as planetary EVAs involve higher DCS risk than microgravity operations due to increased physical workload (Conkin, 2010).

By reducing injury risk, physiological uncertainty, and operational stress, these integrated technologies directly support crew morale. Confidence in suit fit, health protection, and decision-support systems allows astronauts to focus on mission objectives rather than physical discomfort or health concerns. Ultimately, safeguarding physiological well-being through predictive modeling, validated countermeasures, and inclusive design is fundamental to sustaining both performance and morale during future lunar and Martian exploration missions.

4.4.3 Sustainable Life Support Systems (Bioregenerative Applications)

Long-duration human space missions demand life support systems that go beyond simple resupply and recycling technologies. Sustainable life support systems (LSS) increasingly rely on bioregenerative life support systems (BLSS), which integrate biological organisms to regenerate air, water, food, and other essential resources in a closed or semi-closed loop. BLSS are considered a cornerstone for future exploration and habitation of the Moon and Mars, where continuous resupply from Earth is technically and economically unfeasible (De Micco et al., 2023).

Bioregenerative systems are inspired by terrestrial ecosystems, in which matter circulates through interconnected trophic levels. A typical BLSS consists of three primary compartments: producers, consumers, and decomposers. Producers include higher plants, microalgae, and photosynthetic bacteria that convert carbon dioxide into oxygen and biomass through photosynthesis. Consumers are the human crew, who rely on biological outputs for food and oxygen. Decomposers, mainly microbial communities, recycle organic waste into inorganic nutrients that can be reused by plants, thereby closing the loop (De Micco et al., 2023).

Among biological components, higher plants play a central role. Beyond oxygen generation and carbon dioxide removal, plants provide fresh food rich in vitamins, antioxidants, and bioactive compounds that degrade over time in pre-packaged space food. Plant cultivation also contributes to water purification through transpiration and condensation processes and offers psychological benefits to astronauts, mitigating stress and isolation during long missions. The selection of crop species depends on mission duration: short missions favor fast-growing leafy greens, while long-term missions require staple crops such as wheat, rice, or potatoes to supply carbohydrates, proteins, and fats (De Micco et al., 2023).

Microbial biotechnology is equally essential in BLSS design. Microorganisms enable waste degradation, nutrient recycling, and water purification and may support food production through fermentation or microalgal biomass generation. Microalgae, in particular, have been extensively studied for their high efficiency in oxygen production and carbon dioxide fixation, as well as their potential use as protein-rich dietary supplements. Despite promising ground-based results, microbial bioprocesses in space remain underexplored, especially regarding long-term stability and process control under altered gravity and radiation conditions (De Micco et al., 2023).

A major challenge in bioregenerative applications lies in system integration and control. Biological processes are dynamic and sensitive to environmental parameters such as light, temperature, humidity, and atmospheric composition. Effective BLSS, therefore, require advanced monitoring, automation, and predictive modeling to ensure reliability, biosafety, and optimal resource balance. Ground-based demonstrators such as BIOS, Biosphere 2, Lunar Palace 1, and the MELiSSA pilot plant have demonstrated partial loop closure, but full operational BLSS with humans in space remain a future goal (De Micco et al., 2023).

The sustainable life support systems based on bioregenerative principles represent a transformative approach to human space exploration. By mimicking natural ecosystems and integrating plants, microbes, and humans into a closed loop, BLSS enable long-term autonomy while also generating knowledge applicable to sustainable agriculture, waste management, and circular economy solutions on Earth.

4.4.4 Reduction of Launch Mass and Mission Costs

Advanced food systems play a critical role in reducing launch mass and overall mission costs for long-duration space missions. Historically, the high cost of space launch, often exceeding tens of thousands of dollars per kilogram, has strongly constrained spacecraft design, life-support strategies, and mission architectures. As a result, food systems were traditionally optimized for minimum mass through dehydration, pre-packaging, and strict rationing. However, recent reductions in launch costs driven by commercial launch providers have significantly altered this trade space, enabling new approaches to space food production and logistics (Jones, 2018).

Advanced food systems encompass technologies such as regenerative life-support systems, bioregenerative food production, and partial resupply strategies. These systems aim to reduce the mass of consumables launched from Earth by recycling nutrients, water, and carbon dioxide, while potentially producing fresh food during the mission. Although recycling and in-situ food production require additional hardware mass and system complexity, they can substantially reduce the recurring mass of food resupply for long-duration missions, such as lunar bases or Mars expeditions.

Jones (2018) highlights that launch mass has traditionally been minimized because of extremely high launch costs, particularly during the Space Shuttle era, when costs to the International Space Station (ISS) exceeded $90,000 per kilogram. Under such conditions, advanced food systems with high recycling efficiency were viewed as essential, despite their development and operational complexity. However, with the advent of lower-cost commercial launch systems, such as Falcon 9, launch costs to low Earth orbit have dropped by nearly an order of magnitude. This reduction fundamentally changes the economic balance between resupply and recycling.

From a systems engineering perspective, advanced food systems can reduce mission costs by decreasing the number of resupply launches required over the mission lifetime. Even when resupply becomes cheaper, logistics still introduce operational risk, schedule dependence, and vulnerability to launch failures. Regenerative food systems increase mission autonomy and robustness, which is particularly valuable for deep-space missions where resupply is impractical or impossible (Jones, 2018).

Furthermore, advanced food systems enable more flexible spacecraft and habitat design. Reduced reliance on tightly mass-optimized prepackaged food allows for increased dietary variety, improved crew health, and enhanced psychological well-being. These benefits can translate indirectly into cost savings by improving crew performance and reducing medical risks during extended missions. Lower launch costs also allow designers to accept higher system mass in exchange for increased reliability, redundancy, and simplicity key principles for sustainable food production systems in space.

The application of advanced food systems contributes to launch mass and mission cost reduction through a combination of reduced consumable mass, fewer resupply missions, and enhanced mission resilience. While declining launch costs may make resupply more economically attractive for near-Earth missions, advanced food systems remain essential for long-duration exploration. The interaction between lower launch costs and life-support design underscores the need to reassess traditional assumptions and adopt integrated, cost-optimized food system strategies for future human spaceflight (Jones, 2018).

4.4.5 Enhanced Food Safety and Microbial Control

Microbial contamination is an unavoidable challenge in human space missions due to the ubiquity and resilience of microorganisms in closed habitats. In space environments such as the International Space Station (ISS), microorganisms originate primarily from crew, cargo, food, and water systems. Although the ISS has consistently been shown to be a microbiologically safe working and living habitat, the unique risks associated with spaceflight, particularly astronaut immune dysregulation, elevate concerns regarding food safety and microbial control (Lee et al., 2021).

Advanced space food systems are increasingly shifting from Earth-dependent, prepackaged foods toward bio-regenerative food production involving plants and microorganisms. While this transition improves nutritional quality, sustainability, and crew well-being, it introduces new microbial risks that cannot be fully mitigated using terrestrial food safety frameworks. Current ISS food safety protocols, including Hazard Analysis and Critical Control Point (HACCP) systems, sterilized seeds, microbial tolerance limits, and surface sanitization, have successfully prevented foodborne illness to date. However, these strategies represent the strict minimum required to maintain safety and are not scalable for long-duration deep-space missions (Lee et al., 2021).

Microbial monitoring aboard the ISS relies heavily on culture-based methods and post-flight analysis on Earth, resulting in delayed detection and limited situational awareness. This limitation is particularly concerning for space-grown fresh produce and microbially produced foods, where microbial loads may be higher than in terrestrial controls, and pathogenic species may occasionally be detected at low levels. Advanced food systems, therefore, require next-generation microbial monitoring technologies that are automated, rapid, and capable of simultaneous quantification and identification of microorganisms in flight (Lee et al., 2021).

Emerging approaches such as nucleic-acid-based detection, biosensors, hyperspectral imaging, and non-thermal microbial inactivation methods offer promising solutions for enhancing food safety in space. Integrating these technologies into advanced space food systems will be essential to ensure crew health, protect life-support hardware, and enable sustainable human exploration beyond low Earth orbit.

4.4.6 *Use of Synthetic Biology for On-Demand Nutrient Production*

Long-duration human space missions require food systems that are not only safe and palatable but also nutritionally complete over extended periods. Conventional prepackaged space foods suffer from nutrient degradation during long storage, particularly for vitamins such as C, B_1, and carotenoids, which are essential for astronaut health. Synthetic biology offers a transformative solution by enabling on-demand nutrient production using engineered microorganisms within advanced space food systems (Llorente et al., 2022).

Synthetic biology-based food systems leverage genetically engineered microbes to biosynthesize essential nutrients, dietary supplements, and potentially therapeutic compounds during missions. Projects such as NASA's BioNutrients program demonstrate the feasibility of using yeast and bacterial cultures to produce carotenoids and other micronutrients directly in space, reducing reliance on Earth resupply and mitigating nutrient loss associated with long-term storage (Llorente et al., 2022). These microbial bioreactors are compact, lightweight, and compatible with closed-loop life-support systems, making them well suited for deep-space exploration.

The economic viability of synthetic biology-enabled food systems is further strengthened by the recent reduction in space launch costs. Lower launch costs relax mass constraints, allowing additional bioreactor hardware, redundancy, and monitoring systems to be incorporated without prohibitive expense. This shift supports a transition from fully Earth-dependent food logistics toward partially autonomous, regenerative food production architectures (Ball, 2022).

However, the integration of synthetic biology into space food systems introduces new challenges related to microbial containment, food safety, and crew health. Astronaut immune dysregulation and the altered behavior of microorganisms in microgravity necessitate stringent monitoring and control strategies. Current ISS experience shows that microbial contamination is generally well controlled, but existing protocols represent the minimum acceptable standard and must evolve to accommodate intentional use of live microbes for food production (Onofri et al., 2025).

The synthetic biology enables resilient, flexible, and mass-efficient nutrient production for future space missions. When combined with advanced monitoring technologies and improved launch economics, it represents a cornerstone of sustainable human exploration beyond low Earth orbit.

4.4.7 Integration of 3D Food Printing Systems

Advanced space food systems are evolving to address the nutritional, psychological, and logistical challenges of long-duration human space missions. Among emerging technologies, integrated three-dimensional (3D) food printing systems offer a highly promising approach for enhancing food variety, personalization, and resource efficiency in space environments. Recent advances in additive manufacturing have enabled the fabrication of complex food structures using digitally controlled processes, making 3D food printing particularly suitable for confined and resource-limited habitats such as spacecraft and planetary bases (Zhang et al., 2022).

Material extrusion is currently the most widely used 3D food printing technique for space-relevant applications. This process involves depositing fluid or semi-solid food "inks" through a nozzle layer by layer, followed by solidification via cooling, heating, or gelation. An important advantage of this approach is the ability to generate a wide range of food products from a limited number of shelf-stable food inks, reducing storage volume, packaging mass, and waste. Such flexibility is critical for extended missions where resupply is limited and crew morale depends on dietary diversity (Enfield et al., 2023).

Integrated 3D food printing systems also enable nutritional customization, allowing foods to be tailored to individual astronaut needs based on mission phase, health status, or personal preferences. This capability is especially important given the physiological changes experienced during spaceflight, including altered metabolism, bone loss, and immune dysregulation. By precisely controlling ingredient composition and portion size, 3D food printing can support optimized nutrient delivery while maintaining palatability and sensory appeal (Santhoshkumar et al., 2024).

Despite its promise, several challenges remain before widespread deployment in space. These include ensuring food safety, preventing microbial contamination, achieving reliable operation in microgravity, and developing printable food inks with suitable rheological and nutritional properties. Addressing these challenges through system integration, automation, and advanced monitoring will be essential to realizing the full potential of 3D food printing as a core component of future advanced space food systems (Santhoshkumar et al., 2024).

4.4.8 Planetary Habitat Development (Lunar and Martian Food Ecosystems)

The establishment of permanent or semi-permanent human habitats on the Moon and Mars requires a fundamental transformation of space food systems from Earth-dependent logistics to autonomous, regenerative food ecosystems. As mission durations increase and resupply becomes impractical, advanced space food systems must integrate food production, processing, storage, and waste recycling into planetary habitat infrastructure (Pluchon et al., 2024).

Controlled environment agriculture (CEA) represents a cornerstone technology for lunar and Martian food ecosystems. Recent experimental results from modular plant cultivation systems demonstrate that crops can be grown under fully controlled conditions with yields comparable to, or exceeding, high-performance terrestrial agriculture. Modular cultivation chambers allow each crop to be grown in optimized microclimates tailored to its specific light, temperature, humidity, and nutrient requirements, thereby maximizing productivity while minimizing resource use (Duri et al., 2022).

Planetary food ecosystems must provide not only calories but also nutritional diversity and psychological benefits. Studies show that a combination of leafy greens, fruiting crops, root vegetables, legumes, cereals, and fungi can supply a substantial portion of crew caloric and nutritional needs. Projections indicate that integrated cultivation systems occupying approximately 155 m^2 could provide more than 50% of daily caloric requirements for a four-person crew, while achieving a more balanced macronutrient distribution than plant-only systems (Duri et al., 2022; Pluchon et al., 2024). Such diversity is essential for long-term dietary adherence and crew well-being.

Food ecosystems in planetary habitats are tightly coupled with life-support systems. Plants contribute to atmospheric revitalization through CO_2 uptake and oxygen generation, while closed-loop irrigation and nutrient recycling reduce water and fertilizer demands. Waste biomass can be repurposed for composting, biofertilizer production, or further bioprocessing, enhancing system circularity and resilience (Oluwafemi et al., 2018).

Despite these advances, challenges remain, including system reliability, crop pollination, post-harvest processing, and microbial safety. Cereals and legumes, while critical for caloric density, require additional processing infrastructure, underscoring the need for integrated food transformation and cooking systems within habitats. Continued development of modular, scalable food ecosystems is therefore essential for enabling sustainable human presence on the Moon and Mars (Pluchon et al., 2024).

4.4.9 Earth Applications and Spin-Off Technologies

Many consider space the "final frontier," a proving ground for humanity's endurance and technological capability. Since the Apollo Moon landings and decades of human activity in low Earth orbit aboard Skylab, MIR, and the International Space Station (ISS), space agencies worldwide have advanced technologies to sustain human life in extreme, resource-scarce environments. Among these, advanced space food systems developed under stringent closed-loop requirements have generated significant Earth-based applications and spin-off technologies that directly support global sustainability goals (Zeidler & Schubert, 2014).

Living in space requires near-total efficiency: water, nutrients, and energy must be recycled, losses minimized, and food production tightly controlled. These constraints mirror emerging challenges on Earth, where resources are finite and ecosystems are increasingly strained. Sustainability, defined as meeting present needs without compromising future generations (Maiwald et al., 2021), has become central to global development strategies, culminating in the United Nations Sustainable Development Goals (SDGs) (Venturini et al., 2025). Advanced space food systems offer practical pathways to address these goals, particularly food security, water management, and sustainable agriculture.

A key spin-off technology is controlled environment agriculture (CEA), originally developed for bioregenerative life-support systems in long-duration space missions. CEA enables precise control of plant growth parameters such as light spectrum, temperature, humidity, CO_2 concentration, and nutrient delivery. On Earth, these principles underpin vertical farming, an urban agriculture concept that cultivates crops in multi-story buildings using stacked growth layers. By decoupling food production from natural climatic and soil conditions, vertical farms can dramatically increase yield per land area, reduce water consumption, and eliminate the need for pesticides (Maiwald et al., 2021; Venturini et al., 2025).

Research by the German Aerospace Center (DLR) demonstrates the feasibility of such systems through a conceptual 37-floor vertical farm capable of producing approximately 13.3 metric tons of edible biomass per day. Advanced aeroponic nutrient delivery systems, high-efficiency LED lighting, and closed-loop water recycling technologies refined for space missions enable faster crop cycles and higher productivity compared to traditional agriculture (Maiwald et al., 2021).

Beyond food production, space-derived sanitation and water-recycling technologies address another critical global challenge: access to clean water. With nearly two billion people lacking safe drinking water (Zeidler & Schubert, 2014), closed-loop purification systems developed for spacecraft and lunar habitats provide scalable solutions for terrestrial use.

The advanced space food systems exemplify how technologies designed for humanity's expansion beyond Earth can be redirected to sustain life on our home planet. By transferring knowledge from spaceflight to Earth applications, these spin-offs contribute meaningfully to ecological and economic sustainability while reinforcing the interconnectedness of space exploration and global development.

4.4.10 Educational and Scientific Applications

Throughout human history, food and nutrition have been decisive factors in the success or failure of exploration missions. In the context of modern human spaceflight, advanced space food systems have evolved from basic survival provisions into sophisticated platforms for education and scientific research. As space agencies such as NASA prepare for renewed lunar exploration and future missions to Mars, these systems provide unique opportunities to advance knowledge in nutrition science, human physiology, behavioral health, and engineering education (Oluwafemi et al., 2018).

The International Space Station (ISS), approaching more than two decades of continuous human occupation, has served as an unparalleled laboratory for studying the interaction between food systems and human performance in space. The current ISS food system, designed for missions of 4–11 months, relies primarily on shelf-stable foods preserved through dehydration, thermostabilization, or irradiation. Constraints on water availability, storage volume, crew time, and preparation equipment shape not only what astronauts eat but also how food affects their health, morale, and productivity. These operational realities offer rich educational case studies for students in food science, nutrition, aerospace engineering, and human factors disciplines (Allan et al., 2019).

From a scientific perspective, space food systems enable controlled investigations into nutrient stability, dietary adequacy, and the behavioral effects of limited food variety. Astronaut reports highlight the psychological benefits of intermittent deliveries of fresh fruits and vegetables, underscoring the strong link between diet, mental well-being, and mission performance. Studies conducted in closed-chamber environments, such as trials of meal replacement bars designed to reduce system mass, have demonstrated that even nutritionally complete foods may negatively affect intake and behavior if palatability and variety are insufficient. These findings inform both space mission planning and terrestrial research on confined or extreme environments (Pluchon et al., 2024).

Future Mars missions present even greater scientific challenges. Crews will face multi-year missions without resupply, under conditions of isolation, radiation exposure, and prolonged microgravity. Food systems become a central experimental variable, one of the few countermeasures available to mitigate physiological decline and psychological stress. Optimizing food for long-duration missions drives research in food preservation, sensory science, and systems integration, while simultaneously providing educational platforms for training the next generation of scientists and engineers (Duri et al., 2022).

The advanced space food systems extend far beyond sustenance. They function as interdisciplinary educational tools and scientific testbeds, advancing understanding of human nutrition and behavior under extreme conditions while preparing humanity for sustained exploration beyond Earth (Fig. 4.5).

4.5 Conclusion

The evolution of space food systems reflects the broader trajectory of human spaceflight itself from short, survival-focused missions to ambitious plans for long-term habitation beyond Earth. Early programs such as Mercury, Gemini, and Apollo prioritized basic safety, shelf stability, and ease of consumption under microgravity, relying on highly processed foods and innovative packaging solutions. As missions grew longer and more complex during Skylab, the Space Shuttle, and the International Space Station eras, food systems evolved to emphasize nutritional adequacy, variety, psychological well-being, and cultural inclusivity. Advances in food processing, preservation, and packaging transformed space foods from experimental rations into reliable, mission-critical systems. Looking ahead, future lunar, Martian, and deep-space missions demand a fundamental shift in how food is produced, stored, and consumed. Extended mission durations, lack of resupply, radiation exposure, and altered gravity environments require food systems that are adaptive, resilient, and regenerative. Emerging technologies such as hydroponic plant growth, microbial and cellular agriculture, three-dimensional food printing, and bio-regenerative life support systems offer promising solutions by integrating food production with water, air, and waste recycling. These systems not only enhance crew health and performance but also reduce launch mass, mission costs, and dependence on Earth. Beyond their technical function, space food systems play a crucial role in psychological health, social cohesion, and cultural expression, all of which are essential for mission success in isolated and confined environments. Importantly, innovations driven by space exploration have far-reaching implications for Earth, contributing to sustainable agriculture, food security, and resource-efficient production methods. Therefore, advanced space food systems are no longer peripheral technologies but central enablers of human exploration and settlement beyond Earth. By combining engineering, biology, nutrition, and human factors, they provide a blueprint for sustaining life in extreme environments both in space and on our home planet.

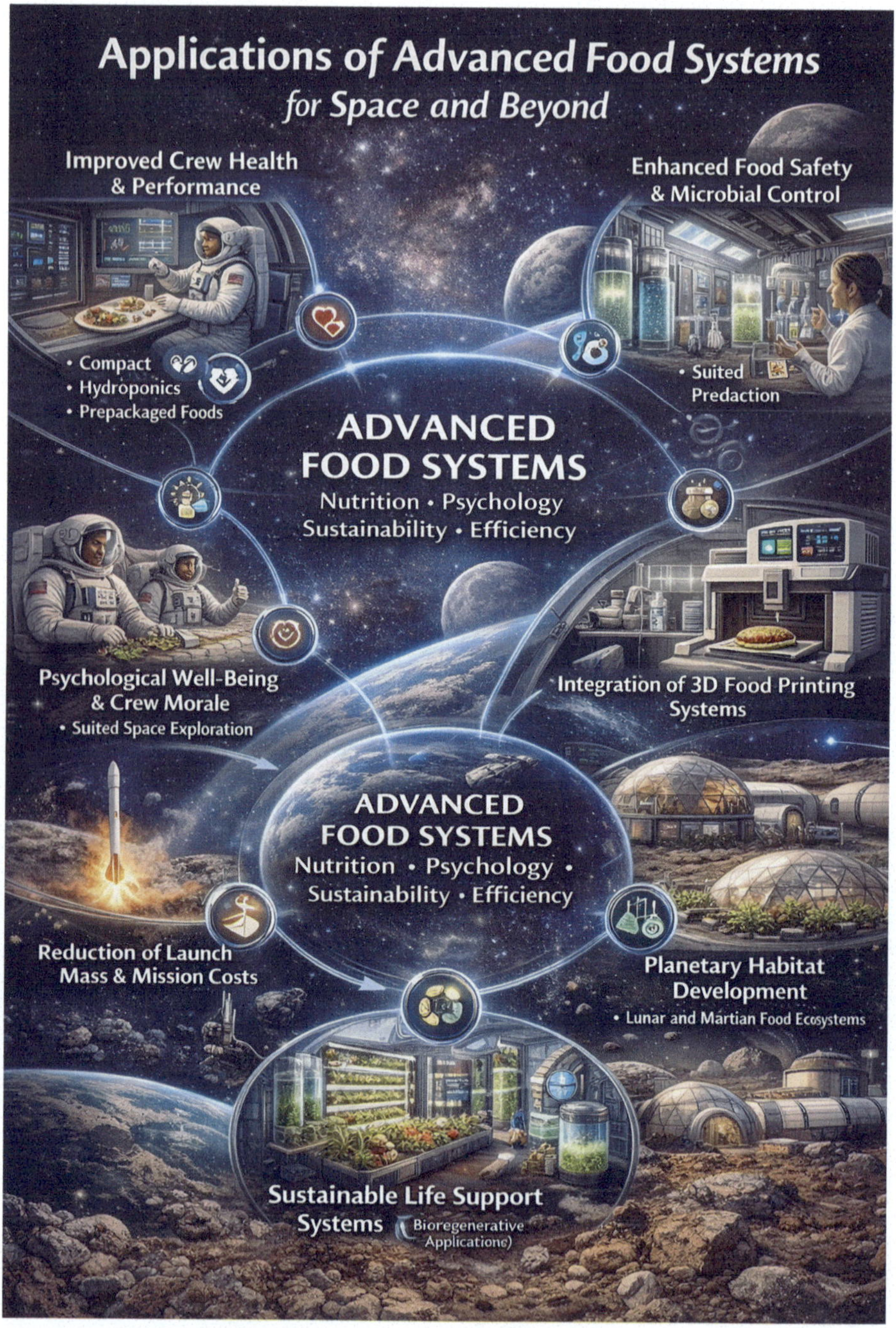

Fig. 4.5 Applications of advanced food system for space

References

Abercromby, A., Douglas, G., Kalogera, K., Somers, J., Suresh, R., Thompson, M., et al. (2022, July). NASA crew health & performance capability development for exploration: 2021 to 2022 overview. In *51st International conference on environmental systems.*

Allan, S. J., De Bank, P. A., & Ellis, M. J. (2019). Bioprocess design considerations for cultured meat production with a focus on the expansion bioreactor. *Frontiers in Sustainable Food Systems, 3*, 44.

Anandharamakrishnan, C., et al. (2022). *3D printing of foods: Applications, challenges, and opportunities*. Academic Press.

Ball, N. (2022). *Development of an on-demand vitamin production system for long-duration space missions* (Doctoral dissertation, NASA Ames Research Center).

Bell, P. J., Paras, F. E., Mandarakas, S., Arcenal, P., Robinson-Cast, S., Grobler, A. S., & Attfield, P. V. (2022). An electro–microbial process to uncouple food production from photosynthesis for application in space exploration. *Life, 12*(7), 1002.

Clément, G., & Slenzka, K. (Eds.). (2006). *Fundamentals of space biology: Research on cells, animals, and plants in space* (Vol. 18). Springer Science & Business Media.

Cooper, M., Douglas, G., & Perchonok, M. (2011). Developing the NASA food system for long-duration missions. *Journal of Food Science, 76*(2), R40–R48.

Conkin, J. (2010). Analysis of NASA decompression sickness and venous gas emboli data and gender. *Women and Pressure,* 41–68.

De Micco, V., Amitrano, C., Mastroleo, F., Aronne, G., Battistelli, A., Carnero-Diaz, E., et al. (2023). Plant and microbial science and technology as cornerstones to Bioregenerative Life Support Systems in space. *npj Microgravity, 9*(1), 69.

Douglas, G. L., Wheeler, R. M., & Fritsche, R. F. (2021). Sustaining astronauts: Resource limitations, technology needs, and parallels between spaceflight food systems and those on Earth. *Sustainability, 13*(16), 9424.

Drysdale, A. E., Maxwell, S., Ewert, M. K., & Hanford, A. J. (2000). *Systems analysis of life support for long-duration missions* (No. 2000-01-2394). SAE Technical Paper.

Duri, L. G., Caporale, A. G., Rouphael, Y., Vingiani, S., Palladino, M., De Pascale, S., & Adamo, P. (2022). The potential for lunar and martian regolith simulants to sustain plant growth: A multidisciplinary overview. *Frontiers in Astronomy and Space Sciences, 8*, 747821.

Enfield, R. E., Pandya, J. K., Lu, J., McClements, D. J., & Kinchla, A. J. (2023). The future of 3D food printing: Opportunities for space applications. *Critical Reviews in Food Science and Nutrition, 63*(29), 10079–10092.

Esfandabadia, M. M., & Bannovab, O. (2019). Designing a martian greenhouse as a habitable space: Feasibility studies and design approach. In *70th International Astronautical Congress (IAC)* (Vol. 10). IAF.

Evans, R. (2023). Space food packaging: A review of its past, present and future materials and technologies. *Packaging Technology and Science, 36*(8), 617–627.

Grover, Y., Bhasin, J., Dhingra, B., Nandi, S., Hansda, M., Sharma, R., et al. (2022). Developments and scope of space food. *Current Nutrition & Food Science, 18*(3), 248–258.

Häder, D. P., Braun, M., & Hemmersbach, R. (2018). Bioregenerative life support systems in space research. In *Gravitational biology I: Gravity sensing and graviorientation in microorganisms and plants* (pp. 113–122). Springer International Publishing.

Jones, H. (2018, July). The recent large reduction in space launch cost. In *48th International conference on environmental systems.*

Kaschubek, D. (2021). Optimized crop growth area composition for long duration spaceflight. *Life Sciences in Space Research, 30*, 55–65.

Kechagias, J., & Chaidas, D. (2023). Additive manufacturing for space applications: A review. *Aerospace Science and Technology, 134*, 108038.

Lane, H. W., & Schoeller, D. A. (Eds.). (1999). *Nutrition in spaceflight and weightlessness models* (Vol. 24). CRC Press.

Lee, J. A., Brecht, J. K., Castro-Wallace, S., Donovan, F. M., Hogan, J. A., Liu, T., et al. (2021). *Microbial food safety in space production systems*. NASA.

Ley, W., Wittmann, K., & Hallmann, W. (Eds.). (2009). *Handbook of space technology*. Wiley.

Llorente, B., Williams, T. C., Goold, H. D., Pretorius, I. S., & Paulsen, I. T. (2022). Harnessing bioengineered microbes as a versatile platform for space nutrition. *Nature Communications, 13*(1), 6177.

Macdonald, M., & Badescu, V. (Eds.). (2014). *The international handbook of space technology*. Springer.

Maiwald, V., Schubert, D., Quantius, D., & Zabel, P. (2021). From space back to Earth: Supporting sustainable development with spaceflight technologies. *Sustainable Earth, 4*(1), 3.

Massa, G. D., Wheeler, R. M., Morrow, R. C., & Levine, H. G. (2017). Growth chambers on the International Space Station for large plants. *Acta Horticulturae, 1134*, 215–222.

Miller, J. (2019). *Literature and meat since 1900*. Palgrave Macmillan.

Miotto, M., Gouveia, R., Abidin, F. Z., Figueiredo, F., & Connon, C. J. (2017). Developing a continuous bioprocessing approach to stromal cell manufacture. *ACS Applied Materials & Interfaces, 9*(47), 41131–41142.

National Aeronautics and Space Administration. (2022). *Crew health & performance capability development for exploration: 2021–2022 overview*. NASA Johnson Space Center.

National Research Council, Division on Earth, Life Studies, Board on Life Sciences, & Committee on Key Challenge Areas for Convergence. (2014). *Convergence: Facilitating transdisciplinary integration of life sciences, physical sciences, engineering, and beyond*. The National Academies Press.

Oluwafemi, F. A., De La Torre, A., Afolayan, E. M., Olalekan-Ajayi, B. M., Dhital, B., Mora-Almanza, J. G., et al. (2018). Space food and nutrition in a long term manned mission. *Advances in Astronautics Science and Technology, 1*(1), 1–21.

Onofri, S., Moeller, R., Billi, D., Balsamo, M., Becker, A., Benvenuto, E., et al. (2025). Synthetic biology for space exploration. *npj Microgravity, 11*(1), 41.

Pandith, J. A., Neekhra, S., Ahmad, S., & Sheikh, R. A. (2023). Recent developments in space food for exploration missions: A review. *Life Sciences in Space Research, 36*, 123–134.

Perchonok, M., & Bourland, C. (2002). NASA food systems: Past, present, and future. *Nutrition, 18*(10), 913–920.

Pluchon, P. F., Aubry, B., Oliver, P. H., & Navarro, G. (2024). Plant cultivation system in space environment: Towards food autonomy for Lunar and Martian missions by 2050. In *53rd International conference on environmental systems* (pp. 21–25). ICES.

Rahul, W. (2023). Evolution of space food, category, challenges and packaging. *Journal of Pharmaceutical Innovation, 12*, 1233–1244.

Rollock, A. E. (2023). *Guiding principles for prioritizing use of emergent technology towards self-reliant deep space habitats* (Doctoral dissertation, University of Colorado at Boulder).

Santhoshkumar, P., Negi, A., & Moses, J. A. (2024). 3D printing for space food applications: Advancements, challenges, and prospects. *Life Sciences in Space Research, 40*, 158–165.

Schmidt, M., & Legato, M. (Eds.). (2025). *Building a space-faring civilization: Advancing the renaissance of science, medicine and human performance in civilian spaceflight*. Elsevier.

Shayanfar, S., & Pillai, S. D. (2015). Future trends in electron beam technology for food processing. In *Electron beam pasteurization and complementary food processing technologies* (pp. 295–309). Woodhead publishing.

Stephens, N., & Ellis, M. (2020). Cellular agriculture in the UK: A review. *Wellcome Open Research, 5*, 12.

Tang, H., Rising, H. H., Majji, M., & Brown, R. D. (2021). Long-term space nutrition: A scoping review. *Nutrients, 14*(1), 194.

Venturini, K., Scatto, M., Pini, B., & Bigliardi, B. (2025). Towards sustainability in space food packaging. *Space Policy*, 101733. https://doi.org/10.1016/j.spacepol.2025.101733

Viejo, C. G., Harris, N., & Fuentes, S. (2024). Assessment of changes in sensory perception, biometrics and emotional response for space exploration by simulating microgravity positions. *Food Research International, 175*, 113827.

Waters, G. C., Olabi, A., Hunter, J. B., Ddkon, M., & Lasseur, C. (2002). Bioregenerative food system cost based on optimized menus for advanced life support. *Life Support & Biosphere Science, 8*(3–4), 199–210.

Williams, D., Kuipers, A., Mukai, C., & Thirsk, R. (2009). Acclimation during space flight: Effects on human physiology. *Canadian Medical Association Journal, 180*(13), 1317–1323.

Zabel, P., Bamsey, M., Schubert, D., & Tajmar, M. (2016). Review and analysis of over 40 years of space plant growth systems. *Life Sciences in Space Research, 10*, 1–16.

Zeidler, C., & Schubert, D. (2014). From bioregenerative life support systems for space to vertical farming on earth – The 100% spin-off. In *Life in space for life on Earth symposium.*

Zhang, L. Z., Dong, H. S., Yu, Y. B., Liu, L. Y., & Zang, P. (2022). Application and challenges of 3D food printing technology in manned spaceflight: A review. *International Journal of Food Science and Technology, 57*(8), 4906–4917.

Chapter 5
Precision Agriculture

Abbreviations

AI	Artificial intelligence
BRD	Bovine respiratory disease
CNNs	Convolutional Neural Networks
DL	Deep learning
DSMs	Digital Surface Models
ECa	Soil electrical conductivity
FDR	Frequency domain reflectometry
GMOT	Grass Measurement Optimization Tool
GNSS	Global Navigation Satellite Systems
GPR	Ground penetrating radar
GPS	Global Positioning System
GRS	Gamma ray spectrometry
IoT	Internet of Things
LiDAR	Light Detection and Ranging
ML	Machine learning
NIRS	Near Infrared Spectroscopy
PA	Precision agriculture
PLF	Precision livestock farming
PRYSM	Precision Robotic Sprayer
SVMs	Support Vector Machines
TDR	Time domain reflectometry
UAVs	Unmanned Aerial Vehicles
VOCs	Volatile organic compounds
VRT	Variable-rate technology

Y. Kumar, S. Dey, *Emerging Technology in Food Processing*,
https://doi.org/10.1007/978-3-032-24505-2_5

5.1 Introduction

Global agriculture is undergoing a fundamental transformation driven by the growing demand for food, heightened environmental concerns, climate variability, and advances in digital technologies. In response, precision agriculture (PA) has emerged as a data-intensive, technology-driven farming approach aimed at optimizing resource use and improving production efficiency. Precision agriculture is broadly defined as the application of modern technologies and data analytics to manage variability in agricultural production, thereby increasing output while minimizing environmental impacts (Monteiro et al., 2021). The concept of PA encompasses both precision crop farming and precision livestock farming (PLF). In crop production systems, PA allows farmers to monitor and respond to intra-field variability using tools such as soil sensors, drones, and GPS-guided equipment. This enables more accurate seeding, irrigation, fertilization, and harvesting, enhancing crop yields while reducing input waste. For example, sensor-based soil assessments and remote sensing platforms have revolutionized field management by providing real-time data on moisture levels, nutrient content, and plant health. The integration of these technologies with machinery allows for variable-rate application (VRA), thereby increasing input efficiency (Praveen et al., 2025). In pasture-based and intensive livestock systems, PLF technologies offer new opportunities to monitor animal behavior, health, reproduction, and productivity in real time. Wearable sensors, camera systems, and automated feeders are now commonly used to track individual animal activity, detect anomalies, and assess welfare parameters. These technologies enable the early detection of disease, improved nutritional management, and labor efficiency through automation. Innovations such as robotic milking systems, automated weight tracking, and edge computing-based behavior analysis exemplify the rapidly expanding field of PLF (Bernabucci et al., 2025). Underlying both crop and livestock precision systems is a robust technological infrastructure that supports sensing, analysis, and actuation. The first stage sensing and data collection involves technologies such as the Internet of Things (IoT), unmanned aerial vehicles (UAVs), and GNSS (Global Navigation Satellite Systems), which generate high-resolution, real-time data. In the data analysis stage, artificial intelligence (AI), machine learning (ML), and deep learning (DL) algorithms process these data streams to uncover patterns, make predictions, and support decision-making. Finally, the implementation phase uses this information to direct automated actions via actuators, variable-rate applicators, and robotics, enabling site-specific responses to field or animal conditions (Cruz et al., 2020). Furthermore, recent advancements are expanding the horizons of PA. Technologies such as cloud computing and big data analytics facilitate the storage and interpretation of large-scale agricultural datasets. Blockchain is increasingly used for ensuring traceability and transparency in the agricultural supply chain. Meanwhile, 6G communication technologies are expected to significantly enhance connectivity and data transmission speed in remote farming areas. Digital twins (DT), simulation modeling, and even quantum computing are being explored to push the boundaries of agricultural forecasting and management (Polymeni et al., 2024). Importantly, precision agriculture also contributes to broader goals of sustainability and environmental conservation. By applying water, fertilizers, and pesticides only where needed, PA minimizes runoff and greenhouse gas (GHG) emissions. In livestock systems, early detection of illness and stress not only improves animal welfare but also reduces the environmental footprint associated with animal production. Additionally, PA supports biodiversity and land stewardship by promoting conservation practices informed by real-time environmental data (Nawar et al., 2017). This chapter aims to provide a comprehensive overview of the current state of precision agriculture, with a particular focus on the integration of sensing technologies, data analytics, and automation across crop and livestock systems. It highlights recent advancements, emerging tools, and future directions that are shaping the digital transformation of agriculture.

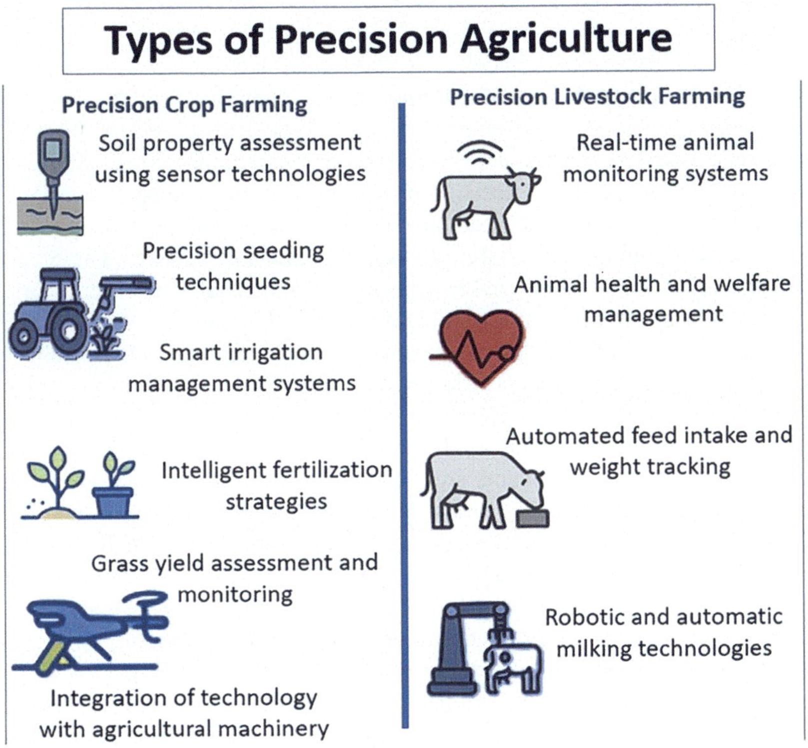

Fig. 5.1 Types of precision agriculture

5.2 Type of Precision Agriculture (Fig. 5.1)

5.2.1 Precision Crop Farming

Precision crop farming leverages advanced technologies such as artificial intelligence (AI), machine learning (ML), and the Internet of Things (IoT) to optimize agricultural practices by analyzing large datasets collected from various sources, including soil sensors, remote sensing, and historical yield records. These tools enhance predictive accuracy in crop yield estimation by considering complex and variable environmental factors like soil quality, climate, and water availability. Machine learning models such as Random Forests, Support Vector Machines (SVMs), and Convolutional Neural Networks (CNNs) are particularly effective in handling non-linear relationships and large-scale data. When integrated with smart systems such as precision seeding, intelligent fertilization, and smart irrigation, these technologies support real-time decision-making and resource optimization. The result is improved productivity, reduced input waste, and lower environmental impact. Precision crop farming thus represents a shift toward data-driven agriculture, promoting sustainability, economic efficiency, and resilience in food production systems through innovative, scalable, and accurate methodologies (Monteiro et al., 2021; Praveen et al., 2025).

5.2.1.1 Soil Property Assessment Using Sensor Technologies

The accurate assessment of soil properties is a foundational aspect of precision agriculture, enabling informed decision-making in crop management. A wide array of soil sensors has been developed to measure various soil parameters such as moisture content, nutrient availability, salinity, and soil texture. These sensor-based approaches significantly reduce the need for manual labor, while offering timely and high-resolution insights into soil health and variability. When accurately calibrated and appropriately interpreted, soil sensors can become indispensable tools for sustainable farm management (Monteiro et al., 2021).

One of the most commonly employed sensor types is the soil electrical conductivity (ECa) sensor, which quantifies the solute concentration in the soil to assess its salinity hazard. ECa measurements are widely utilized to map spatial variability in soil properties across agricultural fields. Mobile ECa scanning systems have gained traction due to their efficiency in delineating management zones, which is critical for site-specific interventions such as variable rate fertilization and irrigation. These spatial maps allow for more targeted resource allocation, ultimately contributing to cost reduction and improved environmental stewardship (Higgins et al., 2019; Ehsani & Sullivan, 2006). Soil water content is another critical parameter, often measured using frequency domain reflectometry (FDR) or time domain reflectometry (TDR) sensors. These sensors assess soil moisture by detecting changes in the dielectric constant, which correlates with volumetric water content. However, accurate measurement depends on the dielectric properties of the specific soil type, necessitating site-specific calibration. These sensors are crucial for irrigation management, helping determine when and how much water should be applied to optimize crop performance and minimize water waste (Lakhankar et al., 2009). In addition to FDR and TDR, tensiometers are widely used to evaluate soil water tension or suction. This metric reflects the effort required by plant roots to extract water from the soil and is valuable for estimating the soil's available water capacity. Tensiometers assist in quantifying the soil's current moisture status and are useful in scheduling irrigation to meet crop water demand without inducing stress or overwatering. Another innovation in this area is the use of optical sensors like photodiodes, which measure light reflectance to estimate soil characteristics such as clay content, organic matter, and moisture levels. These sensors provide a non-destructive and rapid method for assessing soil properties, further facilitating real-time monitoring (Lakhankar et al., 2009).

Recent advances have expanded the capabilities of soil nutrient analysis through portable sensor technologies. Smolka et al. (2017) introduced a mobile sensor platform capable of on-site analysis of soil extracts to detect key macronutrients, including nitrate (NO_3^-), ammonium (NH_4^+), potassium (K^+), and phosphate (PO_4^{3-}). These sensors significantly reduce the cost and time typically associated with laboratory-based soil testing, thereby promoting more frequent and site-specific soil health assessments. Electrochemical sensors and ion-selective electrodes have shown promise for real-time soil nutrient profiling, laying the groundwork for fully automated nutrient management systems. In addition to electrical and chemical sensors, geophysical methods such as ground penetrating radar (GPR) and gamma ray spectrometry (GRS) are also employed to evaluate soil properties. GPR data have demonstrated strong correlations with soil hydrological parameters, while GRS can detect variations in soil nutrient content and texture through vegetation and surface cover. These non-invasive techniques allow for deeper subsurface analysis, making them useful for large-scale soil monitoring (Monteiro et al., 2021).

Optical sensing technologies, including multispectral and hyperspectral imaging, further enhance soil assessment by analyzing light reflectance across different wavelengths. These sensors can detect subtle differences in soil composition and are often mounted on satellites, UAVs, or tractors for continuous field scanning. Their utility in mapping organic matter content, texture, and moisture levels is well documented, making them valuable tools in digital soil mapping (Gaikwad et al., 2015). Moreover, innovations in remote sensing have enabled the integration of Global Navigation Satellite Systems (GNSS) reflectometry into UAV-based platforms for soil moisture mapping. The EU-funded MISTRALE project demonstrated the use of GNSS signals, such as those from Galileo or GPS, to generate high-resolution moisture maps from drones flying at low altitudes. Though still in prototype development, such systems could provide valuable insights into field-scale water distribution, aiding in irrigation planning and flood management (Monteiro et al., 2021). The soil property assessment using sensor technologies has evolved into a multifaceted approach that combines physical, chemical, and remote sensing tools. These innovations support precision agriculture by enabling more efficient, sustainable, and data-informed farming practices. With ongoing advancements in sensor miniaturization, calibration algorithms, and data integration platforms, soil sensors are expected to become even more accessible and accurate, playing a central role in the digital transformation of agriculture.

5.2.1.2 Precision Seeding Techniques

Precision seeding is at the forefront of modern agricultural technologies, aiming to improve seeding efficiency, accuracy, and uniformity while reducing input costs. As agriculture transitions towards automation and sustainability, precision seeding plays a pivotal role in optimizing crop yield and ensuring the uniform emergence of seedlings. It refers to the accurate placement of seeds at controlled depth and spacing, tailored to specific crop requirements and environmental conditions (Monteiro et al., 2021). The primary role of precision seeding is to enhance the uniformity and efficiency of the sowing process. Unlike traditional seeding methods, which often suffer from inconsistencies in seed spacing and depth, precision seeding ensures optimal plant population and spatial distribution. This uniformity in seed placement leads to better competition for sunlight, water, and nutrients among plants, thus reducing variability in crop biomass and improving overall yield. According to data, precision seeding can increase crop output by 10–30% compared to conventional drill methods (Zhai JianBo et al., 2014). Precision seeding also supports sustainable agricultural practices by minimizing seed wastage and enabling site-specific seeding. It forms the foundation for variable-rate technology (VRT) and precision agriculture, where seeds are placed based on soil fertility zones, moisture content, or yield potential, enhancing resource use efficiency (Monteiro et al., 2021).

Modern precision seeding systems rely on advanced control mechanisms to synchronize the seed metering devices with the operating speed of the machinery. Monteiro et al. (2021) introduced a control system based on a single-chip microcomputer that allows real-time coordination between the seeder's movement and the seed metering rate. Experimental results demonstrated that such systems maintain consistent seeding rates, even under varying speed conditions, enhancing uniformity. Further developments include the use of sensors and feedback systems. For instance, Nielsen et al. (2018) proposed a dynamic coulter depth control mechanism that uses coulter position sensors and ultrasonic soil surface sensors to monitor and adjust the sowing depth in real-time. This adaptive system maintains a consistent depth across variable field terrains, ensuring synchronized seed emergence and reducing inter-plant competition during early growth stages. The precision of seed placement depends on the mechanics of the seed metering device and the delivery system. These mechanisms have evolved significantly, incorporating pneumatic, mechanical, and electric systems to improve seed singulation and drop accuracy. For high-speed operations, Gao et al. (2020) developed an air-assisted precision seed metering device that addresses the problem of short filling times and seed accumulation in the venturi tube. They determined that an optimal nozzle convergence angle of 70° and a feeding angle of 45° contributed to enhanced seed flow and placement accuracy, even at higher travel speeds. The seed delivery mechanism generally includes a seed box, metering wheel, and a delivery tube. The air-assisted systems use negative or positive pressure to control the movement of individual seeds, minimizing bounce and clumping during placement. These systems are particularly beneficial when seeding small-sized or lightweight seeds, which are otherwise difficult to handle mechanically. Recent trends point towards the integration of precision seeding with robotics and automation to further enhance precision and reduce labor dependency. Haibo et al. (2015) developed a wheel mobile robot specifically for wheat precision seeding. This four-wheel-drive robot incorporates a kinematic model and precision positioning system, offering consistent seed placement and depth control in varying field conditions. Similarly, Jin et al. (2019) designed an electric seeder tailored for small-sized vegetable seeds. This machine utilizes power-drive and optical fiber detection technology for real-time monitoring of seeding quality. It combines furrowing, seed placement, and soil compaction in one pass, improving field operation efficiency and accuracy. Robotic seeding systems are especially useful for crops that require precise spacing or are sensitive to seed placement errors. These machines can operate autonomously using GPS and sensor guidance, adjusting parameters dynamically based on field conditions and crop requirements. Precision seeding is transforming the landscape of crop cultivation by ensuring accurate seed placement and depth control, which are critical factors for uniform crop emergence and maximum yield potential. The integration of advanced control systems, sensor technologies, air-assisted delivery mechanisms, and robotics has significantly improved the performance of precision seeders. These advancements contribute not only to increased productivity but also to resource conservation, cost reduction, and sustainable farming practices. As the technology continues to evolve, precision seeding will remain a cornerstone of smart agriculture, enabling farmers to meet the growing food demand with greater efficiency and precision.

5.2.1.3 Smart Irrigation Management Systems

Smart irrigation management systems have emerged as a transformative innovation within precision agriculture, offering a strategic response to the increasing demand for water-efficient and sustainable farming practices. In the context of global climate change, growing population pressures, and the competition for freshwater across sectors, the adoption of intelligent irrigation systems is not only a technological advancement but also a necessity for ensuring food security and long-term agricultural productivity (Monteiro et al., 2021). Smart irrigation systems play a critical role in optimizing irrigation water use efficiency (IWUE), ensuring crops receive the right amount of water at the right time. Unlike traditional irrigation methods such as flood or fixed-schedule systems which often result in over irrigation or under irrigation, smart irrigation is knowledge and data driven. It leverages real-time environmental data and automation to fine-tune water application in accordance with actual crop water needs, thereby reducing wastage, preserving water resources, and enhancing crop yield (Rodriguez-Ortega et al., 2017; Shi et al., 2021). One of the primary roles of smart irrigation systems is to support sustainable agriculture through the precise allocation of water, which is increasingly important in arid and semi-arid regions. Additionally, these systems help mitigate the environmental impact of excessive water usage, such as nutrient leaching, salinization, and soil degradation (Rodriguez-Ortega et al., 2017; Goap et al., 2018). Smart irrigation systems function through the integration of sensor networks, automation hardware, communication technologies, and data analytics. These systems are capable of monitoring various environmental and soil parameters, including soil moisture, temperature, humidity, and rainfall, and automatically adjusting irrigation schedules based on current and forecasted conditions (Monteiro et al., 2021). For example, Krishnan et al. (2020) developed a GSM-based smart irrigation system that automates water delivery using feedback from soil moisture, temperature, and rain sensors. The system reduces manual labor and ensures that irrigation decisions are based on real-time data, thus preventing water wastage. Similarly, Goap et al. (2018) introduced an IoT-based smart irrigation prototype that incorporates a hybrid machine learning algorithm to predict soil moisture levels. Their model uses historical sensor data and weather forecasts to make proactive irrigation decisions. This predictive approach significantly enhances the accuracy of irrigation scheduling and ensures better planning of water usage. The core mechanism of a smart irrigation system is the feedback loop between sensors, controllers, and actuators. Sensors deployed in the field collect data on soil and climatic conditions. This data is transmitted often wirelessly to a central controller or server. The controller uses predefined thresholds or intelligent algorithms, such as fuzzy logic or machine learning, to decide when and how much to irrigate (Krishnan et al., 2020). Finally, actuators are triggered to open or close irrigation valves or operate pumps accordingly. Benyezza et al. (2021) designed a zoning-based smart irrigation system enhanced with a fuzzy logic controller (FLC). The system divides the agricultural land into zones based on crop and soil characteristics. Each zone is monitored individually, and the irrigation time is calculated using the FLC. This zoning strategy allows for more targeted water application, minimizing water and energy use while maintaining optimal crop health. Some advanced smart irrigation systems incorporate satellite remote sensing and weather forecasting models to optimize irrigation scheduling on a larger spatial scale. Corbari et al. (2019) demonstrated the use of LANDSAT satellite data in conjunction with meteorological models to update a distributed hydrological model for irrigation management in Southern Italy. This method allows for real-time adaptation of irrigation schedules based on evapotranspiration estimates and soil moisture trends across different plots. To further improve sustainability, Al-Ali et al. (2019) developed an IoT-based smart irrigation system powered by renewable solar energy. This system is especially relevant in off-grid rural regions, where energy availability is limited. It uses a single-board system-on-a-chip controller with built-in Wi-Fi, connected to solar panels, and sensors measuring soil and weather conditions. By operating autonomously, it offers a low-maintenance and eco-friendly solution for precision irrigation.

5.2.1.4 Intelligent Fertilization Strategies

The development and implementation of intelligent fertilization systems represent a transformative step toward sustainable and efficient agricultural practices. These systems leverage advanced technologies such as the Internet of Things (IoT), artificial intelligence (AI), machine learning, remote sensing, and robotics to optimize fertilizer use, reduce environmental impacts, and enhance crop productivity. By integrating various data sources, soil nutrient levels, crop growth status, weather forecasts, and topographic information, intelligent fertilization systems enable precise and site-specific nutrient management. This data-driven approach helps in addressing the key issues associated with traditional fertilization practices, such as over-fertilization, nutrient leaching, and runoff, which lead to environmental degradation and economic inefficiency (Mikula et al., 2020; Monteiro et al., 2021). Variable-rate technology (VRT) is one of the most significant components of intelligent fertilization systems. VRT enables the application of different fertilizer amounts at specific locations within a field, based on real-time data or prescription maps. Song et al. (2021) developed a UAV-based VRT granular fertilizer spreader capable of adjusting the application rate automatically by controlling the fluted roller motor's speed in response to site-specific requirements. The system exhibited high stability, quick response, and minimal error in both fixed and variable-rate modes, making it a promising tool for precision nutrient management (Song et al., 2021). Robotics also plays an essential role in the evolution of intelligent fertilization systems. For example, Ghafar et al. (2023) introduced a low-cost, two-wheeled agricultural robot capable of spraying fertilizers and pesticides while monitoring crop health using onboard cameras. The robot, controlled via a wireless interface, offers a practical and affordable solution for small to medium-sized farms. Its autonomous operation and adaptability to real-world conditions suggest that robotic systems can significantly enhance the efficiency and safety of fertilization tasks. Moreover, fertigation systems integrating irrigation and fertilization are being optimized using intelligent control frameworks. Lin et al. (2020) proposed an IoT-based fertigation system that utilizes an integer linear programming model to maximize both economic return and environmental sustainability. This system accounts for both long- and short-term planning, optimizing the allocation of limited resources across multiple crops. Such integrated decision-making frameworks support sustainable agricultural development (Lin et al., 2020). Finally, coupling crop models with optimization algorithms offers another dimension to intelligent fertilization. Bai and Gao (2021) applied a genetic algorithm alongside the DSSAT model to optimize nitrogen scheduling for maize. After calibrating and validating the model, they demonstrated improved decision-making for nitrogen applications, balancing crop yield with environmental considerations (Bai & Gao, 2021).

5.2.1.5 Grass Yield Assessment and Monitoring

Grass yield assessment and monitoring play a pivotal role in optimizing pasture management, enhancing livestock productivity, and supporting sustainable agricultural practices. Accurate and timely information on above-ground biomass yield enables informed decisions regarding grazing rotations, fertilization schedules, and harvesting, ultimately improving farm profitability and resource use efficiency. According to Beukes et al. (2020), consistent herbage measurements can lead to a 15% increase in farm profitability, underscoring the economic value of reliable grass yield data. Traditionally, grass yield was assessed using destructive sampling methods combined with weighing systems on agricultural equipment such as mowers, trailers, or balers. While accurate, these methods are labor-intensive, time-consuming, and often impractical for large-scale or frequent monitoring (Beukes et al., 2019; Murphy et al., 2020). The advent of non-destructive sensing technologies has revolutionized grass yield monitoring. Optical sensing techniques, including light detection and ranging (LiDAR) and active optical sensors, now enable the estimation of biomass through measurements of pasture canopy height and spectral reflectance. Combining LiDAR-based height data with reflectance signatures allows for more accurate biomass prediction. Furthermore, ultrasonic sensors have shown promising results in predicting herbage mass through canopy height estimation, with enhanced accuracy when paired with spectral reflectance data. These technological innovations provide rapid, repeatable, and non-invasive means of assessing biomass, which are crucial for real-time pasture management (Higgins et al., 2019). Unmanned aerial vehicles (UAVs), or drones, have further expanded the capabilities of grass yield monitoring. Equipped with high-resolution cameras and multispectral or hyperspectral sensors, UAVs can capture detailed image data at various plant phenological stages. These images, processed through photogrammetric software, generate 3D point clouds and digital surface models (DSMs) that enable precise sward height measurements over time. The reliability of dry biomass yield predictions using sward height metrics derived from low-cost UAV imaging across multiple years. The continuous miniaturization and affordability of these sensing platforms, coupled with advancements in data processing algorithms and computing hardware, are paving the way for more accessible and scalable pasture monitoring systems (Lussem et al., 2019, 2020; Schaefer & Lamb, 2016). Remote sensing also offers valuable insights into spatial variability in grass yield within and across fields. Oliveira et al. (2020) highlighted the potential of drone-based spectral remote sensing to assess field heterogeneity and predict yield with adequate accuracy, supporting better decision-making for silage production. These technologies enable site-specific management by identifying areas of low productivity and guiding targeted interventions. On-the-go pasture meters and tools like the Grass Measurement Optimization Tool (GMOT) represent another frontier in automated yield assessment. Pasture meters, using optical sensors and GPS integration, can perform real-time sward-yield mapping. GMOT enhances measurement precision by incorporating geo-spatial data and herbage management history, such as fertilization rates and previous grazing events, to refine herbage mass predictions. These tools help eliminate sampling bias and ensure spatially balanced data collection, further enhancing the reliability of yield assessments (Murphy et al., 2020). Long et al. (2016) collected the accurate yield data for alfalfa/grass and corn silage fields has been challenging, limiting adaptive management and on-farm research. Recent advancements in forage yield monitors on self-propelled forage harvesters (SPFHs) offer a potential solution. A study across 11 farms in 2013 showed that moisture sensors estimated crop moisture content within 3.7% DM for alfalfa/grass, while flow sensors predicted dry yield within ±0.5 Mg DM/ha. When regularly calibrated, these monitors provided precise yield data, supporting their use in large plots with significant treatment effects. This technology enables practical, data-driven decisions for improved pasture and silage management. The integration of advanced sensing technologies, UAV imaging, remote sensing, and smart decision-support tools has significantly enhanced the mechanisms of grass yield assessment. These developments not only improve the accuracy and efficiency of biomass monitoring but also support sustainable and profitable grassland management practices in modern agriculture.

5.2.1.6 Integration of Technology with Agricultural Machinery

The integration of advanced technology with agricultural machinery has revolutionized modern farming by enhancing precision, efficiency, and sustainability across various agricultural operations. Mechanization, once reliant on steam-powered tools and rudimentary engines, has now evolved into a highly automated system driven by intelligent electronics, mechatronics, and information and communication technologies (ICT). This transformation plays a critical role in addressing the growing need for food production while minimizing environmental impacts and reducing resource wastage (Adams, 2019; Monteiro et al., 2021). One of the most impactful advancements in this domain is the adoption of automated steering systems in tractors and field machinery. These systems utilize Global Navigation Satellite Systems (GNSS), along with sensors such as steering angle detectors and cameras, to guide agricultural vehicles along predetermined paths with high precision. This not only minimizes overlaps and misses during field operations like tillage, planting, and harvesting but also reduces fuel consumption, time, labor input, and soil compaction. The automation of steering, often referred to as autopilot systems, integrates hardware (GPS units, actuators, sensors) and software (path planning algorithms and navigation control systems) to enable tractors to operate autonomously with minimal human intervention (Azmi et al., 2020; Romano et al., 2020). Further, the introduction of the ISOBUS standard (ISO 11783) has significantly improved communication between tractors, implements, and on-board computers. This standardized protocol allows various digital devices in agricultural machinery to exchange data seamlessly, thus enhancing operational coordination and enabling real-time decision-making in precision farming. ISOBUS supports automated data acquisition, system diagnostics, and centralized control of multiple implements, streamlining complex field operations and improving management accuracy (Fountas et al., 2015). Mechatronic systems, comprising mechanical, electronic, and software components, are also central to modern intelligent machinery and agricultural robots (agrobots). These autonomous machines perform repetitive tasks such as weeding, seeding, fertilizing, and harvesting with high precision, reducing the dependency on manual labor and minimizing the application of agrochemicals. Lighter than traditional tractors, agrobots reduce soil compaction and are suitable for challenging terrains like sloped vineyards or waterlogged fields (Lowenberg-DeBoer et al., 2020). Variable rate application (VRA) systems further exemplify the integration of technology with machinery. Intelligent VRA equipment can adjust input levels (seeds, fertilizers, pesticides) based on real-time data from prescription maps or onboard sensors. Mechanisms like electronically controlled valves and dynamic feedback control systems ensure accurate dispensing, even when operating under varying environmental conditions or speeds (Rovira-Más et al., 2020). The fusion of technology with agricultural machinery has shifted the role of farmers from manual operators to digital managers, leveraging data-driven systems to enhance productivity, sustainability, and operational control across the agricultural value chain (Bo et al., 2024).

5.2.2 Precision Livestock Farming

Precision livestock farming (PLF) is an emerging paradigm that integrates advanced technologies to enhance animal production, health, welfare, and environmental sustainability. PLF uses real-time animal monitoring systems, including wearable sensors and remote sensing tools, to continuously assess animal behavior, location, and physiological parameters. Automated systems track feed intake, weight gain, and health indicators, enabling early detection of illness, improved feeding strategies, and optimized resource use. In dairy systems, robotic and automatic milking technologies reduce labor demands and improve milk quality by allowing consistent, animal-friendly milking routines. While widely adopted in intensive systems, extending PLF to extensive livestock system poses challenges due to environmental variability. However, innovations such as satellite-based pasture monitoring and edge computing are expanding the reach of PLF. This approach supports informed decision-making, better logistics, and sustainable management, especially in multispecies systems. PLF is pivotal in transforming livestock farming into a data-driven, efficient, and welfare-oriented industry (Bernabucci et al., 2025).

5.2.2.1 Real-Time Animal Monitoring Systems

Real-time animal monitoring systems play a pivotal role in modern livestock management by enabling precise, continuous, and individualized tracking of animal behavior, health, and welfare. These systems have evolved significantly with the integration of advanced sensor technologies, machine learning, and data transmission methods. Their primary function is to enhance decision-making in pasture management, improve animal health and welfare, increase productivity, and reduce labor costs, especially in large-scale and extensive livestock systems (Bernabucci et al., 2025). One of the core mechanisms of real-time animal monitoring is the use of Global Navigation Satellite System (GNSS) technologies, particularly GPS sensors, to record the movement patterns and spatial behavior of animals. GPS collars allow the continuous tracking of livestock, helping farmers understand grazing patterns, walking routes, resting areas, and preferred pasture zones. This geospatial data is critical in optimizing rotational grazing strategies and identifying under- or over-utilized pasture sections. GPS-based monitoring also helps differentiate between animal activities such as grazing, resting, or walking by analyzing movement speed and trajectory (Ungar & Nevo, 2025). In addition to location tracking, accelerometers embedded in collars or tags are used to classify specific behaviors such as feeding, rumination, standing, lying, or signs of lameness. These sensors collect data at high frequencies (typically 4–10 Hz), and when combined with GPS data, allow accurate identification of posture and behavior patterns with over 90% accuracy. For example, changes in head position detected by accelerometers can distinguish feeding-related behaviors, while subtle gait changes can be used to detect lameness, often earlier than visual inspection would allow (Monteiro et al., 2021; Bernabucci et al., 2025). Behavior-monitoring collars, integrated with various sensors, have proven effective in detecting estrus and abnormal behavioral changes, such as reduced rumination or feeding time, which may indicate illness or stress. Studies have shown that such automated systems outperform human observation in consistency and accuracy. Data collected from these devices can be stored locally and transmitted periodically or in real-time to farm management platforms, allowing for prompt intervention and improved animal welfare (Lamanna et al., 2025). Emerging technologies also include machine vision and audio analysis. Image-based systems can monitor posture, movement, and facial expressions, providing insight into stress, discomfort, or health issues. Facial recognition systems, such as Convolutional Neural Networks (CNNs), can now identify individual animals like pigs with high accuracy (up to 96.7%), enabling personalized monitoring. Likewise, sound analysis helps detect behavioral states; differences in vocalization frequencies have been associated with specific activities such as ruminating or stress responses (Monteiro et al., 2021). Furthermore, machine learning algorithms enhance data interpretation by classifying complex behavior patterns and predicting anomalies. These technologies are increasingly capable of integrating multimodal data (location, motion, sound, and visuals) for comprehensive animal monitoring. The integration of GPS with Internet of Things (IoT) frameworks has further enhanced monitoring by enabling seamless data exchange between devices and central systems. Coupled with Low Power Wide Area Networks (LPWAN) technologies like LoRaWAN, Sigfox, NB-IoT, and DASH7, real-time animal tracking becomes feasible over long ranges with low data rates and minimal energy consumption. However, GPS data transmission remains the most power-consuming activity, necessitating optimized battery management and maintenance schedules (Bernabucci et al., 2025). The real-time animal monitoring systems represent a transformative advancement in precision livestock farming (PLF). By providing continuous, objective, and accurate insights into animal behavior and health, these systems empower farmers to make timely and informed management decisions, contributing to more sustainable, efficient, and welfare-conscious livestock production.

5.2.2.2 Animal Health and Welfare Management

Effective animal health and welfare management is central to sustainable livestock production, ensuring productivity, minimizing economic losses, and enhancing animal well-being. Poor animal health impairs growth, increases mortality risk, and can compromise welfare through pain and discomfort. Additionally, it poses significant public health concerns, especially through the transmission of zoonotic diseases and the misuse of antibiotics. The early detection and prevention of disease, therefore, form the cornerstone of precision livestock farming (PLF), where the goal is not only to identify clinical signs but also to anticipate and mitigate health issues before they escalate (Bernabucci et al., 2025). Modern technologies have revolutionized health management through the use of sensors, artificial intelligence (AI), machine learning (ML), and big data analytics. These systems continuously collect data related to key health indicators such as animal movement, feeding and drinking behavior, temperature, and air quality, enabling real-time monitoring and predictive diagnostics. Algorithms analyze the data to detect deviations from baseline behaviors or physiological norms, allowing farmers and veterinarians to intervene early, improving treatment outcomes and reducing reliance on antibiotics (Michelena et al., 2025). Sound-based technologies are particularly promising for disease detection. Tools such as Pig Cough Monitor™, Soundtalks®, and Fancom B.V. systems employ mathematical algorithms to detect abnormal respiratory patterns like coughing in pigs and calves. These systems have demonstrated over 90% precision in detecting signs of bovine respiratory disease (BRD), enabling faster and more accurate diagnoses than human observation alone. Similarly, distress in animals can be vocalized or manifested in altered activity patterns, which can be captured through microphones and surveillance cameras. Interpreting these vocalizations or movements using AI further improves the sensitivity and specificity of disease monitoring (Meen et al., 2015). Advanced air sensors have also been deployed, particularly in the poultry industry, to monitor volatile organic compounds (VOCs) that correlate with disease outbreaks like Coccidiosis. These sensors can detect elevated VOC concentrations well before clinical symptoms become visible, providing a significant head start in disease prevention and control. For instance, in dairy cattle, sensor technologies can predict and manage mastitis by analyzing milk conductivity, udder temperature, and activity levels (Neethirajan, 2020). Image analysis technologies also support welfare management by identifying physical abnormalities. Automated camera systems can assess posture to detect lameness, while ingestible sensor pills can measure core physiological data such as heart rate. These non-invasive methods offer cost-effective, continuous, and accurate health assessments without adding stress to the animals (Bernabucci et al., 2025). Feed and water intake monitoring is another critical aspect of welfare. Reduced consumption is often an early indicator of health issues or environmental stress. The integration of climate control sensors such as temperature, humidity, and CO_2 detectors provides additional insight into thermal comfort and respiratory health, enabling automatic ventilation or cooling system adjustments to mitigate environmental stressors (Van Hertem, 2018). Overall, the integration of PLF technologies into animal health and welfare management systems enables farmers to maintain high standards of care while maximizing efficiency and sustainability. These intelligent systems provide robust tools for proactive, data-driven decisions that support early disease detection, improve treatment outcomes, enhance animal comfort, and ultimately safeguard both livestock productivity and public health.

5.2.2.3 Automated Feed Intake and Weight Tracking

Automated feed intake and weight tracking represent vital components in the broader context of precision livestock farming (PLF). These systems enable individualized animal monitoring and data collection, facilitating optimized nutrition strategies, improved animal welfare, and increased farm profitability. Precision livestock feeding, informed by real-time sensor feedback, aims to deliver the optimal amount of nutrients tailored to individual animals or groups. This optimization helps to maximize nutrient utilization without compromising animal performance, while also minimizing environmental impacts. Notably, it can reduce protein intake by up to 25% and nitrogen excretion by 40%, with potential profitability increases of around 10% (Banhazi et al., 2012; Hendriks et al., 2019). Automated feeding systems (AFS) are increasingly being adopted as a cost-effective alternative to manual feeding regimes. These systems have been developed for various species, including cattle, pigs, and sheep. Modern AFS platforms typically include interfaces that record essential feeding metrics such as the time and frequency of feed access, animal identification via RFID tags, feed weight consumed, and duration of feeding events. Such systems allow for accurate tracking of individual feed consumption, which is crucial in evaluating animal growth, health, and efficiency (Bezen et al., 2020). Several regional innovations have demonstrated the practical benefits of AFS. For instance, in Australia, feed sensors have been developed that precisely measure and control feed delivery at the level of individual feeders (Banhazi et al., 2012). In Canada, next-generation systems are emerging that not only regulate feed quantity but also offer customization of nutrient composition, allowing real-time dietary adjustments based on sensor data reflecting the animals' physiological state (O'Grady & O'Hare, 2017). Such developments pave the way toward truly individualized feeding regimens. Portable near infrared spectroscopy (NIRS) has also been explored for use in dairy farms, enabling rapid and accurate analysis of total mixed ration (TMR) and manure composition. Evangelista et al. (2021) demonstrated the potential of NIRS to provide on-site, real-time data that could inform feeding strategies and nutrient management. RGB-D cameras further contribute to feed intake measurement by visually tracking individual cow feeding behavior, offering non-invasive and continuous monitoring capabilities. Automated weight tracking is equally critical, as average daily gain is a important performance indicator in livestock systems. Systems like the Osborne Weight-Watcher™ and image-based weighing platforms use dimensional analysis to estimate animal body weight with acceptable accuracy (Banhazi et al., 2012). These technologies have proven effective in generating performance records and identifying growth trends in successive animal cohorts. Banhazi et al. (2012) reported the successful application of such systems in pig production, highlighting their reliability and timely data delivery. In combination, automated feed and weight tracking systems contribute significantly to data-driven livestock management. Advanced algorithms and nutritional models are being integrated to calibrate feed rations, assess environmental impacts such as greenhouse gas (GHG) emissions, and optimize economic outcomes. As technology advances, the integration of sensor data, AI-driven analytics, and dynamic modeling will enable even finer control over feed efficiency and animal productivity, marking a substantial shift toward sustainable and profitable livestock farming (Banhazi et al., 2011; O'Grady & O'Hare, 2017).

5.2.2.4 Robotic and Automatic Milking Technologies

Robotic and automatic milking technologies are among the most established and widely recognized applications of precision livestock farming (PLF). The adoption of Automatic Milking Systems (AMS) has accelerated, especially in Western Europe, driven by the need to reduce manual labor, increase milk production per cow, and improve the overall quality of life for dairy farming families. The use of AMS has transformed traditional milking routines by allowing cows the freedom to choose their milking times, thereby enabling a more animal-centric approach to dairy management. Initially designed for indoor systems or farms with access to adjacent grazing areas, AMS technology is now expanding into larger-scale pasture-based dairy operations, further enhancing its adaptability and reach (Rodenburg, 2017; John et al., 2016). Modern AMS units are equipped with a range of advanced features, including sensors, vision systems, and real-time monitoring tools. These upgrades facilitate efficient teat detection, udder health monitoring, and milking process control. The integration of time-of-flight (ToF) depth-sensing cameras has significantly improved the precision of teat localization, achieving accuracy levels within 5 mm when the search area is restricted to a defined region of interest. Such precision is vital, given the anatomical variability in udder shape among cows and across different lactation stages. The capability to accurately identify and track teats using depth imaging, point-cloud data, and intelligent algorithms marks a significant advancement in automated milking technology (O'Mahony et al., 2019). Recent developments have explored various computer vision techniques for improving AMS performance. Rodenburg (2017) demonstrated effective teat detection using point-cloud analysis, while other studies have experimented with structured light cameras such as the Microsoft Kinect and machine learning approaches like Haar

Cascade classifiers. Additionally, the use of thermal imaging and stereovision has been investigated for enhanced teat tracking, especially under challenging lighting or environmental conditions (Ben Azouz et al., 2015). These technologies not only improve the success rate of teat cup attachment but also reduce the cycle time of the milking process, increasing overall system efficiency. Reducing milking cycle time is a critical factor in the efficiency of AMS operations. Intelligent robotic control systems, informed by real-time visual feedback, are essential to navigate the milking cluster quickly and accurately onto the udder. This capability not only boosts throughput but also reduces stress on the animals. Moreover, the ability to accommodate cows with atypical udder shapes or behavioral resistance to conventional AMS units represents a key area of innovation. Designing flexible robotic systems that can adapt to a wider variety of cows can reduce the need to cull animals that do not conform to standard milking system parameters (O'Mahony et al., 2019). The robotic and automatic milking technologies are revolutionizing dairy farming by enhancing productivity, animal welfare, and operational sustainability. As sensor precision and artificial intelligence continue to improve, the potential for AMS to operate effectively in diverse farming environments will expand. The future of AMS lies in greater integration of vision-based automation, adaptive learning systems, and real-time animal behavior analytics, ultimately contributing to more sustainable and efficient dairy production systems worldwide.

5.3 Technology of Precision Agriculture (Fig. 5.2)

5.3.1 Sensing and Data Collection (Perception Stage: Sensors; UAVs (Drones); GNSS; IoT)

Sensors function by detecting physical, chemical, or biological parameters from their environment and converting them into digital signals for analysis. In agriculture, soil sensors measure moisture, temperature, salinity, and nutrient levels using electrical resistance, capacitance, or dielectric properties. Plant health sensors such as leaf chlorophyll meters use optical methods (e.g., light absorption or fluorescence) to assess photosynthetic activity. Environmental sensors monitor ambient temperature, humidity, wind speed, and solar radiation through thermistors, hygrometers, and anemometers. These sensors are often deployed in situ (in the soil or on plants) or mounted on equipment and platforms. Once activated, the sensors collect data at regular intervals and relay it to data loggers, local devices, or cloud platforms. Real-time monitoring helps optimize irrigation, fertilization, and pest control. Advanced sensing systems integrate machine learning algorithms for automated diagnostics and decision-making. Overall, sensors form the foundational layer of data-driven precision agriculture by providing granular, site-specific insights (Khoramshahi et al., 2025).

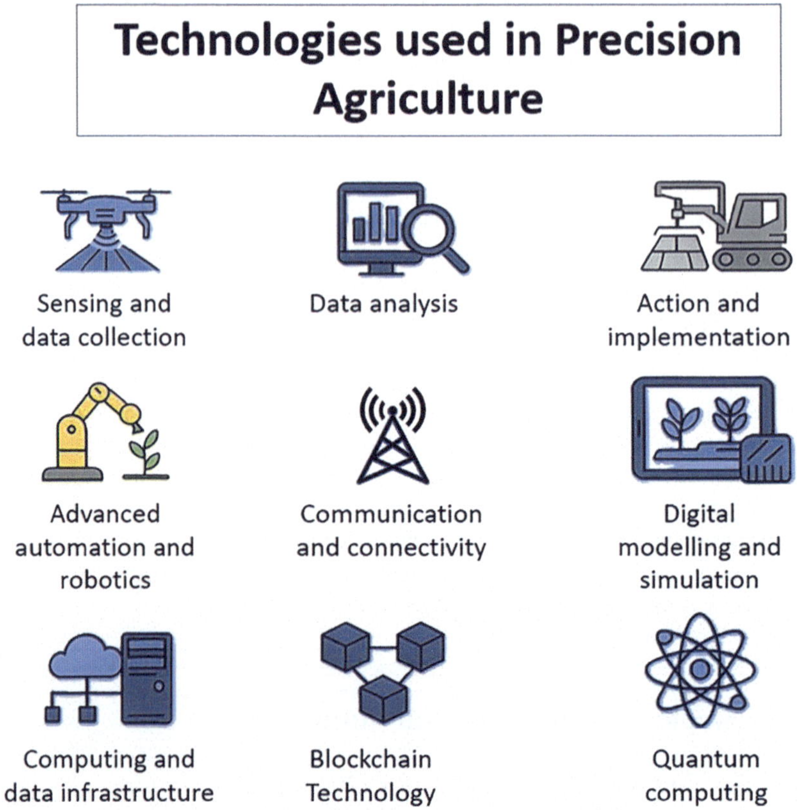

Fig. 5.2 Technologies used in precision agriculture

Unmanned aerial vehicles (UAVs), commonly known as drones, are aerial platforms equipped with various imaging sensors that facilitate high-resolution crop and field monitoring. These systems typically consist of a flight controller, GPS module, inertial sensors (gyroscopes, accelerometers), and a camera payload (RGB, multispectral, thermal, or hyperspectral). The UAV follows pre-programmed flight paths defined by GNSS waypoints. During flight, it captures geo-referenced images of the farmland. Multispectral and hyperspectral sensors measure specific wavelength bands that reveal plant physiological states, such as chlorophyll content, water stress, and disease symptoms. Thermal sensors detect temperature variations that may indicate irrigation inefficiencies or pest infestations. Once the UAV completes its mission, image data is processed using photogrammetry and vegetation index algorithms (e.g., NDVI) to create orthomosaics, 3D terrain models, and prescription maps. These outputs enable precision interventions like variable-rate spraying. UAVs thus offer a rapid, scalable, and non-invasive mechanism for real-time crop assessment and decision support (Khoramshahi et al., 2025).

Global Navigation Satellite Systems (GNSS) operate by using signals from multiple satellite constellations such as GPS, GLONASS, Galileo, and BeiDou to determine the precise location of a receiver on Earth. Each satellite transmits timing signals that the GNSS receiver uses to calculate distances based on signal travel time. When signals from at least four satellites are received, the receiver triangulates its position in three-dimensional space. For agriculture, GNSS-enabled machinery like tractors, UAVs, and sprayers uses these coordinates for autonomous navigation and accurate field operations. Advanced positioning techniques like Real-Time Kinematic (RTK) and Precise Point Positioning (PPP) improve accuracy to the centimeter level by correcting atmospheric delays and clock errors. GNSS supports site-specific management practices such as planting, fertilizing, and harvesting with minimal overlap or gaps. It also allows for field mapping and georeferencing of sensor data. GNSS thus acts as the spatial backbone of precision agriculture, ensuring high positional accuracy for smart farming applications (Khoramshahi et al., 2025). The Internet of Things (IoT) in agriculture connects a network of sensors, actuators, and devices through wireless communication protocols, enabling automated, real-time data collection and control. Each IoT device comprises sensing units (for soil, crop, and weather), a microcontroller, and a communication module (e.g., LoRa, NB-IoT, Wi-Fi, Bluetooth (BT)). These devices transmit data to a central gateway or cloud server, where it is processed and visualized using data analytics tools. Edge computing nodes often perform preliminary processing to reduce latency and bandwidth usage. IoT enables seamless integration of devices like irrigation valves, fertilizer injectors, and UAVs with sensor networks for responsive action. For example, a soil moisture sensor can trigger irrigation via a smart controller based on real-time thresholds. Cloud platforms support mobile alerts, remote monitoring, and AI-based forecasting. This interconnected ecosystem allows for scalable, data-driven farm management, improves decision accuracy, and reduces labor, input costs, and environmental impact (Khoramshahi et al., 2025).

5.3.2 *Data Analysis (Analysis Stage: AI; ML; DL)*

AI in precision agriculture functions as a decision-making engine that integrates data from multiple sources, such as UAVs, IoT sensors, GPS systems, and satellite imagery, to provide automated, optimized recommendations for crop management. AI systems ingest real-time environmental and crop condition data, process it using rule-based systems or intelligent agents, and deliver actions like when and where to irrigate, apply pesticides, or harvest. AI technologies help identify hidden patterns, simulate outcomes, and improve operational efficiency. For example, AI can suggest variable-rate fertilizer application by analyzing past crop yield, weather forecasts, and soil nutrient levels. AI also integrates image recognition and natural language processing for early pest detection or field monitoring. As a supervisory layer, AI orchestrates ML and DL models for high-level insights and strategic planning, supporting farmers with timely, data-driven decisions while reducing waste and improving sustainability (Botero-Valencia et al., 2025).

Machine learning operates by learning patterns from agricultural data to make accurate predictions or classifications. The process begins with data collection from sources like UAVs, soil sensors, weather stations, and satellite imagery. This data undergoes preprocessing such as cleaning, normalization, and feature extraction to prepare it for model training. Supervised ML models (e.g., Decision Trees, SVM, Random Forest) are trained on labelled datasets to predict outcomes like crop yield, disease presence, or optimal irrigation schedules. In unsupervised learning, clustering algorithms can segment fields by soil fertility or crop health. After training, the models are validated and deployed to support ongoing field operations. For instance, ML algorithms have been used to analyze drone imagery to predict nitrogen nutrition index in rice or estimate aboveground biomass in soybean crops. The strength of ML lies in its adaptability to different conditions and its capability to integrate multi-source, multi-temporal data for precision decision-making (Botero-Valencia et al., 2025).

Deep learning, a subset of ML, mimics the human brain using artificial neural networks to interpret complex, high-dimensional agricultural data. Its mechanism relies on architectures like Convolutional Neural Networks (CNNs) to analyze spatial data, such as UAV-captured images, for detecting diseases, classifying crop types, or estimating biomass. The process begins with raw input data (e.g., RGB, thermal, or hyperspectral images), which is passed through multiple layers of the neural network. Each layer extracts progressively abstract features, edges, shapes, and textures that help in identifying patterns like plant stress or disease symptoms. DL models require large datasets and high computational resources but offer superior accuracy. For example, DL has been used in early detection of tomato plant viruses using image-based classification. In combination with IoT and remote sensing, DL enhances real-time monitoring and predictive capabilities, making it vital for automating complex tasks in smart farming environments and improving sustainability (Botero-Valencia et al., 2025).

5.3.3 Action and Implementation (Actuation Stage: Smart Sprayers; Prescription Maps)

The urgent need to modernize spraying operations in steep-slope vineyards arises from the unique combination of labor shortages, harsh terrain, and the inefficiency of current methods. Manual spraying is no longer sustainable due to its physically demanding nature and the declining availability of vineyard workers. Meanwhile, tractor-based sprayers, though somewhat mechanized, struggle with terrain constraints, causing off-target drift, low efficiency, and soil compaction. These limitations highlight a clear need for intelligent, precision spraying systems tailored to the specific challenges of mountainous vineyards (Baltazar et al., 2021). Smart sprayers, such as the Precision Robotic Sprayer (PRYSM), represent a transformational solution. They reduce labor dependency and optimize resource use by targeting application areas more accurately, thereby lowering pesticide usage and environmental impact. Equipped with GNSS and LiDAR-based navigation, these autonomous systems adapt to steep and irregular slopes, narrow row widths, and heterogeneous canopies. Integrated sensors and machine learning algorithms, including SVM-based leaf density classifiers, enable real-time decision-making to dynamically adjust spray parameters. This precision-driven approach ensures uniform coverage, minimizes waste, and enhances operational flexibility, positioning smart sprayers as a cornerstone of sustainable, high-efficiency viticulture in challenging terrains (Baltazar et al., 2021). The smart spraying mechanism implemented in the PRYSM robot follows a closed-loop approach that combines advanced sensing, machine learning, and real-time actuation to optimize fertilizer application in steep-slope vineyards. The process begins with the activation of the robot's core systems, including localization components (GNSS and LiDAR) and the AgIoT 2.0 module, which manages sensor and actuator communication. As the robot navigates autonomously along the vineyard rows, a stereo camera mounted on the front captures high-resolution images of the grapevine canopy ahead of the spraying zone. These images are essential for detecting variability in foliage density and ensuring accurate, data-driven spraying decisions. Once an image is captured, it is divided into three regions of interest (ROIs), each corresponding to one of the PRYSM robot's three centrifugal spray drums. Each ROI is processed independently to extract a set of 14 visual descriptors, which represent the structural and spectral characteristics of the canopy. These descriptors include local binary pattern (LBP) texture features, hue histograms, vegetation indices (RGBVI, GLI, and NGRDI), and average pixel intensity values related to greenness and vegetation health. These features form the input to a trained Support Vector Machine (SVM) classifier that estimates the leaf area index (LAI) for each region. The classifier assigns each ROI to one of four LAI classes: 0%, 33%, 66%, or 100% leaf coverage. These classes correspond to specific control settings for the sprayer, allowing real-time adaptation of the spray parameters. The brushless motor adjusts the airflow, the DC motor regulates the fertilizer concentration (density), and the water pump manages the flow rate. Each spray drum operates independently, spraying only when sufficient leaf density is detected in its corresponding region. This targeted approach avoids overapplication in sparse areas and reduces chemical drift, improving both environmental sustainability and application efficiency. Following this decision-making step, the sprayers are actuated, and the robot continues its autonomous movement down the row. The sensing-analysis-actuation loop operates continuously, enabling the PRYSM system to respond dynamically to canopy variability throughout the vineyard. This smart spraying mechanism exemplifies how automation and artificial intelligence can transform conventional viticulture by reducing labor, increasing precision, and promoting sustainable agricultural practices in even the most challenging terrain (Baltazar et al., 2021). Prescription maps have become a cornerstone of precision agriculture, enabling site-specific management by translating spatial variability into actionable field operations. Derived from UAV-based remote sensing data, these maps incorporate key variables such as vegetation indices (e.g., NDVI, GNDVI), elevation models, and moisture distribution to inform variable-rate applications (VRAs) of fertilizers, pesticides, and defoliants. By linking crop growth status with input requirements, prescription maps ensure that resources are applied where and when they are needed most, improving efficiency and reducing environmental impact. Agricultural drones equipped with VRA systems can execute these prescriptions autonomously, enhancing adaptability across diverse crop types and terrains. Despite advancements, practical implementation in staple crops like wheat remains limited, especially for fine-tuned parameters like protein uniformity. Nonetheless, emerging research shows promise, with UAV-generated prescription maps being used to adjust nitrogen inputs, balance intra-field variability, and ultimately enhance yield and grain quality in a sustainable and economically viable manner (Guan et al., 2025). In the study on the wheat variety "Yukichikara" in the snow-prone Hokuriku region of Japan, utilized UAV-based remote sensing and drone technologies to generate and execute prescription maps tailored to in-field growth variability. UAVs, including the DJI Phantom 4 RTK and P4 Multispectral, captured NDVI imagery prior to heading, which was processed using DJI Terra software to produce spatial distribution maps of crop vigor. These maps were translated into prescription maps that specified fertilization rates across zones of low, medium, and high growth. The Agras T10 agricultural drone, equipped with a spreading system and real-time sensors, executed VRA by adjusting flight speed and fertilizer outlet size according to the prescription. Fields F4v and F5v received precision applications in 2023, while multi-layered treatments were applied in F1v in 2022 due to technical constraints. The prescription maps ensured nitrogen was applied at the anthesis stage—critical for protein formation maximizing grain quality while minimizing resource wastage. This approach demonstrates how UAV-generated prescription maps integrate remote sensing, data processing, and automated machinery to optimize agricultural productivity and sustainability (Guan et al., 2025).

5.3.4 Advanced Automation and Robotics

Advanced automation and the integration of robotics in precision agriculture are revolutionizing traditional farming practices by enhancing efficiency, reducing labor dependency, and promoting sustainability. Robotics ranging from ground-based machines to aerial drones are now applied across various stages of agricultural production, including soil preparation, planting, weeding, plant protection, harvesting, and yield estimation. These systems utilize cutting-edge technologies such as deep learning, machine vision, and RTK-GPS for precise navigation, object detection, and real-time decision-making in complex, unstructured field environments. Autonomous agricultural robots, equipped with stereo cameras, lidar, and internet connectivity, can work independently or collaboratively, even communicating with other machines to ensure continuous operation. As regulatory frameworks evolve, including the upcoming EU Machinery Regulation (2023/1230), safety and certification standards for autonomous machines are becoming more stringent, requiring third-party validation for systems with self-evolving capabilities. Despite the high initial costs, these innovations offer significant benefits, such as optimized nitrogen use, reduced greenhouse gas emissions, and lower pest-related losses. However, barriers to adoption persist, particularly in developing countries, where affordability, infrastructure, and technical expertise remain challenges. To unlock the full potential of robotic technologies, it is essential to integrate them with broader ecological and social systems, aligning with strategies like Farm to Fork to ensure both productivity and sustainability in global agriculture (Spagnuolo et al., 2025). Robotics plays a transformative role in precision agriculture by enhancing efficiency, accuracy, and adaptability in complex and dynamic farming environments. Agricultural robotic arms, tailored for specific field operations such as harvesting, weeding, seeding, spraying, and crop management, are classified based on their function, structure, and operational environment. These systems integrate advanced hardware components, manipulators, end-effectors, sensors, and drivers with sophisticated software frameworks that support perception, motion planning, and control. This integration allows robotic arms to intelligently interact with crops and adapt to real-time environmental changes. Recent innovations, such as high-speed dual-arm harvesters and vision-guided weeding robots, demonstrate their effectiveness in improving productivity and reducing crop damage. Despite challenges like environmental variability, safety concerns, and the need for real-time adaptability, robotic arms offer promising solutions to optimize agricultural operations, reduce labor dependence, and support sustainable farming practices (Jin & Han, 2024). Tomato harvesting, which accounts for 40% of global vegetable production, remains labor-intensive, costly, and inefficient. Robotics offers a sustainable solution by automating the harvesting process, reducing labor costs, minimizing physical strain on workers, and improving consistency and yield quality. The core mechanism involves advanced machine vision and deep learning models, such as a modified Single-Shot Multi-Box Detector (SSD) enhanced with a self-attention mechanism. This enables real-time, accurate detection and localization of ripe tomatoes, even under complex field conditions like occlusion and variable lighting, ensuring efficient picking and supporting the broader goals of precision agriculture and sustainability (Song et al., 2025). The robotic tomato detection mechanism in this study integrates a modified Single-Shot Multi-Box Detector (SSD) model with a self-attention mechanism to optimize object detection accuracy and efficiency. Traditionally, SSD utilizes six feature maps, but this study discards four middle layers (conv7, conv8.2, conv9.2, conv10.2) to reduce computational complexity and focus on detecting large objects like tomatoes. This simplification enhances speed and maintains crucial image features through the retained conv4.3 and conv11.2 layers, which now have broader receptive fields suitable for large-object recognition. To further enhance performance, a self-attention mechanism is embedded into the conv4.3 and conv11.2 layers. This mechanism transforms input vectors into query, key, and value spaces to compute attention weights. It helps the model concentrate on critical regions of the image, boosting detection precision under complex scenarios such as occlusions and variable lighting. The conv4.3 layer refines local focus, while conv11.2 integrates global context, improving feature representation and robustness. For training, 1000 labeled images of tomatoes were collected using a camera setup that simulates a robot's perspective. Labeled using LabelImg (Google Brain, version 1.7), the dataset was processed in Tensor Flow using a high-performance computing system, supporting real-time, accurate tomato detection for robotic harvesting (Song et al., 2025). The modified SSD model with self-attention achieved 95% accuracy and 96.1% recall in tomato detection, outperforming traditional SSD models. It reduced model size and computation time by over 21%, enabling faster, more efficient robotic harvesting (Song et al., 2025).

5.3.5 Communication and Connectivity (6G Technology)

6G technology enables precision agriculture by integrating ultra-reliable communications, sensing, and robotics to enhance productivity and reduce environmental impact. Ambient IoT (A-IoT), a key 6G component, supports low-cost, battery-free sensor networks for real-time monitoring and control in agricultural environments (Arun et al., 2025). 6G technology, particularly through Ambient IoT (A-IoT), plays a transformative role in precision agriculture by enabling dense, energy-efficient sensor networks that support real-time monitoring and data-driven decision-making. A-IoT devices, which operate using ambient energy sources such as electromagnetic, solar, or thermal energy, offer a low-cost and sustainable solution for critical agricultural applications like underground soil sensing. These devices utilize backscatter communication to interface with existing 3GPP architecture, eliminating the need for constant power supplies or bulky batteries, which are often impractical in large-scale, rural deployments. By facilitating seamless integration of communication, sensing, and control technologies, 6G empowers precision agriculture to enhance productivity, optimize resource use, and mitigate environmental impacts (Arun et al., 2025). 6G communication in precision agriculture operates through a highly advanced network infrastructure that integrates hyperconnectivity, ultra-reliable low-latency communication (URLLC), and cutting-edge technologies such as AI, quantum sensing, and edge computing. This mechanism enables seamless, real-time data exchange across a vast number of interconnected devices, sensors, drones, autonomous vehicles, and machinery distributed throughout the agricultural landscape. 6G's ultra-low latency ensures instant response and synchronization, critical for time-sensitive applications like autonomous irrigation, real-time pest control, or equipment maintenance. Edge computing processes sensor data locally, minimizing transmission delays and enabling quick decision-making on-site. Quantum sensing offers ultra-precise monitoring of soil and plant health, while quantum communication secures sensitive agricultural data. Four-dimensional communication further enhances signal stability and data speed, even in remote rural areas. Overall, 6G transforms agricultural operations into intelligent ecosystems where high-resolution data is continuously collected, analyzed, and acted upon, leading to optimized resource use, increased yields, and reduced environmental impact (Polymeni et al., 2024).

5.3.6 Digital Modeling and Simulation

Modeling and simulation play a vital role in advancing precision agriculture (PA) by enabling farmers to visualize, assess, and adopt innovative technologies without real-world risks. The transition to PA involves integrating digital tools such as GNSS, sensors, and georeferenced data, which improve accuracy, efficiency, and environmental outcomes. However, real-world limitations like weather variability and operation timing hinder objective evaluation of PA benefits. A simulation platform offers a solution by generating diverse, data-rich scenarios that quantify the impact of varying PA strategies. Virtual reality (VR), especially immersive systems, enhances this by creating interactive environments that simulate real-life farming experiences. These tools can serve as training aids, increasing farmer's confidence and understanding of PA practices. Additionally, non-immersive or partially immersive VR systems offer accessible options for wider user engagement. Such modeling and simulation environments support informed decision-making, training, and the broader adoption of sustainable and efficient farming technologies (Cutini et al., 2023). The mechanism of modeling and simulation in precision agriculture (PA) integrates virtual environments, physical actuators, and digital twins (DT) to replicate and analyze agricultural operations under diverse scenarios. A core component is the DT, a sensor-based virtual replica of physical systems like tractors and implements. It allows bi-directional data flow between the physical and digital environments, enabling real-time updates and interaction. The simulator includes both DT and physical controls (steering wheel, pedals, and command console), allowing the operator to act within a closed-loop system that mirrors actual farming conditions. The system enables simulation of various factors, such as forward speed, field shape, prescription maps, working width, and machine type conditions, that are difficult or impossible to replicate in real-life settings due to constraints like cost, weather, and resource limitations. Modeling and simulation help assess energy consumption, optimize machine parameters, analyze operator behavior, and integrate multiple PA technologies (e.g., autoguiding, ISOBUS, variable rate application (VRA)). These simulations support technology validation, operator training, and performance assessment. By predicting machinery performance under controlled, repeatable, and risk-free scenarios, the simulator fosters informed decision-making and facilitates the adoption of sustainable, efficient PA practices, supporting the design of hardware and software solutions tailored to modern agriculture (Cutini et al., 2023).

5.3.7 *Computing and Data Infrastructure (Cloud Computing and Big Data Analytics)*

The emergence of precision agriculture marks a paradigm shift in modern farming, driven largely by advancements in computing and data infrastructure. Central to this transformation are cloud computing and big data analytics, which enable real-time, scalable, and intelligent agricultural systems. As agriculture becomes increasingly data-intensive, integrating remote sensing, Internet of Things (IoT) devices, and machine learning models, the need for robust computing platforms that can handle vast volumes of heterogeneous data becomes critical. Cloud computing provides the backbone for such infrastructures, offering elastic storage, powerful computational resources, and real-time accessibility, while big data analytics turns raw agricultural data into actionable insights that drive decision-making (Kharraz & Szabó, 2025). Cloud computing supports precision agriculture through its capacity to manage data from geographically dispersed sensors, satellites, drones, and smart farm machinery. This distributed data, which includes soil moisture, weather patterns, nutrient levels, crop health indices, and pest occurrences, is ingested in real-time via cloud-based ETL (Extract, Transform, Load) pipelines. Frameworks like Apache NiFi have proven effective in streamlining this process by enabling continuous data flow from farm to cloud, reducing latency, and supporting time-sensitive decisions. Moreover, cloud platforms facilitate the integration of predictive models and simulation tools, which aid in forecasting crop yield, optimizing irrigation schedules, and detecting anomalies early (Kharraz & Szabó, 2025).

Big data analytics adds another critical dimension by allowing complex pattern recognition and multi-variable correlations that are beyond the reach of traditional methods. In precision agriculture, machine learning algorithms such as Support Vector Machines (SVM), Random Forests, Gradient Boosting, and Deep Neural Networks (DNN) are employed to interpret historical and real-time sensor data. These models help farmers understand the interactions between environmental variables and crop performance, thereby enabling data-driven interventions. For instance, identifying optimal light intensity or nutrient thresholds can directly enhance growth efficiency and reduce waste (Ramakrishna et al., 2019). A significant innovation in this domain is the development of composite efficiency metrics such as the Integrated Agricultural Efficiency Metric (IAEM). This metric synthesizes performance indicators, including data latency, cloud availability, sensor accuracy, and resource usage, to provide a holistic view of farming efficiency. Unlike traditional yield-based metrics, IAEM reflects the technological maturity and operational responsiveness of a precision agriculture system, emphasizing sustainability and resource optimization (Ramakrishna et al., 2019). Additionally, cloud-based systems promote interoperability and scalability. With modular architecture and API-driven integration, these systems can adapt to various crop types, climatic conditions, and farm sizes. Open-source and commercial cloud platforms support collaborative research and allow stakeholders ranging from agronomists to policymakers to access shared datasets and models, thereby democratizing precision farming technologies (Ramakrishna et al., 2019). The integration of cloud computing and big data analytics has become indispensable for the advancement of precision agriculture. These technologies not only improve crop productivity and reduce resource consumption but also enable a more sustainable and resilient food production system. As the global agricultural sector faces increasing pressures from climate change, resource scarcity, and population growth, cloud-enabled data infrastructure will be vital in shaping the future of farming through smarter, faster, and greener practices.

5.3.8 Blockchain Technology

Precision agriculture, empowered by advanced computing, IoT, and big data analytics, aims to optimize resource usage, improve crop yields, and ensure sustainability in farming. However, as modern agricultural systems increasingly rely on real-time data transmission and automation, new challenges around data integrity, security, traceability, and trust have emerged. Blockchain technology, with its decentralized and tamper-proof architecture, has the potential to address these challenges effectively. Integrating blockchain into precision agriculture can revolutionize the way agricultural data is collected, transmitted, stored, and used for decision-making. The role of blockchain in precision agriculture, focusing on key components such as data collection, data transmission, data preprocessing, the blockchain layer, and crop recommendation models (Patel et al., 2023).

Data Collection—The foundation of any precision agriculture system lies in accurate and timely data collection. IoT devices such as soil sensors, weather stations, drones, and remote imaging tools collect a wide range of data, including soil nutrients (N, P, K), pH levels, temperature, moisture, and light intensity. In a blockchain-integrated environment, data is collected in real-time and embedded with metadata (e.g., timestamp, device ID, location), creating a verifiable trail. The use of immutable digital signatures during data capture ensures that each data point is authenticated and cannot be altered post-collection. This enhances trust in the origin and accuracy of the data, which is critical for downstream analytics and recommendations (Torky & Hassanein, 2020; Patel et al., 2023).

Data Migration/Transmission—The transmission of agricultural data from sensor nodes to central servers or cloud platforms introduces vulnerabilities, especially when using wireless communication channels. Traditional systems are susceptible to cyber threats such as data tampering, man-in-the-middle attacks, and unauthorized access. Blockchain addresses these issues by decentralizing data transmission pathways and encrypting all transactions. Through peer-to-peer networks and cryptographic protocols, data packets are transmitted securely across nodes. Each transaction is verified through consensus mechanisms, and once added to a block, it becomes immutable. This ensures that malicious actors cannot alter data during migration, preserving the integrity and authenticity of transmitted agricultural information (Patel et al., 2023).

Data Preprocessing—Before feeding agricultural data into analytics or recommendation models, it undergoes preprocessing stages such as data cleaning, normalization, and transformation. Integrating blockchain into this process guarantees that preprocessing steps are logged transparently and can be audited. For example, if sensor data is smoothed or interpolated, the operations performed are recorded on the blockchain ledger. This provides traceability and transparency in data refinement procedures, ensuring that decision-making is based on verified, high-quality inputs. Smart contracts can also be used to automate preprocessing tasks, triggering data transformations only when certain preconditions are met (e.g., enough data is collected or a specific threshold is exceeded) (Patel et al., 2023).

Blockchain Layer—The core of the blockchain-based agricultural ecosystem is the blockchain layer, which acts as the decentralized ledger for recording all transactions and data exchanges. Each block contains a cryptographic hash of the previous block, a timestamp, and transaction data, thereby ensuring immutability and chronological integrity. In agriculture, this layer enables stakeholders, farmers, agronomists, supply chain partners, and regulatory bodies to access consistent and verifiable data. It also ensures data provenance, where every event from seed selection to harvest is recorded and can be traced back in real time. Public or consortium blockchains can be deployed based on the level of access and control needed. Permissioned blockchains such as Hyperledger Fabric allow only authorized participants to validate transactions, enhancing privacy in commercial agricultural operations (Patel et al., 2023).

Crop Recommendation Model—One of the most impactful applications of blockchain in precision agriculture is in supporting crop recommendation models. These models analyze environmental and soil parameters to recommend the most suitable crop or farming practice. Blockchain enhances the reliability of these models by ensuring that only accurate, verified, and tamper-proof data is used for training and inference. Furthermore, smart contracts can be employed to govern model logic and automate decision-making. For instance, a smart contract can be designed to automatically recommend a specific crop variety when nitrogen and phosphorus levels fall within an optimal range, and weather forecasts indicate suitable climatic conditions (Bodkhe et al., 2022).

Blockchain-based crop recommendation systems also support explainability and accountability. Each recommendation made by the model can be traced back to the exact data inputs and rules used, thereby building farmer trust and enabling continuous model improvement. Additionally, farmers can share model outputs and associated data with agricultural advisors, suppliers, or certification bodies, fostering collaboration and compliance with sustainable farming standards. Blockchain technology plays a transformative role in precision agriculture by safeguarding data throughout its lifecycle from collection to crop recommendation (Bodkhe et al., 2022). When integrated with IoT and cloud systems, blockchain ensures that agricultural decisions are based on secure, authentic, and traceable data. By addressing key issues such as data security, trust, and transparency, blockchain paves the way for building resilient, efficient, and farmer-centric smart farming ecosystems.

5.3.9 Quantum Computing (Frontier Technology)

Quantum computing, with its unique ability to process information through superposition and entanglement, offers promising capabilities for solving complex computational problems in agriculture and life sciences. Many of these problems involve high-dimensional data, intricate biological models, or global optimization tasks, making them computationally expensive for classical hardware. A number of key computational areas where quantum computing could demonstrate a significant advantage include solving systems of linear equations, identifying global extrema in optimization problems, enhancing machine and deep learning models, simulating physical systems using Hamiltonians, and improving search algorithm efficiency (Pook et al., 2025).

One of the most foundational computational problems in both science and engineering is solving systems of linear equations. In the context of agriculture and life sciences, linear systems are ubiquitous, appearing in modeling gene expression, nutrient flow in ecosystems, climate forecasting, and resource allocation. The Harrow-Hassidim-Lloyd (HHL) algorithm is a quantum algorithm designed to solve certain systems of linear equations exponentially faster than classical algorithms under specific conditions. While the practical implementation of the HHL algorithm is currently limited by qubit noise and error correction overheads, it lays the groundwork for developing quantum solvers for real-world large-scale linear systems in agricultural decision-making and environmental modelling (Pook et al., 2025).

Another important computational challenge in agricultural optimization and life sciences research is the identification of global extrema maximum or minimum values in nonlinear functions. For instance, determining the optimal allocation of resources in a farming system or maximizing crop yield under variable climate conditions often involves navigating complex, multidimensional landscapes with many local optima. Quantum annealing and variational quantum algorithms such as the Quantum Approximate Optimization Algorithm (QAOA) have shown promise in identifying global extrema efficiently. These quantum techniques can outperform classical heuristic methods, especially when classical optimization struggles with non-convexity or high dimensionality. While quantum annealers like those developed by D-Wave have provided some early insights into this space, their practical superiority over classical solvers remains an area of active research (Cravero et al., 2022).

Machine learning (ML) and deep learning (DL) have transformed agricultural monitoring, crop classification, soil health prediction, and livestock genomics by enabling data-driven decision-making. However, these models often require vast amounts of data and computational resources for training and inference. Quantum machine learning (QML) seeks to accelerate this by leveraging quantum states to represent and manipulate data. Quantum support vector machines (QSVMs), quantum principal component analysis (QPCA), and quantum Boltzmann machines are being investigated to reduce training times and improve model accuracy. In agriculture, QML could enable real-time disease detection from drone or satellite imagery, or real-time recommendation systems for precision farming. Similarly, quantum deep learning methods could be applied to genomics data or high-resolution climate models to accelerate discovery and prediction. Although quantum neural networks (QNNs) are still in the experimental stage, hybrid quantum-classical models are emerging as practical transitional solutions. Hamiltonian simulation is another domain with significant implications for life sciences, particularly in molecular biology and soil chemistry. A Hamiltonian represents the total energy of a system, and simulating its time evolution can provide insights into molecular interactions, enzyme functions, or nutrient behavior in soil. Quantum computers can simulate Hamiltonians exponentially faster than classical computers, enabling researchers to study quantum systems that are intractable using traditional methods. For example, simulating how pesticides or fertilizers interact at the molecular level with different soil types could inform the development of more sustainable agricultural chemicals. Similarly, modeling protein folding or gene regulation using quantum Hamiltonian simulations could enhance understanding in plant biology and livestock genetics (Cruz et al., 2020). Search algorithms also represent a crucial application area. In agricultural databases and large-scale genomic repositories, rapid search and pattern matching are essential. Grover's algorithm is a quantum algorithm that provides a quadratic speedup over classical unstructured search algorithms. For large datasets common in satellite imagery, soil sensor data, or genetic records, Grover's algorithm could significantly reduce search times. While this quadratic improvement may seem modest compared to exponential gains, in practice it can result in dramatic time savings when applied to datasets with millions or billions of entries (Boev et al., 2021). Despite these exciting prospects, there are considerable challenges to overcome. Current quantum hardware operates in the Noisy Intermediate-Scale Quantum (NISQ) era, meaning that devices are not yet error-resistant or scalable enough to handle many of the practical applications described. Moreover, not all agricultural or life science problems are amenable to quantum speedup; many must be reformulated using linear algebra or probabilistic structures to be compatible with quantum algorithms. Nevertheless, the growing interest in quantum-inspired algorithms, which borrow principles from quantum computing to optimize classical computations, shows that these insights are already influencing current technological development (Berckmans, 2017). Quantum computing holds transformative potential for addressing some of the most demanding computational challenges in agriculture and life sciences. By enabling exponential or quadratic speedups in linear equation solving, global optimization, machine learning, Hamiltonian simulations, and search operations, quantum computing could redefine how we model biological systems, optimize agricultural productivity, and understand environmental dynamics. As hardware advances and algorithms mature, the integration of quantum computing into agricultural research and life science workflows will likely become a key frontier in high-performance scientific computing.

5.4 Applications of Precision Agriculture for Sustainable Crop and Livestock Production and Environmental Conservation (Table 5.1)

5.4.1 Soil Management Through Precision Technologies

Precision agriculture revolutionizes soil management by enabling farmers to implement site-specific interventions based on high-resolution spatial data. Tools such as

Table 5.1 Applications of precision agriculture

Application area	Focus	References
Soil management	Site-specific soil analysis, fertilizer and water optimization	Júnior et al. (2024) and Surendran et al. (2024)
Water management	Smart irrigation, moisture sensing, water-use efficiency	Bwambale et al. (2022) and Lakhiar et al. (2024)
Crop monitoring	UAVs, sensors, stress detection, remote analytics	Neupane and Baysal-Gurel (2021) and Zidan and Febriyanti (2024)
Nutrient management	VRA fertilization, nutrient stress detection	Nawar et al. (2017) and Verma et al. (2020)
Pest and disease management	Remote sensing, predictive models, VRT pesticide use	Zhang et al. (2019) and Pawase et al. (2023)
Planting and harvesting	GPS-guided seeding/harvesting, yield monitoring	Weiss et al. (2020) and Perez-Ruiz et al. (2021)
Climate adaptation	Forecasting, resilient farming practices	Altieri and Nicholls (2017) and Narvekar and Rao (2024)
Data management	Cloud platforms, ML-based insights, predictive analytics	Khan et al. (2023)
Sustainable land management	Controlled traffic, GIS-based resource mapping	McLennon et al. (2021) and Rajanna et al. (2022)
Carbon and energy management	GHG mitigation, soil carbon, energy-efficient machinery	Al Ahmad (2023) and Nath (2024)
Animal health and nutrition	WOW systems, BCS imaging, feeding behavior sensors	Parsons et al. (2023), Aquilani et al. (2022), Segerkvist et al. (2020), Zhang et al. (2022), Yu et al. (2022) and Martinez-Rau et al. (2024)
Behavioral monitoring	Accelerometers, GPS, video-based behavior analysis	Herlin et al. (2021), Cabezas et al. (2022), Riaboff et al. (2021), Ikurior et al. (2020) and Li et al. (2024)
Reproductive event detection	Oestrus/mating sensors, calving alarms	Chang et al. (2024) and Bernabucci et al. (2025)
Pedigree and bonding ID	RFID systems, BT proximity tracking	Scott and Blore (2019) and Bernabucci et al. (2025)
Environmental impact monitoring	GHG sensors, urine detection, manure analytics	Shorten and Welten (2022) and Bernabucci et al. (2025)

GPS mapping, remote sensing, and soil sensors facilitate detailed analysis of soil characteristics, including pH, nutrient content, and moisture levels. These insights allow for targeted application of fertilizers and water, optimizing inputs and reducing environmental risks like nutrient leaching and runoff. Furthermore, continuous monitoring enables early detection of soil degradation or imbalances, empowering farmers to take timely corrective actions. Predictive analytics and machine learning further enhance soil management by forecasting soil behavior and guiding decision-making. This proactive, data-driven approach not only enhances crop productivity but also maintains long-term soil health. As such, precision agriculture supports both the economic and environmental dimensions of sustainable farming by improving efficiency, reducing costs, and preserving vital soil resources (Júnior et al., 2024; Surendran et al., 2024).

5.4.2 Water Management in Precision Agriculture

Water management is a critical aspect of precision agriculture, particularly in the face of increasing climate variability and water scarcity. By integrating technologies such as soil moisture sensors, weather forecasting tools, and IoT-based irrigation systems, precision agriculture enables highly efficient and localized irrigation strategies. Farmers can accurately determine moisture needs for different field zones and apply water only where and when it is required, significantly reducing waste. Advanced irrigation methods, including drip and variable rate irrigation (VRI), are guided by real-time data to deliver precise volumes of water. This minimizes losses due to evaporation and runoff, promotes water conservation, and improves crop health. Additionally, smart irrigation systems contribute to reducing the incidence of water-related crop diseases. Overall, precision water management enhances productivity, ensures sustainability, and supports the responsible use of one of agriculture's most vital resources (Bwambale et al., 2022; Lakhiar et al., 2024).

5.4.3 Crop Monitoring with Advanced Sensing and Analytics

Precision agriculture has significantly enhanced crop monitoring capabilities through the deployment of advanced technologies such as GPS, UAVs (drones), remote sensing, and IoT-based field sensors. These systems collect high-resolution, real-time data on crop conditions, soil moisture, and canopy development. Multispectral and thermal imaging from drones or satellites helps identify stressors like nutrient deficiencies, pest infestations, or disease outbreaks at early stages. This enables farmers to respond with timely and localized treatments, reducing the need for excessive chemical inputs. Furthermore, integrated data analytics platforms and machine learning algorithms analyze trends and historical data to forecast yields, optimize planting schedules, and support long-term planning. Cloud-based applications facilitate remote monitoring and decision-making, increasing operational efficiency. Thus, precision crop monitoring improves not only productivity and profitability but also supports environmentally responsible farming practices through more judicious use of inputs (Neupane & Baysal-Gurel, 2021; Zidan & Febriyanti, 2024).

5.4.4 Site-Specific Nutrient Management

Nutrient management in precision agriculture involves the strategic application of fertilizers based on spatial and temporal variations in soil and crop nutrient status. Technologies such as yield monitors, soil sensors, remote sensing, and GIS-based mapping are used to assess nutrient deficiencies and create variable rate application (VRA) plans. These technologies ensure that fertilizers and amendments are applied only where needed and at optimal rates, improving nutrient use efficiency and minimizing losses through runoff or leaching. Remote sensing provides rapid, non-invasive assessment of nutrient stress, while historical yield data help in refining long-term nutrient strategies. Additionally, this approach contributes to environmental sustainability by mitigating risks such as eutrophication and greenhouse gas emissions linked to fertilizer overuse. Precision nutrient management also reduces production costs, enhances crop quality, and sustains soil fertility, making it a key component in sustainable and resilient agricultural systems (Nawar et al., 2017; Verma et al., 2020).

5.4.5 Precision Pest and Disease Management

Precision agriculture significantly improves pest and disease management through the integration of GPS, GIS, remote sensing, and data analytics. These technologies facilitate early detection of pest infestations and disease outbreaks by monitoring subtle changes in crop color, vigor, and canopy structure, which often precede visible symptoms. Remote sensing and drone imagery can identify stress patterns across a field, allowing for timely and site-specific interventions. Variable rate technology (VRT) enables targeted pesticide application, thereby reducing chemical inputs, environmental contamination, and operational costs. Predictive models based on weather data, historical trends, and real-time monitoring further aid in anticipating pest dynamics, allowing farmers to plan control measures proactively. Precision agriculture also supports integrated pest management (IPM) strategies by providing insights into beneficial insect populations and ecological interactions. These capabilities foster long-term agricultural resilience, enhance input efficiency, and reduce dependence on conventional blanket pesticide use, thereby aligning pest and disease control with sustainable farming objectives (Zhang et al., 2019; Pawase et al., 2023).

5.4.6 Precision in Planting and Harvesting

Precision agriculture has modernized planting and harvesting operations by incorporating GPS-guided equipment, sensors, and data-driven decision-making. During planting, GPS and variable rate seeding systems ensure optimal seed placement by adjusting depth, spacing, and input based on spatial variability in soil moisture, texture, and nutrient availability. This precision promotes uniform crop emergence, optimizes input use, and enhances yield potential. Harvesting operations are equally transformed through the use of yield monitors, remote sensing, and automated machinery. These tools help farmers determine the optimal harvest time by analyzing crop maturity and health in real time, thereby minimizing losses due to overripening or adverse weather. Automated harvesters equipped with sensors improve operational efficiency, reduce labor dependency, and enhance product quality. Furthermore, integrating planting and harvesting data supports continuous farm improvement by providing valuable insights into performance and resource use. Precision-driven planting and harvesting contribute to higher productivity, lower costs, and more sustainable farming systems (Weiss et al., 2020; Perez-Ruiz et al., 2021).

5.4.7 Climate Adaptation Through Precision Agriculture

Climate change poses significant challenges to agriculture, including unpredictable rainfall, rising temperatures, and increased frequency of extreme weather events. Precision agriculture offers adaptive solutions by enabling real-time monitoring, localized input application, and responsive management. Technologies such as soil moisture sensors, drones, and satellite imagery help track field conditions, allowing farmers to adjust irrigation, fertilization, and crop management practices according to evolving climatic patterns. By using predictive analytics and weather forecasting tools, precision agriculture supports climate-smart decision-making, such as adjusting planting schedules or selecting drought-tolerant cultivars. Moreover, conservation practices like cover cropping and no-till farming, guided by precision tools, improve soil structure and carbon sequestration, enhancing resilience to erosion and heat stress. Precision agriculture also aids in reducing greenhouse gas emissions through efficient input management. Overall, precision farming empowers producers to adapt effectively to climate variability while sustaining productivity, conserving resources, and mitigating agriculture's environmental footprint in a warming world (Altieri & Nicholls, 2017; Narvekar & Rao, 2024).

5.4.8 Data Management and Analysis in Precision Agriculture

Data management is central to precision agriculture, underpinning all decision-making processes. A vast array of data is collected from sources such as soil sensors, UAVs, satellite imagery, and climate models. These data streams are integrated into cloud-based platforms that allow real-time access, monitoring, and decision support. Effective data management ensures data quality, consistency, and usability. Machine learning algorithms and predictive analytics further enhance the value of this data by identifying patterns and correlations that inform input application, pest control, and harvest timing. For instance, yield forecasts or pest outbreak risks can be derived from historical trends coupled with current field data. This data-driven approach enables precision agriculture to optimize resource allocation, reduce waste, and enhance productivity. By turning raw data into actionable insights, farmers can make informed, strategic decisions across all aspects of crop production. Ultimately, robust data systems in precision agriculture improve farm efficiency, profitability, and sustainability in a highly dynamic environment (Khan et al., 2023).

5.4.9 Sustainable Land Management Through Precision Agriculture

Precision agriculture plays a critical role in promoting sustainable land management by aligning resource use with environmental stewardship. Technologies such as GPS, sensors, drones, and GIS allow for precise monitoring and spatial mapping of field variability. This enables farmers to tailor management practices such as irrigation, fertilization, and tillage to specific zones based on soil type, fertility, and topography. These site-specific strategies reduce resource wastage and prevent land degradation. Practices like conservation tillage, controlled traffic farming, and variable rate input application are supported by data collected through precision tools, contributing to soil conservation, reduced erosion, and improved water retention. Furthermore, continuous monitoring of key soil health indicators helps implement regenerative practices that enhance long-term soil fertility and productivity. Precision agriculture thus enables sustainable intensification, balancing the need for high yields with the preservation of natural resources. It supports ecological resilience, reduces input-related emissions, and enhances the long-term viability of agricultural land (McLennon et al., 2021; Rajanna et al., 2022).

5.4.10 Carbon and Energy Management in Precision Agriculture

Precision agriculture contributes to carbon and energy management by enhancing efficiency and reducing the environmental footprint of farming operations. Technologies such as GPS-guided equipment, remote sensing, and data analytics ensure precise input application, minimizing the overuse of fertilizers, pesticides, and fuel. This reduces greenhouse gas emissions associated with the production and application of agricultural inputs. Additionally, practices like conservation tillage and cover cropping enabled by detailed field monitoring promote soil carbon sequestration and improve energy efficiency. Precision systems also support carbon accounting, allowing farmers to track emissions, evaluate sustainability metrics, and participate in carbon credit programs. By using automated machinery and optimizing field passes, energy consumption is significantly reduced. Moreover, data collected through precision tools can inform strategies to further cut energy costs and adopt renewable energy solutions on farms. Overall, precision agriculture enables more sustainable production systems by aligning crop management with climate goals, contributing to both economic efficiency and environmental protection (Al Ahmad, 2023; Nath, 2024).

5.4.11 Precision Monitoring of Animal Health and Nutrition

Liveweight is a key indicator in evaluating animal health, nutrition, and performance. However, traditional weighing is labor-intensive and stressful in ELF. Automated Walk-Over-Weigh (WOW) systems with solar power and RFID identification allow real-time, remote weight monitoring (Parsons et al., 2023). Data filtering algorithms enhance reliability, but a training period of 2–3 weeks is necessary to ensure animals use the system correctly (Aquilani et al., 2022). These systems also enable targeted selective treatment (TST) by detecting abnormal weight gain in parasite-infected animals (Segerkvist et al., 2020). BCS assessment using 2D and 3D image analysis has potential but remains limited in extensive systems due to variable outdoor conditions affecting data quality (Aquilani et al., 2022). Feeding behavior reflects both health status and environmental conditions. Embedded accelerometers and microphones help classify feeding and jaw movement patterns (Zhang et al., 2022). Edge computing solutions like DRN-YOLO (Yu et al., 2022) and low-power acoustic systems (Martinez-Rau et al., 2024) have improved real-time monitoring of grazing and rumination, though initial deployment has mostly occurred in controlled indoor settings. Low-energy designs and LoRa transmission show promise for grazing systems.

5.4.12 *Real-Time Behavioral Monitoring and Health Diagnostics*

Animal behavior is traditionally observed visually, which is impractical in ELF. Accelerometers, placed on legs, ears, or necks, and combined with GPS units, can accurately detect behaviors such as grazing, walking, and resting (Herlin et al., 2021). Studies report over 90% accuracy in detecting key behaviors in both cattle and sheep (Cabezas et al., 2022). Tri-axial accelerometers also help identify health disorders. For example, lameness can be inferred from abnormal gait and increased resting time (Riaboff et al., 2021). Similarly, gastrointestinal nematode (GIN) infection reduces activity, as recorded by these sensors (Ikurior et al., 2020). Recent innovations include video analysis algorithms such as YOLOv8n_BiF_DSC, which achieved a 93.6% behavior recognition accuracy in beef cattle, with near-perfect performance for lying postures (Li et al., 2024). However, deployment in ELF remains limited due to power constraints and connectivity issues in remote areas.

5.4.13 *Reproductive Event Detection and Management*

Automated oestrus detection employs pedometers, accelerometers, GPS systems, and mounting sensors. The UNETracker II GNSS system and Alpha-Detector (RFID-based mounting sensor) have demonstrated success in extensive sheep systems by identifying elevated movement and mounting behavior. These systems help optimize mating and breeding schedules. Timely calving detection is critical in ELF to reduce calf mortality. Technologies include GPS-based calving alarms (GPS-CAL) and accelerometers measuring tail raising, rumination drop, or isolation behavior. Although commercial systems like MooCall and SmartVel exist, their suitability for ELF is constrained by environmental challenges and energy limitations. Integrating multi-sensor data may improve detection reliability, but cost and data transmission remain barriers to wider adoption (Chang et al., 2024; Bernabucci et al., 2025).

5.4.14 *Pedigree and Maternal Bonding Identification*

In extensive grazing systems, accurate maternal pedigree identification is difficult. The Pedigree Matchmaker (PMM) system uses RFID-tagged ewes and lambs to infer dam-lamb associations through repeated proximity detection. Integration with WOW systems enhances functionality. Additionally, Bluetooth (BT) systems (e.g., ActiGraph) allow maternal bonding detection with high accuracy by measuring signal proximity between lambs and ewes. However, RFID must still be used for animal identity retention after BT device removal (Scott & Blore, 2019; Bernabucci et al., 2025).

5.4.15 Environmental Impact Monitoring

Monitoring manure deposition, urine patterns, and GHG emissions in ELF is essential for sustainable management but remains underdeveloped due to field difficulties. Open-path laser systems can estimate methane emissions directly from grazing cattle, showing feasibility at the herd level. GPS combined with urine sensors (thermistors detecting temperature changes) has shown high accuracy for urination detection in sheep and cattle (Bernabucci et al., 2025). Urination frequency correlated well with grazing time, especially in sheep. Acoustic sensors, as tested by Shorten and Welten (2022), can detect individual urination events and differentiate cows with higher N loads. These insights could support selection for animals that reduce environmental N leaching.

5.5 Conclusion

Precision agriculture stands at the intersection of digital innovation and agricultural sustainability, offering a transformative approach to modern farming. By integrating cutting-edge technologies with traditional practices, PA enables data-driven decision-making across both crop and livestock production systems. This review has demonstrated how tools such as soil sensors, GNSS, UAVs, and smart irrigation systems enhance precision crop farming by improving yield prediction, input application, and environmental management. In parallel, advancements in wearable sensors, behavior recognition algorithms, and automated feeders have significantly improved the monitoring, health, and welfare of livestock in real-time.

A major insight from this synthesis is that PA operates through a triadic framework: data perception (via sensors and IoT), data analysis (through AI, ML, and DL), and implementation (via actuators, robotics, and smart devices). Supporting technologies such as cloud infrastructure, blockchain for traceability, and 6G for connectivity are essential for creating robust, scalable PA systems. Furthermore, emerging innovations such as digital simulation, quantum computing, and autonomous robotics promise to further increase the adaptability and intelligence of agricultural systems. The environmental and economic benefits of precision agriculture are clear: reduced resource waste, improved productivity, and enhanced animal welfare. However, widespread adoption requires addressing challenges such as high initial investment, data privacy concerns, infrastructure limitations in rural areas, and the need for farmer training. Future research and policy must aim to democratize access to these technologies and promote their integration into diverse agro-ecological contexts. In doing so, precision agriculture can fulfill its potential as a cornerstone of sustainable, resilient, and intelligent global food systems.

References

Adams, B. T. (2019). Farm machinery automation for tillage, planting cultivation, and harvesting. In *Handbook of farm, dairy and food machinery engineering* (pp. 115–131). Academic.

Al Ahmad, A. J. (2023). Harnessing precision agriculture technologies for eco-friendly crop management: A synthesis of environmental biology and agriculture perspectives. *Journal Siplieria Sciences, 4*, 1–10.

Al-Ali, A. R., Al Nabulsi, A., Mukhopadhyay, S., Awal, M. S., Fernandes, S., & Ailabouni, K. (2019). IoT-solar energy powered smart farm irrigation system. *Journal of Electronic Science and Technology, 17*(4), 100017.

Altieri, M. A., & Nicholls, C. I. (2017). The adaptation and mitigation potential of traditional agriculture in a changing climate. *Climatic Change, 140*(1), 33–45. https://doi.org/10.1007/s10584-013-0909-z

Aquilani, C., Confessore, A., Bozzi, R., Sirtori, F., & Pugliese, C. (2022). Review: Precision livestock farming technologies in pasture-based livestock systems. *Animal, 16*(1), 100429. https://doi.org/10.1016/j.animal.2021.100429

Arun, A. N., Lee, B., Castiblanco, F. A., Buckmaster, D. R., Wang, C. C., Love, D. J., & Ghosh, A. (2025). Ambient IoT: Communications enabling precision agriculture. *IEEE Communications Magazine, 63*(4), 137–143.

Azmi, M. A., Mohammad, R., & Pebrian, D. (2020). Evaluation of soil EC mapping driven by manual and autopilot-automated steering systems of tractor on oil palm plantation terrain. *Food Research, 4*, 62–69.

Bai, Y., & Gao, J. (2021). Optimization of the nitrogen fertilizer schedule of maize under drip irrigation in Jilin, China, based on DSSAT and GA. *Agricultural Water Management, 244*, 106555.

Baltazar, A. R., Santos, F. N. D., Moreira, A. P., Valente, A., & Cunha, J. B. (2021). Smarter robotic sprayer system for precision agriculture. *Electronics, 10*(17), 2061.

Banhazi, T. M., Tscharke, M., Ferdous, W. M., Saunders, C., & Lee, S. H. (2011). Improved image analysis based system to reliably predict the live weight of pigs on farm: Preliminary results. *Australian Journal of Multi-Disciplinary Engineering, 8*(2), 107–119. https://doi.org/10.1080/14488388.2011.11464135

Banhazi, T. M., Babinszky, L., Halas, V., & Tscharke, M. (2012). Precision livestock farming: Precision feeding technologies and sustainable livestock production. *International Journal of Agricultural and Biological Engineering, 5*(4), 54–61. https://doi.org/10.3965/j.ijabe.20120504.00x

Ben Azouz, A., Esmonde, H., Corcoran, B., & O'Callaghan, E. (2015). Development of a teat sensing system for robotic milking by combining thermal imaging and stereovision technique. *Computers and Electronics in Agriculture, 110*, 162–170. https://doi.org/10.1016/j.compag.2014.11.016

Benyezza, H., Bouhedda, M., & Rebouh, S. (2021). Zoning irrigation smart system based on fuzzy control technology and IoT for water and energy saving. *Journal of Cleaner Production, 302*, 127001.

Berckmans, D. (2017). General introduction to precision livestock farming. *Animal Frontiers, 7*(1), 6–11.

Bernabucci, G., Evangelista, C., Girotti, P., Viola, P., Spina, R., Ronchi, B., & Primi, R. (2025). Precision livestock farming: An overview on the application in extensive systems. *Italian Journal of Animal Science, 24*(1), 859–884.

Beukes, P. C., McCarthy, S., Wims, C. M., Gregorini, P., & Romera, A. J. (2019). Regular estimates of herbage mass can improve profitability of pasture-based dairy systems. *Animal Production Science, 59*(2), 359–367.

Bezen, R., Edan, Y., & Halachmi, I. (2020). Computer vision system for measuring individual cow feed intake using RGB-D camera and deep learning algorithms. *Computers and Electronics in Agriculture, 172*, 105345. https://doi.org/10.1016/j.compag.2020.105345

Bo, W. A. N. G., Xiaofei, M. A., Fengyuan, W. A. N. G., Zhuoran, C. H. E. N., & Bo, L. I. N. (2024). Integration of agricultural machinery and agronomy promoting high-quality and efficient development of agricultural mechanization. *Agricultural Engineering, 14*(8), 5–9.

Bodkhe, U., Tanwar, S., Bhattacharya, P., & Kumar, N. (2022). Blockchain for precision irrigation: Opportunities and challenges. *Transactions on Emerging Telecommunications Technologies, 33*(10), e4059.

Boev, A. S., Rakitko, A. S., Usmanov, S. R., Kobzeva, A. N., Popov, I. V., Ilinsky, V. V., & Fedorov, A. K. (2021). Genome assembly using quantum and quantum-inspired annealing. *Scientific Reports, 11*(1), 13183.

Botero-Valencia, J., García-Pineda, V., Valencia-Arias, A., Valencia, J., Reyes-Vera, E., Mejia-Herrera, M., & Hernández-García, R. (2025). Machine learning in sustainable agriculture: Systematic review and research perspectives. *Agriculture, 15*(4), 377.

Bwambale, E., Abagale, F. K., & Anornu, G. K. (2022). Smart irrigation monitoring and control strategies for improving water use efficiency in precision agriculture: A review. *Agricultural Water Management, 260*, 107324. https://doi.org/10.1016/j.agwat.2021.107324

Cabezas, J., Yubero, R., Visitación, B., Navarro-García, J., Algar, M. J., Cano, E. L., & Ortega, F. (2022). Analysis of accelerometer and GPS data for cattle behaviour identification and anomalous events detection. *Entropy, 24*(3), 336. https://doi.org/10.3390/e24030336

Chang, A. Z., Swain, D. L., & Trotter, M. G. (2024). A multi-sensor approach to calving detection. *Information Processing in Agriculture, 11*(1), 45–64. https://doi.org/10.1016/j.inpa.2022.07.002

Corbari, C., Salerno, R., Ceppi, A., Telesca, V., & Mancini, M. (2019). Smart irrigation forecast using satellite LANDSAT data and meteo-hydrological modeling. *Agricultural Water Management, 212*, 283–294.

Cravero, A., Pardo, S., Sepúlveda, S., & Muñoz, L. (2022). Challenges to use machine learning in agricultural big data: A systematic literature review. *Agronomy, 12*(3), 748.

Cruz, P. M., Catarina, G., Gautier, R., & Fernández-Rossier, J. (2020). Optimizing quantum phase estimation for the simulation of Hamiltonian eigenstates. *Quantum Science and Technology, 5*(4), 044005.

Cutini, M., Bisaglia, C., Brambilla, M., Bragaglio, A., Pallottino, F., Assirelli, A., & Menesatti, P. (2023). A co-simulation virtual reality machinery simulator for advanced precision agriculture applications. *Agriculture, 13*(8), 1603.

Ehsani, R., & Sullivan, M. (2006). Soil electrical conductivity (EC) sensors, extension factsheet. AEX-565-02, 2002. Available online: http://www.nurserycropscience.info/nutrition/equipment-to-monitor-fertility/soil/measuring-ec-in-foeld-soils-by-tractor.pdf/view. Accessed on 13 Feb 2021.

Evangelista, C., Basiricò, L., & Bernabucci, U. (2021). An overview on the use of near infrared spectroscopy (NIRS) on farms for the management of dairy cows. *Agriculture, 11*(4), 296. https://doi.org/10.3390/agriculture11040296

Fountas, S., Sorensen, C. G., Tsiropoulos, Z., Cavalaris, C., Liakos, V., & Gemtos, T. (2015). Farm machinery management information system. *Computers and Electronics in Agriculture, 110*, 131–138.

Gaikwad, P., Devendrachari, M. C., Thimmappa, R., Paswan, B., Raja Kottaichamy, A., Makri Nimbegondi Kotresh, H., & Thotiyl, M. O. (2015). Galvanic cell type sensor for soil moisture analysis. *Analytical Chemistry, 87*(14), 7439–7445.

Gao, X., Zhou, Z., Xu, Y., Yu, Y., Su, Y., & Cui, T. (2020). Numerical simulation of particle motion characteristics in quantitative seed feeding system. *Powder Technology, 367*, 643–658.

Ghafar, A. S. A., Hajjaj, S. S. H., Gsangaya, K. R., Sultan, M. T. H., Mail, M. F., & Hua, L. S. (2023). Design and development of a robot for spraying fertilizers and pesticides for agriculture. *Materials Today: Proceedings, 81*, 242–248.

Goap, A., Sharma, D., Shukla, A. K., & Krishna, C. R. (2018). An IoT based smart irrigation management system using machine learning and open source technologies. *Computers and Electronics in Agriculture, 155*, 41–49.

Guan, S., Shimazaki, Y., Takahashi, K., Kato, H., Fukami, K., & Watanabe, S. (2025). Agricultural drone-based variable-rate N application for regulating wheat protein content. *Drones, 9*(4), 310.

Haibo, L., Shuliang, D., Zunmin, L., & Chuijie, Y. (2015). Study and experiment on a wheat precision seeding robot. *Journal Of Robotics, 2015*(1), 696301.

Hendriks, W. H., Verstegen, M. W. A., & Babinszky, L. (Eds.). (2019). *Poultry and pig nutrition*. Wageningen Academic Publishers.

Herlin, A., Brunberg, E., Hultgren, J., Högberg, N., Rydberg, A., & Skarin, A. (2021). Animal welfare implications of digital tools for monitoring and management of cattle and sheep on pasture. *Animals, 11*(3), 829. https://doi.org/10.3390/ani11030829

Higgins, S., Schellberg, J., & Bailey, J. S. (2019). Improving productivity and increasing the efficiency of soil nutrient management on grassland farms in the UK and Ireland using precision agriculture technology. *European Journal of Agronomy, 106*, 67–74.

Ikurior, S. J., Pomroy, W. E., Scott, I., Corner-Thomas, R., Marquetoux, N., & Leu, S. T. (2020). Gastrointestinal nematode infection affects overall activity in young sheep monitored with tri-axial accelerometers. *Veterinary Parasitology, 283*, 109188. https://doi.org/10.1016/j.vetpar.2020.109188

Jin, T., & Han, X. (2024). Robotic arms in precision agriculture: A comprehensive review of the technologies, applications, challenges, and future prospects. *Computers and Electronics in Agriculture, 221*, 108938.

Jin, X., Li, Q., Zhao, K., Zhao, B., He, Z., & Qiu, Z. (2019). Development and test of an electric precision seeder for small-size vegetable seeds. *International Journal of Agricultural and Biological Engineering, 12*(2), 75–81.

John, A. J., Clark, C. E. F., Freeman, M. J., Kerrisk, K. L., Garcia, S. C., & Halachmi, I. (2016). Review: Milking robot utilization, a successful precision livestock farming evolution. *Animal, 10*(9), 1484–1492. https://doi.org/10.1017/S1751731116000721

Júnior, M. R. B., de Almeida Moreira, B. R., dos Santos Carreira, V., de Brito Filho, A. L., Trentin, C., & de Souza, F. L. P. (2024). Precision agriculture in the United States: A comprehensive meta-review inspiring further research, innovation, and adoption. *Computers and Electronics in Agriculture, 221*, 108993. https://doi.org/10.1016/j.compag.2024.108993

Khan, A., Hassan, M., & Shahriyar, A. K. (2023). Optimizing onion crop management: A smart agriculture framework with IoT sensors and cloud technology. *Applied Research in Artificial Intelligence and Cloud Computing, 6*, 49–67.

Kharraz, N., & Szabó, I. (2025). *Cloud-driven data analytics for growing plants indoor.*

Khoramshahi, E., Nezami, S., Pellikka, P., Honkavaara, E., Chen, Y., & Habib, A. (2025). A taxonomy of sensors, calibration and computational methods, and applications of mobile mapping systems: A comprehensive review. *Remote Sensing, 17*(9), 1502.

Krishnan, R. S., Julie, E. G., Robinson, Y. H., Raja, S., Kumar, R., Thong, P. H., & Son, L. H. (2020). Fuzzy logic based smart irrigation system using internet of things. *Journal of Cleaner Production, 252*, 119902.

Lakhankar, T., Ghedira, H., Temimi, M., Sengupta, M., Khanbilvardi, R., & Blake, R. (2009). Non-parametric methods for soil moisture retrieval from satellite remote sensing data. *Remote Sensing, 1*(1), 3–21.

Lakhiar, I. A., Yan, H., & Zhang, C. (2024). A review of precision irrigation water-saving technology under changing climate for enhancing water use efficiency, crop yield, and environmental footprints. *Agriculture, 14*(6), 1141. https://doi.org/10.3390/agriculture14061141

Lamanna, M., Bovo, M., & Cavallini, D. (2025). Wearable collar technologies for dairy cows: A systematized review of the current applications and future innovations in precision livestock farming. *Animals: an Open Access Journal from MDPI, 15*(3), 458.
Li, G., Shi, G., & Zhu, C. (2024). Dynamic serpentine convolution with attention mechanism enhancement for beef cattle behavior recognition. *Animals, 14*(3), 466. https://doi.org/10.3390/ani14030466
Lin, N., Wang, X., Zhang, Y., Hu, X., & Ruan, J. (2020). Fertigation management for sustainable precision agriculture based on Internet of Things. *Journal of Cleaner Production, 277*, 124119.
Long, E. A., Ketterings, Q. M., Russell, D., Vermeylen, F., & DeGloria, S. (2016). Assessment of yield monitoring equipment for dry matter and yield of corn silage and alfalfa/grass. *Precision Agriculture, 17*(5), 546–563.
Lowenberg-DeBoer, J., Huang, I. Y., Grigoriadis, V., & Blackmore, S. (2020). Economics of robots and automation in field crop production. *Precision Agriculture, 21*(2), 278–299.
Lussem, U., Bolten, A., Menne, J., Gnyp, M. L., Schellberg, J., & Bareth, G. (2019). Estimating biomass in temperate grassland with high resolution canopy surface models from UAV-based RGB images and vegetation indices. *Journal of Applied Remote Sensing, 13*(3), 034525–034525.
Lussem, U., Schellberg, J., & Bareth, G. (2020). Monitoring forage mass with low-cost UAV data: Case study at the Rengen grassland experiment. *PFG–Journal of Photogrammetry, Remote Sensing and Geoinformation Science, 88*(5), 407–422.
Martinez-Rau, L. S., Chelotti, J. O., Giovanini, L. L., Adin, V., Oelmann, B., & Bader, S. (2024). On-device feeding behaviour analysis of grazing cattle. *IEEE Transactions on Instrumentation and Measurement, 73*, 1–13. https://doi.org/10.1109/TIM.2024.3376013
McLennon, E., Dari, B., Jha, G., Sihi, D., & Kankarla, V. (2021). Regenerative agriculture and integrative permaculture for sustainable and technology driven global food production and security. *Agronomy Journal, 113*(6), 4541–4559.
Meen, G. H., Schellekens, M. A., Slegers, M. H. M., Leenders, N. L. G., van Erp-van der Kooij, E., & Noldus, L. P. J. J. (2015). Sound analysis in dairy cattle vocalisation as a potential welfare monitor. *Computers and Electronics in Agriculture, 118*, 111–115. https://doi.org/10.1016/j.compag.2015.08.018
Michelena, Á., Fontenla-Romero, Ó., & Luis Calvo-Rolle, J. (2025). A review and future trends of precision livestock over dairy and beef cow cattle with artificial intelligence. *Logic Journal of the IGPL, 33*(4), jzae111.
Mikula, K., Izydorczyk, G., Skrzypczak, D., Mironiuk, M., Moustakas, K., Witek-Krowiak, A., & Chojnacka, K. (2020). Controlled release micronutrient fertilizers for precision agriculture – A review. *Science of the Total Environment, 712*, 136365.
Monteiro, A., Santos, S., & Gonçalves, P. (2021). Precision agriculture for crop and livestock farming – Brief review. *Animals, 11*(8), 2345.
Murphy, D. J., O'Brien, B., & Murphy, M. D. (2020). Development of a grass measurement optimisation tool to efficiently measure herbage mass on grazed pastures. *Computers and Electronics in Agriculture, 178*, 105799.
Narvekar, C., & Rao, M. (2024). Agricultural productivity improvement – Role of AI and yield prediction using machine learning. In *Precision agriculture for sustainability* (pp. 83–97). Apple Academic Press.
Nath, S. (2024). A vision of precision agriculture: Balance between agricultural sustainability and environmental stewardship. *Agronomy Journal, 116*, 1126–1143.
Nawar, S., Corstanje, R., Halcro, G., Mulla, D., & Mouazen, A. M. (2017). Delineation of soil management zones for variable-rate fertilization: A review. *Advances in Agronomy, 143*, 175–245. https://doi.org/10.1016/bs.agron.2017.01.003
Neethirajan, S. (2020). The role of sensors, big data and machine learning in modern animal farming. *Sensors and Bio-Sensing Research, 29*, 100367. https://doi.org/10.1016/j.sbsr.2020.100367

Neupane, K., & Baysal-Gurel, F. (2021). Automatic identification and monitoring of plant diseases using unmanned aerial vehicles: A review. *Remote Sensing, 13*(19), 3841. https://doi.org/10.3390/rs13193841

Nielsen, S. K., Munkholm, L. J., Lamandé, M., Nørremark, M., Edwards, G. T., & Green, O. (2018). Seed drill depth control system for precision seeding. *Computers and Electronics in Agriculture, 144*, 174–180.

O'Grady, M. J., & O'Hare, G. M. P. (2017). Modelling the smart farm. *Information Processing in Agriculture, 4*(3), 179–187. https://doi.org/10.1016/j.inpa.2017.06.001

O'Mahony, N., Campbell, S., Carvalho, A., Krpalkova, L., Riordan, D., & Walsh, J. (2019). 3D vision for precision dairy farming. *IFAC-PapersOnLine, 52*(30), 312–317. https://doi.org/10.1016/j.ifacol.2019.12.473

Oliveira, R. A., Näsi, R., Niemeläinen, O., Nyholm, L., Alhonoja, K., Kaivosoja, J., & Honkavaara, E. (2020). Machine learning estimators for the quantity and quality of grass swards used for silage production using drone-based imaging spectrometry and photogrammetry. *Remote Sensing of Environment, 246*, 111830.

Parsons, I. L., Norman, D. A., Karisch, B. B., Webb, S. L., Stone, A. E., Proctor, M. D., & Street, G. M. (2023). Automated walk-over-weigh system to track daily body mass and growth in grazing steers. *Computers and Electronics in Agriculture, 212*, 108113. https://doi.org/10.1016/j.compag.2023.108113

Patel, D. H., Shah, K. P., Gupta, R., Jadav, N. K., Tanwar, S., Neagu, B. C., & Tolba, A. (2023). Blockchain-based crop recommendation system for precision farming in IoT environment. *Agronomy, 13*(10), 2642.

Pawase, P. P., Nalawade, S. M., Bhanage, G. B., Wagh, D. P., & Borkar, D. S. (2023). Variable rate fertilizer application technology for nutrient management: A review. *International Journal of Agricultural and Biological Engineering, 16*(3), 11–19. https://doi.org/10.25165/j.ijabe.20231603.7483

Perez-Ruiz, M., Martínez-Guanter, J., & Upadhyaya, S. K. (2021). High-precision GNSS for agricultural operations. In *GPS and GNSS technology in geosciences* (pp. 299–335). Elsevier.

Polymeni, S., Skoutas, D. N., Sarigiannidis, P., Kormentzas, G., & Skianis, C. (2024). Smart agriculture and greenhouse gas emission mitigation: A 6G-IoT perspective. *Electronics, 13*(8), 1480.

Pook, T., Vandenplas, J., Boschero, J. C., Aguilera, E., Leijnse, K., Chauhan, A., & Aldridge, M. (2025). Assessing the potential of quantum computing in agriculture. *Computers and Electronics in Agriculture, 235*, 110332.

Praveen, R. V. S., Maindola, M., Bhavana, M., Nijhawan, G., Raju, H., & Bansal, S. (2025, February). AI for smart farming: Machine learning models for precision crop yield prediction. In *2025 International Conference on Computational, Communication and Information Technology (ICCCIT)* (pp. 797–802). IEEE.

Rajanna, G., Dass, A., Suman, A., et al. (2022). Co-implementation of tillage, irrigation, and fertilizers in soybean: Impact on crop productivity, soil moisture, and soil microbial dynamics. *Field Crops Research, 288*, 108672. https://doi.org/10.1016/j.fcr.2022.108672

Ramakrishna, C., Venkateshwarlu, B., Srinivas, J., & Srinivas, S. (2019). IoT based smart farming using cloud computing and machine learning. *International Journal of Innovative Technology and Exploring Engineering, 9*(1), 3455–3458.

Riaboff, L., Relun, A., Petiot, C. E., Feuilloy, M., Couvreur, S., & Madouasse, A. (2021). Identification of discriminating behavioural and movement variables in lameness scores of dairy cows at pasture from accelerometer and GPS sensors using a partial least squares discriminant analysis. *Preventive Veterinary Medicine, 193*, 105383. https://doi.org/10.1016/j.prevetmed.2021.105383

Rodenburg, J. (2017). Robotic milking: Technology, farm design, and effects on work flow. *Journal of Dairy Science, 100*(9), 7729–7738. https://doi.org/10.3168/jds.2016-11522

Rodriguez-Ortega, W. M., Martinez, V., Rivero, R. M., Camara-Zapata, J. M., Mestre, T., & Garcia-Sanchez, F. (2017). Use of a smart irrigation system to study the effects of irrigation

management on the agronomic and physiological responses of tomato plants grown under different temperatures regimes. *Agricultural Water Management, 183*, 158–168.
Romano, E., Bisaglia, C., Calcante, A., Oberti, R., Zani, A., Vinnikov, D., & Rapisarda, V. (2020). Assessment of comfort variation among different types of driving agricultural tractors: Traditional, satellite-assisted and semi-automatic. *International Journal of Environmental Research and Public Health, 17*(23), 8836.
Rovira-Más, F., Zhang, Q., & Saiz-Rubio, V. (2020). Mechatronics and intelligent systems in agricultural machinery. In N. M. Holden, M. L. Wolfe, J. A. Ogejo, & E. J. Cummins (Eds.), *Introduction to biosystems engineering*. Virginia Tech Publishing.
Schaefer, M. T., & Lamb, D. W. (2016). A combination of plant NDVI and LiDAR measurements improve the estimation of pasture biomass in tall fescue (Festuca arundinacea var. Fletcher). *Remote Sensing, 8*(2), 109.
Scott, N., & Blore, C. (2019). *Final report – Pedigree MatchMaker for beef*. Meat & Livestock Australia.
Segerkvist, K. A., Höglund, J., Österlund, H., Wik, C., Högberg, N., & Hessle, A. (2020). Automatic weighing as an animal health monitoring tool on pasture. *Livestock Science, 240*, 104157. https://doi.org/10.1016/j.livsci.2020.104157
Shi, J., Wu, X., Zhang, M., Wang, X., Zuo, Q., Wu, X., & Ben-Gal, A. (2021). Numerically scheduling plant water deficit index-based smart irrigation to optimize crop yield and water use efficiency. *Agricultural Water Management, 248*, 106774.
Shorten, P. R., & Welten, B. G. (2022). An acoustic sensor technology to detect urine excretion. *Biosystems Engineering, 214*, 90–106. https://doi.org/10.1016/j.biosystemseng.2021.12.004
Smolka, M., Puchberger-Enengl, D., Bipoun, M., Klasa, A., Kiczkajlo, M., Śmiechowski, W., & Vellekoop, M. J. (2017). A mobile lab-on-a-chip device for on-site soil nutrient analysis. *Precision Agriculture, 18*(2), 152–168.
Song, C., Zhou, Z., Zang, Y., Zhao, L., Yang, W., Luo, X., & Zhu, Q. (2021). Variable-rate control system for UAV-based granular fertilizer spreader. *Computers and Electronics in Agriculture, 180*, 105832.
Song, K., Chen, S., Wang, G., Qi, J., Gao, X., Xiang, M., & Zhou, Z. (2025). Research on high-precision target detection technology for tomato-picking robots in sustainable agriculture. *Sustainability, 17*(7), 2885.
Spagnuolo, M., Todde, G., Caria, M., Furnitto, N., Schillaci, G., & Failla, S. (2025). Agricultural robotics: A technical review addressing challenges in sustainable crop production. *Robotics, 14*(2), 9.
Surendran, U., Nagakumar, K. C. V., & Samuel, M. P. (2024). Remote sensing in precision agriculture. In *Digital agriculture: A solution for sustainable food and nutritional security* (pp. 201–223). Springer.
Torky, M., & Hassanein, A. E. (2020). Integrating blockchain and the internet of things in precision agriculture: Analysis, opportunities, and challenges. *Computers and Electronics in Agriculture, 178*, 105476.
Ungar, E. D., & Nevo, Y. (2025). Rhythms, patterns and styles in the jaw movement activity of beef cattle on rangeland as revealed by acoustic monitoring. *Sensors, 25*(4), 1210.
Van Hertem, T. (2018). Objective sustainability assessment by precision livestock farming. *Agricultural and Food Policy and Environmental Economics Policy*.
Verma, P., Chauhan, A., & Ladon, T. (2020). Site specific nutrient management: A review. *Journal of Pharmacognosy and Phytochemistry, 9*(1), 233–236.
Weiss, M., Jacob, F., & Duveiller, G. (2020). Remote sensing for agricultural applications: A meta-review. *Remote Sensing of Environment, 236*, 111402. https://doi.org/10.1016/j.rse.2019.111402
Yu, Z., Liu, Y., Yu, S., Wang, R., Song, Z., Yan, Y., Li, F., Wang, Z., & Tian, F. (2022). Automatic detection method of dairy cow feeding behaviour based on YOLO improved model and edge computing. *Sensors (Basel), 22*(9), 3271. https://doi.org/10.3390/s22093271

Zhai JianBo, Z. J., Xia JunFang, X. J., Zhou Yong, Z. Y., & Zhang Shun, Z. S. (2014). *Design and experimental study of the control system for precision seed-metering device*.

Zhang, J., Huang, Y., Pu, R., Gonzalez-Moreno, P., Yuan, L., Wu, K., & Huang, W. (2019). Monitoring plant diseases and pests through remote sensing technology: A review. *Computers and Electronics in Agriculture, 165*, 104943. https://doi.org/10.1016/j.compag.2019.104943

Zhang, J., Khoshbakht, M., Liu, J., Gou, Z., Xiong, J., & Jiang, M. (2022). A clustering review of vegetation-indicating parameters in urban thermal environment studies towards various factors. *Journal of Thermal Biology, 110*, 103340. https://doi.org/10.1016/j.jtherbio.2022.103340

Zidan, F., & Febriyanti, D. E. (2024). Optimizing agricultural yields with artificial intelligence-based climate adaptation strategies. *IAIC Transactions on Sustainable Digital Innovation, 5*, 136–147.

Chapter 6
3D Food Printing

Abbreviations

3D	Three-Dimensional
3D FP/3D	Food Printing Three-Dimensional Food Printing
AM	Additive Manufacturing
CAD	Computer-Aided Design
G′	Storage Modulus (Elastic Modulus)
G″	Loss Modulus (Viscous Modulus)
G-code	Geometric Code (Machine Instruction Code)
IDDSI	International Dysphagia Diet Standardisation Initiative
MIT	Massachusetts Institute of Technology
NASA	National Aeronautics and Space Administration
R&D	Research and Development
SFF	Solid Freeform Fabrication
SLS	Selective Laser Sintering
STL	Standard Tessellation Language
TNO	Netherlands Organisation for Applied Scientific Research

6.1 Introduction

Three-dimensional (3D) food printing, also known as additive manufacturing, rapid prototyping, or food-layered manufacturing, is an emerging technology that is redefining the boundaries of conventional food processing and product design. By applying the principles of additive manufacturing to edible materials, 3D food printing enables the fabrication of food products through a digitally controlled, layer-by-layer deposition process. This technology allows precise control over food shape, structure, texture, composition, and nutritional content, making it one

Y. Kumar, S. Dey, *Emerging Technology in Food Processing*,
https://doi.org/10.1007/978-3-032-24505-2_6

of the most innovative developments in modern food engineering and digital gastronomy (Sun et al., 2015a; Derossi et al., 2020).

Unlike traditional food manufacturing methods that rely on molding, extrusion, cutting, or shaping, 3D food printing follows an additive approach in which materials are deposited incrementally according to a computer-aided design (CAD) model. This approach enables the creation of complex geometries, internal structures, and multi-material foods that are difficult or impossible to produce using conventional techniques. As a result, 3D food printing offers unprecedented flexibility in food design and customization, supporting the development of personalized, functional, and sustainable food systems (Dankar et al., 2018).

The increasing interest in 3D food printing is driven by several global challenges faced by modern food systems. These include the rising prevalence of diet-related chronic diseases, the need for personalized nutrition, aging populations, food waste reduction, and the demand for sustainable protein sources. Through digital control of ingredient deposition, 3D food printing allows precise adjustment of portion size, calorie content, macronutrient ratios, and micronutrient fortification. This capability makes it possible to design foods tailored to individual health conditions such as obesity, diabetes, malnutrition, and dysphagia, marking a shift from mass food production toward consumer-centered nutrition solutions (Dankar et al., 2018; Varvara et al., 2021).

Sustainability is another important motivation behind the development of 3D food printing technologies. Conventional livestock-based food production systems are resource-intensive and contribute significantly to environmental degradation. In contrast, 3D food printing facilitates the incorporation of alternative and underutilized food resources such as plant-based proteins, algae, insect protein, mycoprotein, food-processing by-products, and cultured meat. By transforming these materials into nutritionally balanced and visually appealing foods, 3D food printing supports circular economy principles and promotes more sustainable and resilient food systems (Severini & Derossi, 2016; Liu et al., 2019).

The origins of 3D food printing can be traced back to early experimental research in the early 2000s, when scientists began exploring the feasibility of using modified 3D printers to fabricate edible materials. A significant milestone was achieved in 2007 with the development of the Fab@Home project at Cornell University. Fab@Home was an open-source 3D printing platform capable of extruding semi-solid edible materials such as chocolate, cake frosting, Nutella, and processed cheese. This project demonstrated that food materials could be digitally fabricated with reasonable accuracy and repeatability, laying the foundation for subsequent research in food additive manufacturing (Holland et al., 2018).

Further advancements were made in 2010 by researchers at the Massachusetts Institute of Technology (MIT), who introduced the concept of digital gastronomy. Through experimental systems such as the Virtuoso Mixer, the Digital Fabricator, and the Robotic Chef, MIT researchers highlighted the potential of integrating digital design, multi-material deposition, and automated cooking processes. These concepts emphasized the fusion of culinary creativity with digital fabrication technologies, expanding the scope of food design and preparation (Zoran & Coelho, 2011).

The relevance of 3D food printing gained global attention when the National Aeronautics and Space Administration (NASA) began investigating its application for long-duration space missions in 2013. NASA collaborated with industry partners such as BeeHex and Systems and Materials Research Corporation to develop printable food formulations, including doughs, sauces, and nutrient pastes, capable of meeting astronauts' nutritional requirements during extended space travel. This initiative highlighted the potential of 3D food printing for food security in extreme and resource-limited environments (Lipton et al., 2015).

The transition of 3D food printing from research laboratories to commercial and domestic settings began around 2014 with the introduction of consumer-oriented printers such as Foodini by Natural Machines. Foodini enabled users to print a variety of foods, including pizza bases, pasta, and vegetable purées, using fresh ingredients. Around the same time, BeeHex gained attention for its automated pizza-printing technology, capable of producing customized pizzas with precise ingredient layering. These developments marked an important step toward the commercialization and broader adoption of 3D food printing technologies.

In Europe, research organizations such as the Netherlands Organisation for Applied Scientific Research (TNO) and Biozoon Food Innovations in Germany focused on healthcare and elderly nutrition applications of 3D food printing. Their work emphasized the development of texture-modified foods for elderly individuals and patients with dysphagia, a condition characterized by difficulty swallowing. By producing visually appealing foods with controlled texture and nutritional composition, these initiatives demonstrated the potential of 3D food printing to improve quality of life and nutritional intake in clinical and assisted-living settings (Dankar et al., 2018).

From a scientific perspective, 3D food printing represents an interdisciplinary convergence of food science, material science, mechanical engineering, and digital design. Successful printing requires a deep understanding of food material properties, particularly rheological behavior, viscoelasticity, and structural stability. Food formulations, often referred to as "food inks," must exhibit suitable flow behavior during extrusion while maintaining shape and integrity after deposition. These requirements have driven extensive research into food ink formulation, hydrocolloid modification, and post-printing stabilization techniques (Sun et al., 2015b; Mantihal et al., 2020; Oliveira et al., 2020).

Beyond technological novelty, the societal implications of 3D food printing are significant. The technology offers solutions for personalized nutrition, healthcare feeding, sustainable food production, and innovative culinary experiences. It also opens new opportunities for food product innovation, digital food design, and decentralized food manufacturing (Oliveira et al., 2020). However, challenges related to processing speed, equipment cost, material limitations, food safety, and regulatory frameworks remain barriers to large-scale adoption.

The 3D food printing is a rapidly evolving technology with the potential to transform how food is designed, produced, and consumed. By enabling customization, nutritional precision, sustainability, and functional food development, it addresses many of the pressing challenges faced by modern food systems. This book chapter aims to provide a comprehensive understanding of the principles, materials, processes, nutritional aspects, and applications of 3D food printing, equipping students and professionals with the knowledge required to engage with this emerging field.

6.2 Fundamentals of 3D Food Printing

Three-dimensional (3D) food printing is an emerging food manufacturing technology derived from additive manufacturing, where edible materials are deposited layer by layer to create complex food structures directly from digital designs. Unlike conventional food processing methods that depend on molding, extrusion dies, or manual shaping, 3D food printing enables the fabrication of customized food products without object-specific tooling (Sun et al., 2015c). The fundamentals of 3D food printing lie at the intersection of solid freeform fabrication principles, digital design technologies, and the physicochemical behavior of printable food materials.

The core objective of 3D food printing is to achieve precise control over food shape, texture, composition, and nutritional content through digitally controlled deposition. This capability has gained significant attention due to increasing demand for personalized nutrition, texture-modified foods for the elderly, and innovative food designs for culinary and industrial applications (Sher & Tutó, 2015). Understanding the fundamental principles governing this technology is essential for food technologists to develop printable formulations and optimize processing conditions (Table 6.1).

6.2.1 Solid Freeform Fabrication (SFF) Principle

Solid freeform fabrication (SFF) is the foundational principle underlying all 3D printing technologies, including food printing. SFF refers to a group of additive manufacturing processes in which three-dimensional objects are constructed layer by layer directly from digital models, without the use of molds, dies, or subtractive machining (Sun et al., 2015b). In 3D food printing, edible materials are selectively deposited or solidified to form self-supporting structures based on predefined geometries.

The SFF process begins with the creation of a digital three-dimensional model, which is subsequently sliced into multiple two-dimensional layers. Each layer represents a cross-section of the final object and is fabricated sequentially. Food materials are deposited point-by-point or line-by-line according to the toolpath generated by the slicing software. The deposited layers must bond adequately to previous layers to ensure mechanical stability and shape retention.

A critical aspect of SFF in food printing is the ability of the deposited food material to maintain its shape under gravitational and mechanical forces. Unlike

Table 6.1 Fundamental principles and components of 3D food printing

Aspect	Description	Relevance to 3D food printing	References
Three-dimensional (3D) food printing	Additive manufacturing-based food fabrication method where edible materials are deposited layer by layer from digital designs	Enables customized food shapes, textures, and nutritional profiles without molds or dies	Sun et al. (2015c) and Sher and Tutó (2015)
Solid freeform fabrication (SFF)	Layer-by-layer construction of objects directly from digital models without subtractive machining	Forms the core manufacturing principle, allowing mass customization, reduced waste, and complex internal geometries	Sun et al. (2015b)
Digital 3D model creation	Development of a three-dimensional digital model representing the final food product	Serves as the foundation for slicing, toolpath generation, and precise fabrication	Sher and Tutó (2015)
Slicing and toolpath generation	Conversion of digital models into multiple 2D layers and printer instructions (G-code)	Controls deposition sequence, layer thickness, extrusion rate, and printing accuracy	Pitayachaval et al. (2018)
Computer-aided design (CAD)	Use of CAD software to design geometry, dimensions, and internal structure of foods	Allows control over porosity, infill patterns, texture, portion size, and ingredient distribution	Sher and Tutó (2015) and Sun et al. (2015c)
Food ink	Printable edible material used for deposition during printing	Determines print quality, shape fidelity, texture, nutritional value, and safety	Sun et al. (2015c)
Rheological behavior	Flow and deformation characteristics such as viscosity, yield stress, and shear-thinning behavior	Ensures smooth extrusion during printing and rapid shape recovery after deposition	Pitayachaval et al. (2018) and Phuhongsung et al. (2020)
Viscoelastic properties	Combination of viscous and elastic responses of food ink	Supports layer stacking, interlayer adhesion, and structural stability	Sun et al. (2015b) and Liu et al. (2019)
Common food inks	Chocolate, dough, cheese, starch pastes, fruit/vegetable purees, protein-rich formulations	Provide versatility for bakery, confectionery, functional, and personalized food applications	Sher and Tutó (2015) and Pant et al. (2021)
Use of additives	Incorporation of hydrocolloids, emulsifiers, and gelling agents	Improves printability, structural stability, and rheological performance	Sher and Tutó (2015) and Pant et al. (2021)
Temperature sensitivity	Dependence of material behavior on temperature during extrusion and solidification	Critical for thermoreversible materials such as chocolate and fat-based inks	Chua et al. (2018) and Pant et al. (2021)
Nutritional and microbiological considerations	Retention of nutrients and control of microbial risks during printing and post-processing	Ensures safety, quality, and functionality of printed foods	Phuhongsung et al. (2020)

engineering materials such as plastics or metals, food materials are often soft, viscoelastic, and sensitive to temperature and moisture. Therefore, successful application of SFF to food requires careful control of deposition parameters and material properties to prevent collapse or deformation of printed structures (Sun et al., 2015b).

SFF-based food printing offers several advantages over traditional food structuring techniques. It enables mass customization, reduces material waste, and allows the creation of complex internal geometries that influence texture and mouthfeel. As a result, SFF has become a promising approach for producing customized foods for medical nutrition, space missions, and high-value culinary products.

6.2.2 Digital Design and Computer-Aided Modeling (CAD)

Digital design and computer-aided modeling (CAD) form the backbone of 3D food printing. CAD software is used to design the geometry, dimensions, and internal structure of food products before fabrication. These digital models provide precise control over shape complexity, portion size, and spatial distribution of ingredients, which is not feasible with conventional food processing methods (Sher & Tutó, 2015).

Once a CAD model is created, it is converted into a standard tessellation language (STL) file or a similar format. This file is then processed by slicing software, which divides the model into thin horizontal layers and generates machine-readable instructions, commonly referred to as G-code. These instructions control the movement of the printer's axes, extrusion rate, printing speed, and layer thickness during fabrication (Pitayachaval et al., 2018).

Digital design also enables the integration of culinary creativity with engineering precision. By adjusting design parameters, food technologists can manipulate porosity, infill patterns, and internal architecture to influence texture, cooking behavior, and sensory attributes of the final product. For example, varying infill density can alter crispness or softness after baking, while layered ingredient placement can enhance flavor release during consumption (Sun et al., 2015c).

Furthermore, CAD-based food printing supports personalization and digital food libraries, where consumers or manufacturers can store, modify, and share food designs. This digital workflow allows rapid prototyping and on-demand production, making 3D food printing a flexible and efficient approach for customized food fabrication.

6.2.3 Role of Food Ink and Material Properties

Food ink refers to the printable edible material used in 3D food printing, and its properties play a decisive role in determining print quality and structural stability.

Unlike conventional inks, food inks must satisfy not only mechanical and rheological requirements but also nutritional, sensory, and microbiological criteria (Sun et al., 2015c).

From a material science perspective, food inks must possess appropriate flow behavior to enable smooth extrusion or deposition through a nozzle. Most successful food inks exhibit pseudoplastic or shear-thinning behavior, meaning their viscosity decreases under applied shear stress and recovers quickly once the stress is removed. This behavior allows the material to flow during printing while maintaining shape after deposition (Pitayachaval et al., 2018).

Viscoelasticity is another critical property, as it determines the ability of the printed layers to support subsequent layers without deformation. Materials with insufficient elasticity may spread or collapse, whereas excessively stiff materials may clog nozzles or result in poor layer adhesion. Therefore, achieving a balance between viscosity, yield stress, and elasticity is essential for printability.

Common food inks include chocolate, dough, cheese, pureed fruits and vegetables, starch-based pastes, and protein-rich formulations. In many cases, additives such as hydrocolloids, emulsifiers, or gelling agents are incorporated to modify rheological behavior and improve structural stability (Sher & Tutó, 2015; Pant et al., 2021). Temperature sensitivity is also important, particularly for materials like chocolate that require controlled melting and solidification during printing (Chua et al., 2018; Pant et al., 2021).

In addition to mechanical properties, food inks must ensure nutritional quality and biological safety. The formulation should retain essential nutrients during printing and post-processing, while minimizing microbial risks associated with high moisture and ambient temperature processing. As a result, food scientists must integrate knowledge of rheology, nutrition, and food microbiology when developing printable food inks (Phuhongsung et al., 2020).

6.3 Food Ink Formulation

Food ink formulation is one of the most critical aspects of 3D food printing, as it directly determines printability, structural stability, sensory quality, and nutritional value of the final printed product (Fig. 6.1). Food ink refers to any edible material that can be extruded, deposited, or fused in a controlled manner to form a three-dimensional food structure. Unlike conventional food formulations, food inks must meet strict rheological and mechanical requirements while remaining safe, nutritious, and organoleptically acceptable (Sun et al., 2015b).

The success of 3D food printing depends largely on the ability of food ink to flow smoothly through the printing nozzle and rapidly stabilize after deposition. This dual requirement presents a challenge, as many natural food materials lack the ideal flow and recovery behavior needed for additive manufacturing. Consequently, food ink formulation often involves modification of raw ingredients through mechanical

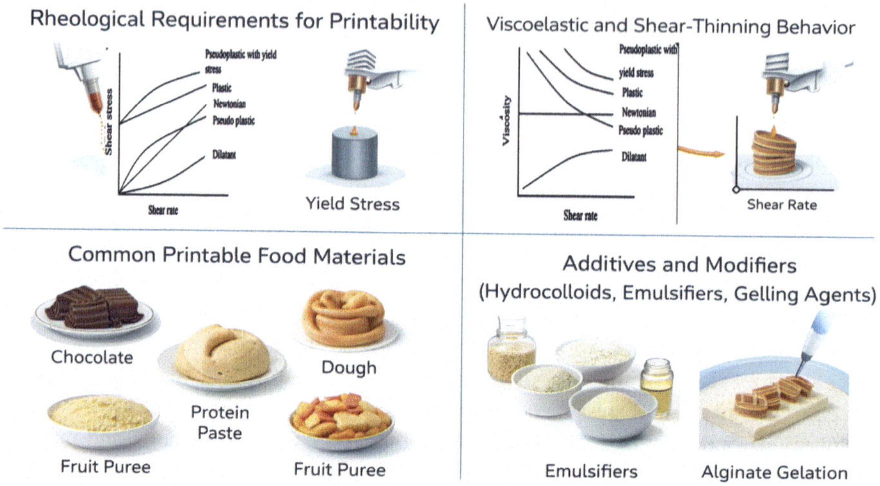

Fig. 6.1 Aspects of 3D food printing ink formulation

processing, thermal treatment, or the incorporation of functional additives such as hydrocolloids, emulsifiers, and gelling agents (Sher & Tutó, 2015).

Food technologists must therefore integrate knowledge of rheology, material science, and food chemistry to develop inks suitable for different 3D printing techniques. This section discusses the rheological requirements for printability, viscoelastic and shear-thinning behavior, commonly used printable food materials, and the role of formulation additives (Phuhongsung et al., 2020).

6.3.1 Rheological Requirements for Printability

Rheology, the study of flow and deformation of materials, plays a decisive role in determining the printability of food inks. During extrusion-based 3D food printing, the ink must flow under applied pressure while maintaining dimensional accuracy once deposited. This requires precise control over viscosity, yield stress, and flow behavior (Derossi et al., 2020; Phuhongsung et al., 2020).

An ideal food ink should possess sufficient viscosity to prevent spreading or collapsing after deposition, yet remain fluid enough to pass through the nozzle without excessive pressure. If viscosity is too high, nozzle clogging and irregular extrusion may occur; if too low, the printed structure may lose its shape (Sun et al., 2015b). Yield stress is equally important, as it defines the minimum stress required to initiate flow. Materials with appropriate yield stress can support subsequent layers, enhancing vertical buildability.

In addition, food inks must exhibit rapid structural recovery after extrusion. This property ensures that the printed layers maintain their geometry and bond effectively with previously deposited layers. Rheological measurements such as shear rate–viscosity curves, oscillatory tests, and creep recovery analyses are commonly used to evaluate the printability of food inks (Derossi et al., 2020).

6.3.2 *Viscoelastic and Shear-Thinning Behavior*

Most successful food inks used in 3D printing exhibit viscoelastic and shear-thinning (pseudoplastic) behavior. Viscoelasticity refers to a material's ability to exhibit both viscous (flow) and elastic (solid-like) properties. This dual behavior is essential for maintaining shape fidelity during layer-by-layer deposition (Liu et al., 2019).

Shear-thinning behavior allows food inks to decrease in viscosity when subjected to shear stress during extrusion, enabling smooth flow through the nozzle. Once the shear stress is removed after deposition, the viscosity rapidly increases, allowing the material to retain its printed shape. This reversible behavior is particularly desirable in extrusion-based food printing systems (Derossi et al., 2020).

Viscoelastic behavior also contributes to interlayer adhesion, which is crucial for structural integrity. If the elastic component is too low, the material may spread excessively; if too high, the ink may resist flow and cause printing defects. Therefore, an optimal balance between elastic modulus (G') and viscous modulus (G'') is necessary to ensure both extrudability and shape retention (Liu et al., 2019).

Many natural food materials do not inherently exhibit ideal viscoelastic and shear-thinning properties, making formulation adjustments essential. These adjustments often involve modifying particle size, moisture content, or incorporating functional additives that enhance rheological performance.

6.3.3 *Common Printable Food Materials*

A wide range of food materials has been explored as printable inks in 3D food printing, depending on the printing technique and intended application. Commonly used materials include carbohydrate-based, protein-based, fat-based, and fruit or vegetable-based formulations (Sun et al., 2015b).

Carbohydrate-based materials, such as starch pastes, doughs, and sugar-based formulations, are among the most widely used food inks. These materials are relatively easy to process and exhibit favorable flow behavior when properly hydrated. Starch gelatinization and retrogradation can further enhance structural stability after printing.

Protein-based materials, including meat pastes, fish surimi, dairy proteins, and plant proteins, are increasingly used due to their high nutritional value. However, protein inks often require modification to improve printability, as they may exhibit poor flow behavior or weak structural recovery. Enzymatic cross-linking and hydrocolloid addition are commonly employed to improve their performance (Sher & Tutó, 2015).

Fat-based materials, such as chocolate, are particularly suitable for heat-assisted 3D printing. Chocolate exhibits thermoreversible behavior, allowing controlled melting during extrusion and rapid solidification upon cooling. This property makes it ideal for producing complex decorative structures with high dimensional accuracy (Sun et al., 2015c).

Fruit and vegetable purees are also used as food inks, especially for personalized nutrition and functional foods. However, their high-moisture content and low structural strength often necessitate the addition of thickeners or gelling agents to improve printability and stability.

6.3.4 Additives and Modifiers (Hydrocolloids, Emulsifiers, Gelling Agents)

To overcome the limitations of natural food materials, various additives and modifiers are incorporated into food ink formulations. These additives play a crucial role in adjusting rheological properties, improving stability, and enhancing printability.

Hydrocolloids are widely used in 3D food printing due to their ability to modify viscosity and improve water binding. Common hydrocolloids include xanthan gum, guar gum, carrageenan, agar, and gelatin. These substances enhance shear-thinning behavior and increase yield stress, enabling better shape retention after printing (Sun et al., 2015c).

Gelling agents such as agar, gelatin, and alginate are particularly useful for post-printing stabilization. Gelation can be induced through cooling, ionic cross-linking, or enzymatic reactions, resulting in stronger and more stable printed structures. For example, calcium-induced alginate gelation is often used to stabilize printed food constructs (Liu et al., 2019).

Emulsifiers improve the stability of multiphase food inks containing fat and water. Ingredients such as lecithin and mono- and diglycerides help maintain homogeneity, prevent phase separation, and improve extrusion consistency. Emulsifiers also contribute to improved mouthfeel and sensory quality of the printed product (Sher & Tutó, 2015).

The selection and concentration of additives must be carefully optimized, as excessive use may negatively affect taste, texture, or nutritional value. Therefore, formulation strategies aim to achieve functional performance while maintaining consumer acceptability and regulatory compliance.

6.4 Types of 3D Food Printing

Three-dimensional (3D) food printing encompasses a group of additive manufacturing techniques that differ in their working principles, material requirements, and end-use applications (Fig. 6.2). While all 3D food printing methods rely on the layer-by-layer fabrication of edible materials from digital designs, the mechanism of material deposition or solidification varies significantly among different technologies. The selection of a suitable 3D food printing method depends on the physical properties of the food ink, desired structural complexity, nutritional objectives, and postprocessing requirements (Sun et al., 2015b).

Based on the method of material delivery and solidification, 3D food printing technologies can be broadly classified into extrusion-based printing, inkjet-based printing, binder jet printing, and heat-fused or selective sintering printing (Table 6.2). Each of these technologies has distinct advantages and limitations and is suitable for specific categories of food materials and applications. Understanding these technologies is essential for food technologists to select appropriate printing systems and develop compatible food formulations.

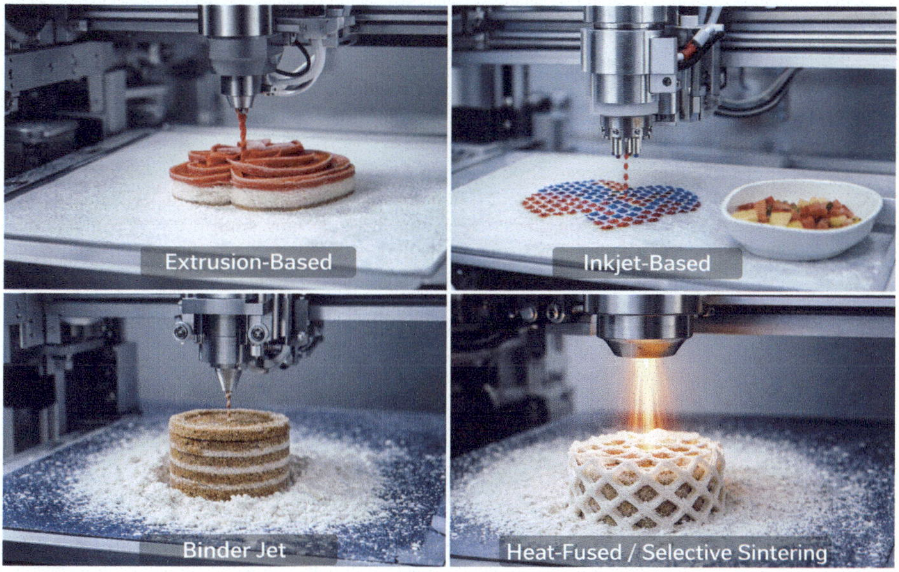

Fig. 6.2 Type of 3D food printing

Table 6.2 Types of 3D food printing technologies, principles, materials, and applications

3D Food printing technique	Working principle	Typical food materials	Advantages	Limitations	References
Extrusion-based 3D food printing	Continuous extrusion of semi-solid food ink through a nozzle using mechanical, pneumatic, or screw-driven force; material deposited layer by layer	Dough, mashed vegetables, meat pastes, cheese, chocolate, starch-based formulations	High versatility; compatible with a wide range of food inks; enables control over internal structure, texture, and porosity; suitable for texture-modified foods	Slow printing speed; requires post-printing stabilization (baking, cooling, gelation); nozzle clogging risk at high viscosity	Sun et al. (2015b), Godoi et al. (2016), Liu et al. (2017), and Mantihal et al. (2020)
Inkjet-based 3D food printing	Selective deposition of micro-droplets of low-viscosity food inks through microscopic nozzles (continuous or drop-on-demand)	Food colorants, flavors, edible inks, sugar solutions, nutrient-enriched liquids	Very high printing resolution; precise dosing of micronutrients and bioactives; minimal material waste	Limited to low-viscosity inks; weak structural strength; nozzle clogging; mainly surface decoration	Sun et al. (2015b), Sher and Tutó (2015), Liu et al. (2017), and Severini and Derossi (2016)
Binder jet 3D food printing	Liquid binder is selectively deposited onto the powdered food bed, bonding particles layer by layer	Sugar, starch, cocoa powder, dehydrated food powders	Ability to create complex geometries and overhangs; no need for support structures; high design freedom	Low mechanical strength after printing; requires post-processing; hygiene and dust-handling concerns	Sun et al. (2015b), Godoi et al. (2016), and Sher and Tutó (2015)
Heat-fused/ Selective sintering 3D food printing	Selective thermal energy is used to melt or sinter-powdered food materials; unheated powder acts as support	Sugar, chocolate, fat-based powders	High resolution and surface finish; immediate solidification; no binders or extrusion required	Nutrient degradation due to heat, high energy consumption, and a limited material range	Sun et al. (2015c), Yang et al. (2017), Liu et al. (2017), and Severini and Derossi (2016)

6.4.1 Extrusion-Based 3D Food Printing

Extrusion-based 3D food printing is the most widely researched and commercially applied technique in the food sector. This method involves the continuous extrusion of food ink through a nozzle using mechanical force, pneumatic pressure, or screw-based systems. The material is deposited layer by layer onto a build platform according to a predefined digital design (Godoi et al., 2016).

The popularity of extrusion-based food printing stems from its versatility and compatibility with a wide range of food materials. Semi-solid and paste-like materials such as dough, mashed vegetables, meat pastes, cheese, chocolate, and starch-based formulations are particularly well suited for this technique. These materials can be easily modified to exhibit shear-thinning and viscoelastic behavior, which are essential for smooth extrusion and shape retention (Mantihal et al., 2020).

From a processing perspective, extrusion-based printers consist of a material reservoir, extrusion mechanism, nozzle, motion control system, and temperature control unit. The extrusion mechanism may operate using a syringe-based piston, compressed air, or rotating screws, depending on the viscosity and particle size of the food ink. Nozzle diameter plays a critical role in determining printing resolution and layer thickness. Smaller nozzles enable higher resolution but require greater extrusion pressure, increasing the risk of clogging.

One of the major advantages of extrusion-based 3D food printing is its ability to fabricate complex geometries with controlled internal structures. By adjusting infill patterns and deposition paths, food technologists can manipulate the texture, porosity, and mechanical strength of printed foods. This capability is particularly useful in designing texture-modified foods for elderly and dysphagic patients, where controlled softness and cohesiveness are required (Liu et al., 2017).

Despite its advantages, extrusion-based printing has limitations. The technique often requires post-printing stabilization, such as baking, cooling, or gelation, to improve structural integrity. Additionally, printing speed is relatively slow compared to conventional food processing methods, which may limit large-scale industrial adoption. Nevertheless, due to its simplicity and adaptability, extrusion-based printing remains the cornerstone of 3D food printing research and development (Fig. 6.2).

6.4.2 Inkjet-Based 3D Food Printing

Inkjet-based 3D food printing operates on the principle of controlled droplet deposition. In this technique, small droplets of liquid food ink are selectively deposited onto a substrate through microscopic nozzles, similar to traditional inkjet printing technology. The droplets are placed precisely according to digital instructions, allowing high-resolution patterning and decoration (Sun et al., 2015b).

Inkjet food printing is primarily suitable for low-viscosity food materials, such as food-grade colorants, flavors, edible inks, sugar solutions, and nutrient-enriched liquids. Due to the limited viscosity range, this technique is often used for surface decoration, flavor enhancement, and nutritional fortification rather than for building self-supporting three-dimensional structures (Sher & Tutó, 2015).

Inkjet printers can operate in continuous or drop-on-demand modes. In drop-on-demand systems, droplets are generated only when required, providing precise control over material deposition and minimizing waste. Thermal or piezoelectric mechanisms are commonly used to generate droplets. In food applications, piezoelectric systems are preferred, as they avoid high temperatures that could degrade sensitive nutrients or bioactive compounds.

One of the main advantages of inkjet-based food printing is its exceptional printing resolution. This makes it ideal for creating intricate patterns, logos, and customized designs on confectionery and bakery products. Additionally, inkjet printing enables precise dosing of micronutrients, probiotics, or functional ingredients, supporting personalized nutrition strategies (Liu et al., 2017; Severini & Derossi, 2016).

However, inkjet-based printing has several limitations. The narrow range of printable viscosities restricts the types of food materials that can be used. Nozzle clogging can occur due to particle aggregation or crystallization of dissolved solids. Moreover, the printed structures often lack mechanical strength and require a supporting substrate or post-processing. As a result, inkjet-based printing is generally used as a complementary technology rather than a standalone food manufacturing method (Fig. 6.2).

6.4.3 Binder Jet 3D Food Printing

Binder jet 3D food printing is a powder-based additive manufacturing technique in which a liquid binder is selectively deposited onto a bed of powdered food material. The binder causes localized bonding of powder particles, forming a solid structure layer by layer. After printing, unbound powder is removed, and the printed object may undergo further processing to enhance strength and stability (Godoi et al., 2016).

In food applications, binder jet printing typically uses powdered materials such as sugar, starch, cocoa powder, or dehydrated food ingredients. The liquid binder may consist of water, sugar solutions, or food-grade adhesives. This technique is particularly suitable for producing complex and fragile structures that would be difficult to fabricate using extrusion-based methods.

One of the primary advantages of binder jet food printing is its ability to create highly detailed and intricate geometries without the need for support structures. The surrounding powder bed provides natural support during printing, allowing the fabrication of overhangs and hollow structures. This capability has made binder jet printing popular in confectionery applications, particularly for decorative sugar sculptures (Sun et al., 2015b).

Despite its advantages, binder jet printing poses several challenges in food applications. The printed objects often have low mechanical strength immediately after printing and may require post-processing steps such as infiltration, baking, or coating. Additionally, the use of powdered materials raises concerns regarding dust generation, material handling, and microbial safety. Careful control of hygiene and storage conditions is essential to ensure food safety (Sher & Tutó, 2015).

Furthermore, binder jet printing is less suitable for high-moisture or protein-rich foods, limiting its application range. As a result, its use is currently confined to niche applications rather than mainstream food production (Fig. 6.2).

6.4.4 Heat-Fused/Selective Sintering 3D Food Printing

Heat-fused or selective sintering 3D food printing involves the use of thermal energy to fuse powdered food materials layer by layer. This technique is adapted from selective laser sintering (SLS) used in polymer and metal additive manufacturing. In food printing, heat is applied selectively to melt or sinter food particles, forming a cohesive structure upon cooling (Sun et al., 2015c).

Common materials used in heat-fused food printing include sugar, chocolate, and fat-based powders due to their favorable thermal properties. When heated, these materials partially melt and bond to adjacent particles, creating solid layers. The surrounding unheated powder acts as a support material, enabling the fabrication of complex geometries (Yang et al., 2017).

A main advantage of heat-fused food printing is the elimination of liquid binders or extrusion mechanisms, reducing formulation complexity. The technique also allows high-resolution printing and excellent surface finish, making it attractive for decorative and artistic food applications. Additionally, thermal fusion provides immediate solidification, reducing the need for extensive post-processing (Liu et al., 2017).

However, the use of high temperatures poses challenges related to nutrient degradation and flavor changes. Heat-sensitive vitamins and bioactive compounds may be lost during sintering. Energy consumption and equipment cost are also relatively high, limiting widespread adoption. Moreover, the range of printable food materials is restricted to those with suitable melting or sintering behavior.

Despite these limitations, heat-fused printing represents a promising technology for specialized applications, particularly in confectionery and customized food design (Severini & Derossi, 2016) (Fig. 6.2).

6.5 Processing of 3D Food Printing and Equipment

The processing of 3D food printing involves a coordinated interaction between digital design, food material formulation, printer hardware, and post-processing operations. Unlike conventional food processing methods, where shaping and structuring occur through molds, extrusion dies, or manual operations, 3D food printing relies on digitally controlled deposition of edible materials to build structures layer by layer. The success of this process depends not only on the formulation of food ink but also on printer components, operational parameters, and stabilization techniques applied after printing (Sun et al., 2015b).

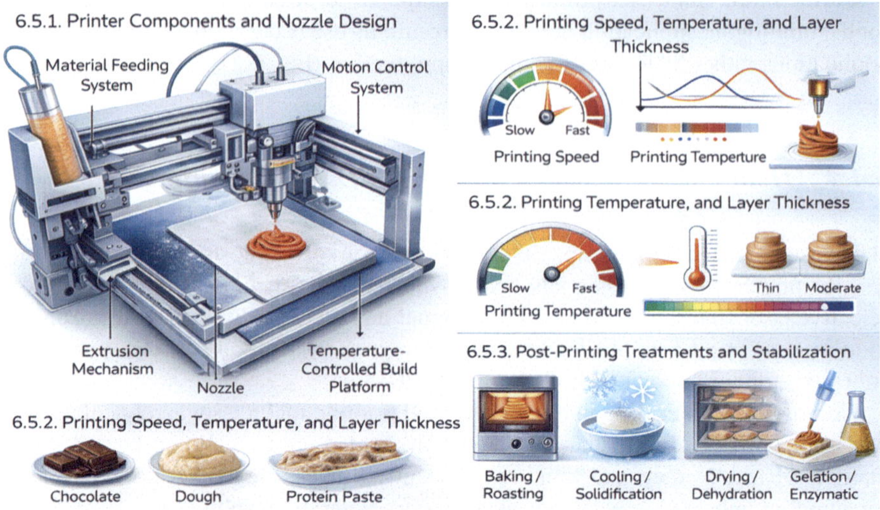

Fig. 6.3 Overview of 3D food printing processing steps and equipment

Processing parameters such as printing speed, temperature, and layer thickness strongly influence dimensional accuracy, textural properties, and structural stability of printed foods. Furthermore, post-printing treatments are often essential to convert freshly printed constructs into safe, stable, and consumer-acceptable food products. This section discusses the major equipment components used in 3D food printing, critical processing parameters, and post-printing stabilization techniques (Fig. 6.3).

6.5.1 Printer Components and Nozzle Design

A typical 3D food printer consists of several main components, including a material feeding system, extrusion or deposition unit, nozzle, motion control system, build platform, and temperature control unit. Each component plays a crucial role in ensuring accurate deposition of food ink and overall printing performance (Godoi et al., 2016).

Material Feeding and Extrusion System

The material feeding system stores and delivers food ink to the extrusion unit. Depending on the printer design and material properties, extrusion may be achieved using syringe-based pistons, pneumatic pressure, or screw-driven mechanisms. Syringe-based systems are commonly used in laboratory-scale printers due to their simplicity and ease of cleaning, whereas screw-based extruders are more suitable for high-viscosity or particle-containing food inks (Mantihal et al., 2020).

Nozzle Design

The nozzle is one of the most critical components in 3D food printing, as it directly influences printing resolution, layer thickness, and flow stability. Nozzle diameter typically ranges from 0.2 to 2.0 mm, depending on the desired resolution and material viscosity. Smaller nozzles produce finer details but require higher extrusion pressure, increasing the risk of clogging, especially when printing particulate or fibrous materials (Liu et al., 2019).

Nozzle geometry, including length, taper angle, and internal surface finish, affects shear stress distribution and flow behavior of food ink. Smooth internal surfaces reduce friction and prevent material accumulation. In some systems, interchangeable nozzles are used to accommodate different food inks and printing resolutions.

Motion Control and Build Platform

The motion control system governs the movement of the nozzle or build platform along the X-, Y-, and Z-axes. Precise control of movement ensures accurate deposition according to the digital design. The build platform may be stationary or movable and is often temperature-controlled to assist in material solidification, particularly for chocolate and fat-based inks (Sun et al., 2015c).

6.5.2 Printing Speed, Temperature, and Layer Thickness

Printing parameters play a decisive role in determining print quality, structural integrity, and processing efficiency in 3D food printing. Among these, printing speed, temperature, and layer thickness are considered the most influential variables.

Printing Speed

Printing speed refers to the rate at which the nozzle moves during material deposition. High printing speeds reduce processing time but may compromise accuracy and layer adhesion, particularly for viscous food inks. Conversely, very low speeds may result in over-deposition, leading to deformation or spreading of printed layers (Mantihal et al., 2020).

Optimizing printing speed requires balancing extrusion rate with material flow behavior. For shear-thinning food inks, moderate speeds are preferred to maintain smooth extrusion and shape retention. Printing speed is often adjusted in conjunction with nozzle diameter and extrusion pressure to achieve consistent filament width (Godoi et al., 2019).

Printing Temperature

Temperature control is especially important for thermally sensitive food inks such as chocolate, cheese, and fat-based formulations. In extrusion-based printing, the ink temperature must be high enough to ensure flow but low enough to prevent degradation of nutrients and sensory attributes. For example, chocolate printing typically requires controlled melting at temperatures slightly above its melting point, followed by rapid cooling to ensure proper crystallization (Sun et al., 2015c).

Temperature also influences microbial stability during printing. Prolonged processing at ambient temperatures may increase the risk of microbial growth in high-moisture food inks. Therefore, temperature control must consider both processing efficiency and food safety.

Layer Thickness
Layer thickness determines vertical resolution and mechanical strength of printed structures. Thin layers improve surface finish and dimensional accuracy but increase printing time. Thicker layers reduce processing time but may compromise detail and layer bonding. Typical layer thickness values in food printing range from 0.5 to 2.0 mm, depending on material properties and application (Liu et al., 2019).

Layer thickness also affects the texture and mouthfeel of the final product. By adjusting layer height and infill patterns, food technologists can manipulate hardness, porosity, and bite characteristics, enabling the development of texture-modified foods for specific consumer groups (Godoi et al., 2019).

6.5.3 Post-Printing Treatments and Stabilization

Freshly printed food structures often lack sufficient mechanical strength and stability for handling, storage, or consumption. Therefore, post-printing treatments are commonly applied to stabilize printed foods and achieve the desired texture, safety, and shelf life.

Thermal Treatments
Thermal processing methods such as baking, cooking, steaming, or roasting are frequently used to stabilize printed foods. For example, printed dough-based products may be baked to induce starch gelatinization and protein denaturation, resulting in improved structural integrity and sensory quality. Thermal treatments also contribute to microbial inactivation, enhancing food safety (Sun et al., 2015c).

Cooling and Solidification
Cooling is an essential post-processing step for materials like chocolate and gelatin-based inks. Controlled cooling promotes crystallization or gelation, leading to improved shape retention and surface finish. Rapid cooling may be achieved using refrigerated build platforms or external cooling systems.

Drying and Dehydration
Drying is commonly used for printed foods with high moisture content, such as fruit or vegetable-based formulations. Drying reduces water activity, improving shelf stability and preventing microbial growth. However, excessive drying may result in shrinkage or texture changes, requiring careful optimization (Sher & Tutó, 2015).

Chemical and Enzymatic Stabilization
Chemical cross-linking and enzymatic treatments are also employed to enhance structural stability. Calcium-induced alginate gelation is widely used in 3D food

printing to stabilize printed constructs immediately after deposition. Enzymes such as transglutaminase are used to cross-link proteins, improving the firmness and elasticity of protein-based printed foods (Liu et al., 2019).

Importance of Post-Processing
The choice of post-printing treatment depends on the type of food ink, intended application, and desired shelf life. Post-processing not only stabilizes printed structures but also influences nutritional quality, sensory properties, and consumer acceptance (Godoi et al., 2019). Therefore, post-printing treatments are considered an integral part of the overall 3D food printing process rather than optional finishing steps.

6.6 Nutritional and Functional Aspects

Three-dimensional food printing (3DFP) represents a transformative approach in modern food processing by integrating digital manufacturing with nutritional science. Beyond its ability to create visually appealing and structurally complex foods, 3DFP provides unprecedented control over nutrient composition, bioavailability, and textural properties. These characteristics position the technology as a powerful tool for addressing personalized nutrition, functional food development, and dietary management for vulnerable populations (Varvara et al., 2021; Dankar et al., 2018).

Unlike conventional food processing methods, 3DFP allows the spatial and quantitative control of ingredients within a food matrix. This enables the development of foods with targeted nutritional profiles, improved nutrient retention, and modified textures suitable for specific physiological needs. The nutritional and functional aspects of 3DFP are therefore central to its adoption in healthcare, geriatric nutrition, sports nutrition, and therapeutic diets.

6.6.1 Nutrient Retention and Bioavailability

Nutrient retention and bioavailability are critical determinants of food quality and functionality. Bioavailability refers to the fraction of ingested nutrients that are absorbed and utilized for normal physiological functions. In 3DFP, nutrient retention is influenced by ingredient selection, printing method, processing parameters, and post-printing treatments such as baking or steaming (Varvara et al., 2021).

One major advantage of 3DFP is its relatively mild processing conditions compared to traditional extrusion or thermal processing. Many extrusion-based 3D printing systems operate at ambient or moderately elevated temperatures, which helps preserve heat-sensitive nutrients such as vitamins, antioxidants, and probiotics (Liu et al., 2017). Studies have shown that functional compounds such as polyphenols, carotenoids, and anthocyanins can be successfully incorporated into

3D-printed foods with minimal degradation when optimized formulations are used (Dankar et al., 2018; Godoi et al., 2019).

Furthermore, 3DFP allows the design of internal food structures that influence digestion kinetics. By controlling infill patterns, porosity, and density, the release and absorption of nutrients can be modulated. For example, foods with higher porosity may enhance enzymatic accessibility, thereby improving nutrient bioavailability, while denser structures may slow digestion and promote satiety (Mantihal et al., 2020).

Encapsulation techniques combined with 3DFP further enhance nutrient stability. Bioactive compounds such as probiotics and antioxidants can be encapsulated within biopolymer matrices, protecting them during printing and storage (Vancauwenberghe et al., 2019). Varvara et al. (2021) reported that probiotics incorporated into 3D-printed mashed potatoes maintained high viability during refrigerated storage, demonstrating the potential of 3DFP in functional food fortification.

6.6.2 Personalized and Customized Nutrition

Personalized nutrition is one of the most significant applications of 3DFP. The technology enables precise control over ingredient composition, portion size, and nutrient distribution, allowing foods to be tailored to individual dietary requirements. This capability aligns closely with the growing demand for precision nutrition and health-oriented food systems (Severini & Derossi, 2016).

Through digital formulation and computer-aided design, 3DFP can customize macronutrient ratios (proteins, carbohydrates, and fats) and micronutrient levels (vitamins and minerals) for specific populations such as athletes, children, pregnant women, and patients with chronic diseases (Dankar et al., 2018). For instance, high-protein foods can be designed for individuals with sarcopenia, while low-potassium or low-sodium formulations can be developed for patients with renal or cardiovascular disorders (Varvara et al., 2021).

Another advantage of personalized 3DFP is its ability to integrate alternative and sustainable nutrient sources. Proteins derived from insects, algae, legumes, and food-processing by-products can be incorporated into printable food inks, contributing to both nutritional adequacy and environmental sustainability (Portanguen et al., 2019). By masking these ingredients within attractive shapes and familiar food formats, 3DFP may also improve consumer acceptance of novel foods.

Children represent another main target group for personalized 3D-printed foods. Nutrient-dense snacks with customized shapes and colors can be designed to improve dietary intake while maintaining sensory appeal. The visual attractiveness of 3D-printed foods plays a crucial role in increasing acceptance among picky eaters and individuals with limited appetite (Severini & Derossi, 2016; Varvara et al., 2021).

6.6.3 *Texture Modification for Special Populations*

Texture is a critical quality attribute influencing food acceptance, safety, and nutritional intake. One of the most promising applications of 3DFP is texture modification for populations with chewing and swallowing difficulties, such as elderly individuals and patients suffering from dysphagia or neurological disorders.

Dysphagia affects a significant proportion of the aging population and is associated with increased risk of malnutrition, dehydration, and aspiration pneumonia. Conventional texture-modified foods often lack visual appeal and sensory quality, leading to reduced food intake. 3DFP offers a solution by enabling the production of texture-modified foods that resemble conventional food shapes while maintaining appropriate softness and cohesiveness (Nachal et al., 2019; Varvara et al., 2021).

By adjusting formulation rheology and internal structure, 3DFP can produce foods that comply with the International Dysphagia Diet Standardisation Initiative (IDDSI) guidelines. Pureed or soft foods can be printed into recognizable shapes such as vegetables or meat portions, improving both dignity and appetite among patients (Mantihal et al., 2020; Phuhongsung et al., 2020).

Texture modification through 3DFP also benefits individuals with dental problems, post-surgical conditions, and pediatric feeding difficulties. The technology enables consistent texture production, which is difficult to achieve with manual preparation methods. Moreover, textural properties can be digitally controlled and replicated, ensuring standardized quality across servings (Liu et al., 2017).

In addition to healthcare applications, texture engineering through 3DFP can influence sensory perception and satiety in the general population. By manipulating infill density and structural geometry, foods with tailored hardness and chewiness can be developed, contributing to novel eating experiences and functional outcomes such as controlled energy intake.

6.7 Advantages and Applications of 3D Food Printing

Three-dimensional (3D) food printing has emerged as a transformative technology in the food processing sector, offering unprecedented flexibility in food design, formulation, and production. By combining digital manufacturing with food science principles, 3D food printing enables the creation of customized food products with controlled shape, texture, and nutritional composition. The ability to fabricate foods layer by layer from digitally designed models has opened new opportunities in bakery, confectionery, alternative protein products, and functional foods (Sun et al., 2015a; Severini & Derossi, 2016).

The applications of 3D food printing extend beyond novelty and aesthetic appeal, addressing important challenges such as personalized nutrition, texture modification for special populations, and sustainable food production. At the same time, the technology presents both technological and economic advantages, as well as certain limitations that must be overcome for large-scale commercialization. This section discusses major application areas and critically evaluates the advantages and constraints of 3D food printing.

6.7.1 Application: Bakery and Confectionery Products

Bakery and confectionery products represent one of the earliest and most successful application areas of 3D food printing. The rheological properties of doughs, chocolate, and sugar-based materials make them highly suitable for extrusion-based and heat-fused printing techniques. Additionally, consumer demand for visually appealing and customized bakery products has driven adoption of this technology in both artisanal and industrial settings (Godoi et al., 2016; Severini & Derossi, 2016).

In bakery applications, 3D printing allows precise control over the shape, size, and internal structure of products such as cookies, biscuits, bread, and pastries. Digital design enables the creation of intricate geometries that are difficult or impossible to achieve using conventional molding or cutting methods. By adjusting infill density and layer thickness, the texture and porosity of baked products can be modified, influencing crispness, softness, and mouthfeel (Liu et al., 2019; Yang et al., 2018).

Confectionery products such as chocolate, candies, and sugar sculptures are particularly well suited for 3D printing due to their thermoreversible behavior. Chocolate printing, for example, involves controlled melting during extrusion followed by rapid solidification, resulting in high-dimensional accuracy and smooth surface finish. Binder jet and selective sintering techniques have also been successfully used to fabricate complex sugar-based structures for decorative and celebratory purposes (Sun et al., 2015a; Yang et al., 2018).

Furthermore, 3D food printing enables mass customization in bakery and confectionery production. Individualized designs, personalized messages, and customized nutrient content can be produced on demand without changing production hardware. This capability aligns with emerging consumer trends toward personalization and experiential food consumption.

6.7.2 Application: Meat and Plant-Based Alternatives

The development of meat and plant-based alternatives is another significant application area for 3D food printing. Growing concerns about environmental sustainability, animal welfare, and human health have increased demand for alternative protein sources. 3D food printing offers a promising approach to structure and texture plant-based and cultured protein materials to mimic the sensory properties of conventional meat (Liu et al., 2019; Shahbazi et al., 2021).

Extrusion-based 3D printing is commonly used to process meat pastes, fish surimi, and plant protein formulations derived from soy, pea, or wheat gluten. Through layer-by-layer deposition, fibrous and anisotropic structures resembling muscle fibers can be created. Digital control over deposition paths allows the formation of complex internal architectures, improving texture and bite characteristics of printed meat analogues (Godoi et al., 2016; Shahbazi et al., 2021).

In addition to plant-based alternatives, 3D food printing has been explored for restructuring low-value meat by-products into visually appealing and nutritionally balanced foods. This approach enhances resource utilization and reduces food waste. Protein cross-linking enzymes and hydrocolloids are often incorporated into formulations to improve printability and structural stability (Shahbazi et al., 2021).

Although the technology is still in the research and development phase for large-scale meat analogue production, 3D food printing demonstrates strong potential for producing customized, sustainable, and nutritionally optimized protein foods.

6.7.3 Application: Functional and Therapeutic Foods

One of the most socially impactful applications of 3D food printing lies in the development of functional and therapeutic foods. Functional foods are designed to provide health benefits beyond basic nutrition, while therapeutic foods are formulated to meet specific medical or dietary requirements. The digital and material flexibility of 3D food printing makes it particularly suitable for these applications (Sun et al., 2015b).

A notable application is the production of texture-modified foods for elderly individuals and patients with dysphagia. By controlling texture, hardness, and cohesiveness through digital design and formulation, 3D printed foods can improve swallowing safety while maintaining visual appeal. This approach addresses a major limitation of conventional pureed diets, which are often nutritionally adequate but visually unappealing (Liu et al., 2019).

3D food printing also enables precise fortification with vitamins, minerals, probiotics, and bioactive compounds. Nutrients can be spatially distributed within a food matrix, allowing controlled release during consumption. This capability supports personalized nutrition strategies tailored to age, health status, and lifestyle.

In clinical and hospital settings, 3D food printing has the potential to produce patient-specific meals with controlled nutrient composition and portion size. Although widespread adoption is still limited by cost and processing speed, the technology holds promise for future healthcare nutrition systems.

6.7.4 Application: Sports and Performance Nutrition

One important application of 3DFP lies in sports and performance nutrition. Athletes require precisely balanced macronutrient ratios, controlled energy intake, and timely nutrient delivery to support training, recovery, and performance. 3DFP allows the formulation of customized high-protein, high-carbohydrate, or electrolyte-fortified foods tailored to individual athletic needs (Dankar et al., 2018; Baiano, 2022).

Layer-by-layer deposition enables controlled release of nutrients during digestion. For example, energy bars or snacks with compartmentalized carbohydrate and protein zones can be designed to provide rapid energy release followed by sustained nutrient availability. Such structural control is difficult to achieve using conventional food-processing methods (Mantihal et al., 2020).

6.7.5 Application: Clinical and Hospital Nutrition

3DFP has significant potential in clinical nutrition, particularly in hospitals and long-term care facilities. Patients suffering from malnutrition, metabolic disorders, cancer, or post-surgical conditions often require foods with precise nutrient composition and modified texture. 3DFP enables the production of patient-specific meals that meet prescribed dietary requirements while improving visual appeal and palatability (Varvara et al., 2021; Baiano, 2022).

For example, nutritionally dense foods enriched with proteins, omega-3 fatty acids, or micronutrients can be printed in soft textures suitable for patients with reduced appetite or swallowing difficulties. Digital reproducibility ensures consistency in nutrient delivery across meals, improving dietary compliance and clinical outcomes (Liu et al., 2017).

6.7.6 Application: Geriatric Nutrition and Aging Populations

Aging populations face challenges such as sarcopenia, reduced appetite, impaired mastication, and dysphagia. 3DFP provides an effective solution by enabling the creation of texture-modified foods with enhanced nutritional density. Foods can be fortified with proteins, calcium, vitamin D, and dietary fiber while maintaining softness and structural integrity (Dick et al., 2019).

Additionally, printing foods into recognizable shapes improves meal attractiveness and dignity among elderly individuals. This has been shown to increase food intake and reduce the risk of malnutrition in institutionalized elderly populations (Varvara et al., 2021; Baiano, 2022).

6.7.7 Application: Pediatric and Infant Nutrition

3DFP also shows promise in pediatric nutrition, where precise nutrient delivery and sensory appeal are critical. Children often exhibit selective eating behavior, leading to nutrient deficiencies. By creating visually engaging shapes and incorporating essential nutrients such as iron, zinc, and vitamins, 3DFP can improve dietary intake among children (Portanguen et al., 2019; Baiano, 2022).

For infants and young children requiring texture-appropriate foods, 3DFP allows controlled modification of hardness and cohesiveness, supporting safe feeding practices while maintaining nutritional adequacy.

6.7.8 Application: Sustainable and Alternative Nutrition

Sustainability is an increasingly important consideration in food systems. 3DFP facilitates the incorporation of alternative nutrient sources such as insect protein, algae, food-processing by-products, and plant-based proteins into acceptable food formats. These ingredients can be nutritionally rich but often face consumer resistance due to unfamiliar appearance or texture (Dankar et al., 2018; Baiano, 2022).

By transforming these materials into familiar shapes and textures, 3DFP enhances consumer acceptance while supporting sustainable protein production. This application aligns with global efforts to reduce the environmental footprint of food systems.

6.7.9 Advantages: Technological and Nutritional Benefits

3D food printing offers several technological advantages over conventional food-processing methods. One of the most significant benefits is design flexibility. Complex geometries, internal structures, and customized shapes can be created directly from digital models without additional tooling. This reduces production complexity and enables rapid prototyping (Godoi et al., 2016; Baiano, 2022).

From a nutritional perspective, 3D food printing supports precise control over ingredient composition and portion size. This enables the formulation of nutritionally balanced foods tailored to individual needs, such as high-protein foods for athletes, low-calorie foods for weight management, or fortified foods for children and elderly populations (Sun et al., 2015c).

Another important advantage is waste reduction. Additive manufacturing deposits only the required amount of material, minimizing trimming and offcuts common in conventional processing. This contributes to sustainable food production and efficient use of raw materials.

Furthermore, the integration of digital technologies allows automation, reproducibility, and scalability in food design. Digital food libraries can store and replicate recipes and designs, enabling consistent quality across production batches.

6.7.10 Advantages: Economic and Technical Constraints

Despite its numerous advantages, 3D food printing faces several economic and technical constraints that limit its widespread industrial adoption. One of the primary challenges is the high initial cost of equipment and maintenance. Specialized printers, temperature control systems, and food-grade components increase capital investment compared to conventional food processing machinery (Sher & Tutó, 2015; Baiano, 2022).

Processing speed is another major limitation. Layer-by-layer fabrication is inherently slower than mass production methods such as extrusion or molding, making it less suitable for high-volume food manufacturing. As a result, current applications are mainly focused on customized, high-value products rather than bulk production (Baiano, 2022).

Material limitations also pose challenges. Not all food materials are suitable for printing, and extensive formulation modification may be required to achieve the desired rheological properties. Additionally, ensuring food safety and microbial stability during printing, particularly for high-moisture foods, requires careful process control (Baiano, 2022).

Regulatory and consumer acceptance issues must also be addressed. As a relatively new technology, 3D food printing raises questions regarding labeling, safety standards, and consumer perception. Educating consumers and establishing clear regulatory frameworks will be essential for broader acceptance.

6.8 Conclusion

Three-dimensional (3D) food printing has emerged as a groundbreaking technology with the potential to fundamentally reshape the way food is designed, manufactured, and consumed. By applying the principles of additive manufacturing to edible materials, this technology introduces a new paradigm in food processing that emphasizes precision, customization, and digital control. Throughout this chapter, the scientific foundations, material requirements, processing technologies, nutritional implications, and diverse applications of 3D food printing have been explored, highlighting its significance in modern food engineering and its relevance to future food systems.

One of the most important contributions of 3D food printing is its ability to move beyond conventional mass production toward personalized and consumer-centered food solutions. The integration of computer-aided design and digitally controlled deposition enables precise control over food structure, portion size, and nutrient composition. This capability supports the development of customized foods tailored to individual dietary needs, health conditions, and sensory preferences. In an era marked by the increasing prevalence of diet-related chronic diseases and growing awareness of nutrition–health relationships, precision nutrition offers a powerful tool for targeted dietary interventions. From a technological perspective, the chapter has demonstrated that the success of 3D food printing relies heavily on a thorough understanding of food material properties and processing parameters. Food ink formulation, rheological behavior, and viscoelastic properties are critical determinants of printability and structural stability. Different printing techniques, such as extrusion-based, inkjet-based, binder jet, and heat-fused or selective sintering methods, offer distinct advantages and limitations depending on the nature of the food material and intended application. Mastery of these technologies allows food technologists to select appropriate methods and optimize processing conditions for high-quality printed foods.

The nutritional and functional aspects of 3D food printing represent one of its most promising contributions to the food sector. The technology enables improved nutrient retention through relatively mild processing conditions and offers opportunities to enhance bioavailability by controlling internal food structure and digestion behavior. Moreover, the ability to spatially distribute nutrients and bioactive compounds within a food matrix opens new possibilities for functional foods with controlled nutrient release. These characteristics make 3D food printing particularly valuable in healthcare nutrition, where precise nutrient delivery and texture modification are essential.

The applications discussed in this chapter demonstrate the broad and multidisciplinary impact of 3D food printing. In bakery and confectionery products, the technology supports creative design, mass customization, and product differentiation. In the development of meat and plant-based alternatives, 3D food printing offers innovative approaches to structuring proteins and improving sensory properties while supporting sustainability goals (Shahbazi et al., 2021). Functional and therapeutic foods, sports nutrition, clinical diets, geriatric and pediatric nutrition represent application areas where the technology can directly improve health outcomes and quality of life. Additionally, the use of alternative and sustainable raw materials highlights the role of 3D food printing in addressing environmental and resource-related challenges. Despite these significant advantages, the chapter has also highlighted several economic and technical constraints that currently limit the widespread adoption of 3D food printing. High equipment costs, relatively slow production rates, limited availability of suitable printable food materials, and challenges related to food safety and microbial stability remain key barriers. Regulatory frameworks and consumer acceptance are additional factors that must be addressed as the technology moves from niche applications toward broader commercialization. Overcoming these challenges will require continued interdisciplinary research, technological innovation, and collaboration between academia, industry, and regulatory authorities.

Looking toward the future, 3D food printing holds strong potential to become an integral component of next-generation food systems. Advances in printer design, material science, automation, and artificial intelligence are expected to improve processing speed, scalability, and reliability. The integration of 3D food printing with digital nutrition platforms, smart sensors, and data-driven health monitoring could further enhance its role in personalized nutrition and precision health. As sustainability and health continue to shape global food priorities, the ability of 3D food printing to deliver customized, functional, and resource-efficient foods positions it as a valuable tool for addressing complex food system challenges.

In conclusion, 3D food printing represents a convergence of food science, engineering, and digital technology that offers transformative possibilities for the future of food production. While still in an evolving stage, its demonstrated benefits in customization, nutrition, functionality, and sustainability underscore its importance as a research and innovation frontier in food technology. For students, researchers, and professionals in the field, understanding the principles, applications, and limitations of 3D food printing is essential for engaging with the future of food engineering and contributing to the development of smarter, healthier, and more sustainable food systems.

References

Baiano, A. (2022). 3D printed foods: A comprehensive review on technologies, nutritional value, safety, consumer attitude, regulatory framework, and economic and sustainability issues. *Food Reviews International, 38*(5), 986–1016.

Chua, C. K., Tan, C., Li, L., & Wong, G. (2018). Enhancing 3D printability of pureed food by addition of hydrocolloids. In *Proceedings of the 3rd international conference on progress in additive manufacturing (Pro-AM 2018)* (pp. 662–666). https://doi.org/10.25341/D42C7V

Dankar, I., Haddarah, A., Omar, F. E., Sepulcre, F., & Pujolà, M. (2018). 3D printing technology: The new era for food customization and elaboration. *Trends in Food Science & Technology, 75*, 231–242.

Derossi, A., Paolillo, M., Caporizzi, R., & Severini, C. (2020). Extending the 3D food printing tests at high speed. Material deposition and the effect of non-printing movements on the final quality of printed structures. *Journal of Food Engineering, 275*, 109865.

Dick, A., Bhandari, B., & Prakash, S. (2019). 3D printing of meat. *Meat Science, 153*, 35–44.

Godoi, F. C., Prakash, S., & Bhandari, B. R. (2016). 3d printing technologies applied for food design: Status and prospects. *Journal of Food Engineering, 179*, 44–54.

Godoi, F. C., Bhandari, B. R., Prakash, S., & Zhang, M. (2019). An introduction to the principles of 3D food printing. In *Fundamentals of 3D food printing and applications* (pp. 1–18). Academic Press.

Holland, S., Foster, T., MacNaughtan, W., & Tuck, C. (2018). Design and characterisation of food-grade powders and inks for microstructure control using 3D printing. *Journal of Food Engineering, 220*, 12–19.

Lipton, J. I., Cutler, M., Nigl, F., Cohen, D., & Lipson, H. (2015). Additive manufacturing for the food industry. *Trends in Food Science & Technology, 43*(1), 114–123.

Liu, Z., Zhang, M., Bhandari, B., & Wang, Y. (2017). 3D printing: Printing precision and application in food sector. *Trends in Food Science & Technology, 69*, 83–94.

Liu, Y., Zhang, W., Wang, K., Bao, Y., Regenstein, J. M., & Zhou, P. (2019). Fabrication of gel-like emulsions with whey protein isolate using microfluidization: Rheological properties and 3D printing performance. *Food and Bioprocess Technology, 12*(12), 1967–1979.
Mantihal, S., Kobun, R., & Lee, B. B. (2020). 3D food printing of as the new way of preparing food: A review. *International Journal of Gastronomy and Food Science, 22*, 100260.
Nachal, N., Moses, J. A., Karthik, P., & Anandharamakrishnan, C. (2019). Applications of 3D printing in food processing. *Food Engineering Reviews, 11*(3), 123–141.
Oliveira, S. M., Fasolin, L. H., Vicente, A. A., Fuciños, P., & Pastrana, L. M. (2020). Printability, microstructure, and flow dynamics of phase-separated edible 3D inks. *Food Hydrocolloids, 109*, 106120.
Pant, A., Lee, A. Y., Karyappa, R., Lee, C. P., An, J., Hashimoto, M., et al. (2021). 3D food printing of fresh vegetables using food hydrocolloids for dysphagic patients. *Food Hydrocolloids, 114*, 106546.
Phuhongsung, P., Zhang, M., & Devahastin, S. (2020). Investigation on 3D printing ability of soybean protein isolate gels and correlations with their rheological and textural properties via LF-NMR spectroscopic characteristics. *Lwt, 122*, 109019.
Pitayachaval, P., Sanklong, N., & Thongrak, A. (2018). A review of 3D food printing technology. In *MATEC web of conferences* (Vol. 213, p. 01012). EDP Sciences.
Portanguen, S., Tournayre, P., Sicard, J., Astruc, T., & Mirade, P. S. (2019). Toward the design of functional foods and biobased products by 3D printing: A review. *Trends in Food Science & Technology, 86*, 188–198.
Severini, C., & Derossi, A. (2016). Could the 3D printing technology be a useful strategy to obtain customized nutrition? *Journal of Clinical Gastroenterology, 50*, S175–S178.
Shahbazi, M., Jäger, H., Ettelaie, R., & Chen, J. (2021). Construction of 3D printed reduced-fat meat analogue by emulsion gels. Part I: Flow behavior, thixotropic feature, and network structure of soy protein-based inks. *Food Hydrocolloids, 120*, 106967.
Sher, D., & Tutó, X. (2015). Review of 3D food printing. *Temes de disseny, 31*, 104–117.
Sun, J., Peng, Z., Yan, L., Fuh, J. H., & Hong, G. S. (2015a). 3D food printing—An innovative way of mass customization in food fabrication. *International Journal of Bioprinting, 1*(1), 27–38.
Sun, J., Peng, Z., Zhou, W., Fuh, J. Y., Hong, G. S., & Chiu, A. (2015b). A review on 3D printing for customized food fabrication. *Procedia Manufacturing, 1*, 308–319.
Sun, J., Zhou, W., Huang, D., Fuh, J. Y., & Hong, G. S. (2015c). An overview of 3D printing technologies for food fabrication. *Food and Bioprocess Technology, 8*(8), 1605–1615.
Vancauwenberghe, V., Mbong, V. B. M., Vanstreels, E., Verboven, P., Lammertyn, J., & Nicolai, B. (2019). 3D printing of plant tissue for innovative food manufacturing: Encapsulation of alive plant cells into pectin based bio-ink. *Journal of Food Engineering, 263*, 454–464.
Varvara, R. A., Szabo, K., & Vodnar, D. C. (2021). 3D food printing: Principles of obtaining digitally-designed nourishment. *Nutrients, 13*(10), 3617.
Yang, F., Zhang, M., & Bhandari, B. (2017). Recent development in 3D food printing. *Critical Reviews in Food Science and Nutrition, 57*(14), 3145–3153.
Yang, F., Zhang, M., Prakash, S., & Liu, Y. (2018). Physical properties of 3D printed baking dough as affected by different compositions. *Innovative Food Science & Emerging Technologies, 49*, 202–210.
Zoran, A., & Coelho, M. (2011). Cornucopia: The concept of digital gastronomy. *Leonardo, 44*(5), 425–431.

Chapter 7
Personalized Nutrition

Abbreviations

AI	Artificial Intelligence
APOA2	Apolipoprotein A2
APOE	Apolipoprotein E
B12	Vitamin B12
BMI	Body Mass Index
CRP	C-Reactive Protein
DASH	Dietary Approaches to Stop Hypertension
DNA	Deoxyribonucleic Acid
GC	Group Specific Component (Vitamin D Binding Protein Gene)
HbA1c	Glycated Hemoglobin
HDL-C	High-Density Lipoprotein Cholesterol
HFE	High Iron Fe Gene (Homeostatic Iron Regulator)
LCT	Lactase Gene
LDL-C	Low-Density Lipoprotein Cholesterol
LDLR	Low-Density Lipoprotein Receptor
MTHFR	Methylenetetrahydrofolate Reductase
NCDs	Non-Communicable Diseases
NF-κB	Nuclear Factor Kappa B
PPARs	Peroxisome Proliferator Activated Receptors
PPARα	Peroxisome Proliferator Activated Receptor Alpha
RNA	Ribonucleic Acid
SCFAs	Short Chain Fatty Acids
SNPs	Single Nucleotide Polymorphisms
TAS2R38	Taste Receptor Type 2 Member 38

Y. Kumar, S. Dey, *Emerging Technology in Food Processing*,
https://doi.org/10.1007/978-3-032-24505-2_7

7.1 Introduction

Nutrition science has historically been guided by population-based dietary recommendations aimed at promoting general health and preventing nutrient deficiencies. These guidelines, while instrumental in improving public health outcomes, are primarily designed to address the average needs of populations rather than individual variability (Verma et al., 2018). However, growing scientific evidence demonstrates that individuals respond differently to the same foods and dietary patterns due to variations in genetics, metabolism, gut microbiota, lifestyle, and environmental exposures (Bush et al., 2020). This recognition has led to the emergence of personalized nutrition, an approach that tailors dietary advice and interventions to the unique biological and behavioral characteristics of individuals (Verma et al., 2018).

Personalized nutrition represents a paradigm shift away from the traditional "one size fits all" dietary framework toward a more precise and individualized model of nutritional guidance (Boland et al., 2019). It integrates advances in genomics, nutrigenomics, metabolomics, microbiome research, and digital health technologies to better understand inter-individual differences in dietary responses. By aligning nutritional recommendations with individual-specific data, personalized nutrition aims to optimize health outcomes, improve dietary adherence, and reduce the risk of chronic diseases (Mathers, 2017; Bush et al., 2020). The increasing relevance of personalized nutrition is closely linked to the global rise in non-communicable diseases (NCDs), including obesity, type 2 diabetes mellitus, cardiovascular diseases, and certain cancers (Verma et al., 2018). These conditions are strongly influenced by dietary factors, yet their incidence, progression, and response to dietary interventions vary considerably among individuals. For example, individuals consuming identical meals may exhibit markedly different postprandial glucose responses, highlighting the limitations of generalized dietary advice (Verma et al., 2018). Personalized nutrition seeks to address this heterogeneity by identifying individual dietary triggers and protective factors, thereby enabling more effective disease prevention and management strategies (Boland et al., 2019).

A central component of personalized nutrition is nutrigenetics, which examines how genetic variations influence an individual's response to nutrients and dietary components. Single-nucleotide polymorphisms (SNPs) in genes involved in lipid metabolism, carbohydrate utilization, vitamin absorption, and inflammatory pathways can significantly alter nutrient requirements and disease susceptibility (Bush et al., 2020; Boland et al., 2019). For instance, polymorphisms in the APOE gene are associated with differential lipid responses to dietary fat, while variations in the MTHFR gene affect folate metabolism and homocysteine levels (Fenech et al., 2011). Incorporating genetic information into dietary planning allows for the development of genotype-informed nutritional recommendations that may enhance metabolic health and reduce disease risk (Fenech et al., 2011).

Closely related to nutrigenetics is nutrigenomics, which focuses on how nutrients and bioactive food compounds influence gene expression, protein synthesis, and cellular function. Nutrients can act as signaling molecules that modulate transcription factors, epigenetic mechanisms, and metabolic pathways, thereby influencing physiological processes linked to health and disease (Mathers, 2017). This bidirectional interaction between diet and the genome underscores the dynamic nature of nutrition-health relationships and provides mechanistic insights into the potential benefits of personalized dietary interventions (Adams et al., 2020).

Another critical pillar of personalized nutrition is the gut microbiome, a diverse and complex community of microorganisms residing in the human gastrointestinal tract (Kviatcovsky et al., 2021). The gut microbiota plays a vital role in digestion, nutrient metabolism, immune regulation, and energy balance. Inter-individual differences in microbial composition and functionality have been shown to influence responses to dietary fiber, fats, and polyphenols, as well as susceptibility to metabolic disorders (Zmora et al., 2019). Research has demonstrated that incorporating microbiome data into personalized nutrition models can improve the prediction of metabolic responses and support the development of targeted dietary strategies that promote beneficial microbial profiles (Kviatcovsky et al., 2021).

Advances in metabolomics and nutritional phenotyping have further strengthened the scientific foundation of personalized nutrition (Laddu & Hauser, 2019). Metabolomics enables the comprehensive analysis of small molecule metabolites in biological samples such as blood, urine, or saliva, providing insights into metabolic status, nutrient utilization, and disease risk (Gibney et al., 2020). These objective biomarkers complement traditional dietary assessment tools, which are often limited by recall bias and reporting errors. By capturing real-time metabolic responses, metabolomics enhances the precision and relevance of individualized dietary recommendations (Gibney et al., 2020).

The practical implementation of personalized nutrition has been accelerated by the rapid development of digital health technologies (Adams et al., 2020). Mobile health applications, wearable devices, continuous glucose monitoring systems, and artificial intelligence-based platforms facilitate continuous data collection and personalized feedback (Kviatcovsky et al., 2021). These tools enable dynamic dietary guidance that adapts to changes in physiological status, lifestyle behaviors, and environmental conditions, thereby improving user engagement and long-term adherence to dietary interventions (Ronteltap et al., 2013). Digital platforms also offer scalable solutions for delivering personalized nutrition services across diverse populations (Kviatcovsky et al., 2021).

Despite its considerable promise, personalized nutrition faces several scientific, ethical, and practical challenges. The integration and interpretation of multiomics data require robust analytical frameworks and interdisciplinary expertise. Additionally, issues related to data privacy, ethical use of genetic information, cost, and accessibility must be carefully addressed to ensure equitable implementation (Kviatcovsky et al., 2021). Translating complex biological data into clear, actionable, and culturally appropriate dietary advice also remains a significant challenge, particularly in low- and middle-income settings (Adams et al., 2020). Nevertheless, personalized nutrition holds substantial potential to transform nutrition science, healthcare delivery, and public health policy. By embracing individual variability and leveraging advances in biological and digital sciences, personalized nutrition aligns closely with the broader goals of precision health and preventive medicine (Adams et al., 2020). As scientific evidence continues to expand and technologies become more accessible, personalized nutrition is poised to play a pivotal role in shaping future dietary guidelines, clinical nutrition practice, and food system innovation (Bush et al., 2020).

7.2 Type of Personalized Nutrition

Personalized nutrition encompasses a spectrum of approaches that tailor dietary recommendations based on individual characteristics rather than relying on generalized population-level guidelines (Bashiardes et al., 2019). These approaches differ in complexity, data requirements, and level of biological specificity, ranging from basic personalization based on demographic and lifestyle factors to advanced strategies integrating multiomics data and digital health technologies (Bashiardes et al., 2019). Understanding the different types of personalized nutrition is essential for appreciating its scientific foundations, practical applications, and potential impact on health promotion and disease prevention (Bush et al., 2020).

Broadly, personalized nutrition can be classified into phenotype-based, genotype-based, microbiome-based, metabolomics-driven, behavioral- and lifestyle-based, and digital and artificial-intelligence–enabled approaches (Bashiardes et al., 2019). These categories are not mutually exclusive; rather, they often overlap and are increasingly integrated into comprehensive personalized nutrition frameworks (Bashiardes et al., 2019; Mathers, 2017). Table 7.1 presents a consolidated overview of these major types of personalized nutrition, highlighting their basis of personalization, key parameters assessed, applications, strengths, and limitations.

7.2.1 Phenotype-Based Personalized Nutrition

In phenotype-based personalized nutrition, the assessment of physical and biochemical indicators allows practitioners to stratify individuals into metabolically relevant categories and design targeted dietary interventions (Kussmann & Fay, 2008). Anthropometric measures such as BMI, waist circumference, and waist-to-hip ratio provide insights into adiposity distribution, which is strongly associated with cardiometabolic risk (Verma et al., 2018). Central obesity, for instance, is linked to insulin resistance and systemic inflammation, necessitating dietary strategies that emphasize energy balance, improved fat quality, and increased intake of anti-inflammatory foods (Verma et al., 2018). Similarly, body composition analysis differentiates between fat mass and lean mass, enabling nutrition plans that preserve muscle mass while reducing excess adiposity, particularly in aging populations (Laddu & Hauser, 2019).

Table 7.1 Types of personalized nutrition and their key characteristics

Type of personalized nutrition	Basis of personalization	Parameters assessed	Primary applications	Strengths	limitations
Phenotype-based personalized nutrition	Observable physical, physiological, and biochemical traits	Age, sex, BMI, body composition, blood pressure, lipid profile, glycemic markers, and inflammatory biomarkers	Clinical nutrition therapy, weight management, chronic disease prevention, and management	Widely accessible; cost-effective; based on routine clinical measurements; strong evidence base	Reflects downstream metabolic outcomes; limited predictive ability; less sensitive to early metabolic changes
Genotype-based personalized nutrition	Genetic variations influencing nutrient metabolism and disease risk	Single-nucleotide polymorphisms (SNPs), gene nutrient interactions, nutrigenetic markers	Disease risk reduction, precision dietary guidance, long-term health optimization	Enables predictive and preventive strategies; stable genetic markers; high personalization potential	High cost; ethical and privacy concerns; limited translation into clear dietary advice
Microbiome-based personalized nutrition	Composition and functional activity of gut microbiota	Microbial diversity, dominant bacterial species, short-chain fatty acid production, dysbiosis markers	Gut health improvement, metabolic disease management, immune modulation	Captures diet microbiota interactions; dynamic and responsive to diet; emerging therapeutic potential	High inter-individual variability; methodological complexity; lack of standardized guidelines
Metabolome-based personalized nutrition	Individual metabolic responses to dietary intake	Metabolites in blood, urine, or tissues; metabolic pathways; nutrient utilization markers	Precision nutrition research, metabolic disorder management, biomarker-driven dietary interventions	Reflects real-time metabolic status; high sensitivity; integrates genetic and environmental influences	Expensive analytical techniques; limited clinical availability; complex data interpretation
Behavior-based personalized nutrition	Individual lifestyle habits and dietary behaviors	Food preferences, eating patterns, physical activity, sleep, and stress levels	Dietary adherence improvement, lifestyle modification programs	Enhances compliance; supports sustainable dietary change; easy to implement	Subjective reporting; limited biological insight; behavior may change over time
Integrated (Multiomics) personalized nutrition	Combination of phenotype, genotype, microbiome, metabolome, and behavioral data	Clinical biomarkers, genetic data, microbial profiles, metabolic markers, lifestyle factors	Advanced precision nutrition, preventive healthcare, personalized dietary algorithms	Comprehensive and highly accurate; maximizes intervention efficacy	High cost; technical complexity; limited scalability

Biochemical phenotyping plays a pivotal role in refining dietary recommendations. Blood lipid profiles, including total cholesterol, low-density lipoprotein cholesterol (LDL-C), high-density lipoprotein cholesterol (HDL-C), and triglycerides, inform individualized fat intake recommendations (Laddu & Hauser, 2019). Individuals with elevated LDL-C may benefit from reduced saturated fat intake and increased consumption of plant sterols, soluble fiber, and omega-3 fatty acids, whereas those with hypertriglyceridemia may require moderation of refined carbohydrates and alcohol intake (Laddu & Hauser, 2019). Glycemic markers such as fasting plasma glucose, postprandial glucose, and glycated hemoglobin (HbA1c) are widely used to personalize carbohydrate quantity, quality, and distribution throughout the day (Verma et al., 2018). Inflammatory biomarkers, including C-reactive protein (CRP) and interleukin-6, further enhance phenotypic characterization by identifying individuals with low-grade chronic inflammation. Dietary patterns rich in fruits, vegetables, whole grains, and unsaturated fats have been shown to reduce inflammatory burden, making such biomarkers valuable tools for tailoring anti-inflammatory dietary interventions (Mathers, 2017). Similarly, micronutrient status indicators, such as serum iron, vitamin D, and vitamin B12 levels, allow targeted correction of deficiencies without unnecessary supplementation (Verma et al., 2018).

Phenotype-based personalized nutrition is particularly effective in clinical nutrition therapy, where dietary interventions must align closely with disease-specific metabolic alterations (Verma et al., 2018). In individuals with hypertension, for example, sodium sensitivity varies considerably, and phenotypic indicators such as blood pressure response to sodium intake can guide personalized sodium restriction strategies (Verma et al., 2018). The Dietary Approaches to Stop Hypertension (DASH) framework is often adapted based on individual blood pressure profiles and electrolyte balance (Laddu & Hauser, 2019). In patients with metabolic syndrome, phenotype-driven interventions focus on improving insulin sensitivity, lipid metabolism, and body weight through tailored macronutrient distribution and energy intake (Laddu & Hauser, 2019).

Weight management programs frequently rely on phenotype-based personalization to improve outcomes. Differences in resting metabolic rate, physical activity level, and metabolic flexibility influence weight loss responses to dietary interventions. Personalized calorie prescriptions and macronutrient ratios based on phenotypic characteristics have demonstrated greater effectiveness than standardized weight loss diets (Gibney et al., 2020). Moreover, phenotype-based approaches are particularly valuable in resource-limited settings, as they rely on widely available clinical and anthropometric data rather than advanced molecular testing (Laddu & Hauser, 2019).

Despite its strengths, phenotype-based personalized nutrition has inherent limitations. Phenotypic markers often represent the cumulative outcome of long-term genetic, environmental, and lifestyle influences rather than underlying causal mechanisms (Laddu & Hauser, 2019). As a result, dietary interventions based solely on phenotypic data may be reactive rather than predictive, addressing existing metabolic disturbances rather than preventing their development (Mathers, 2017).

Additionally, phenotypic measures may lack sensitivity to early metabolic changes, limiting their ability to detect subclinical risk.

Nevertheless, phenotype-based personalization serves as a critical entry point for individualized nutrition care and provides a practical foundation for more advanced personalized nutrition strategies. When combined with behavioral assessment and lifestyle data, phenotypic information enhances the relevance, feasibility, and acceptability of dietary recommendations (Laddu & Hauser, 2019). Furthermore, phenotype-based approaches can be progressively integrated with genetic, metabolomics, and microbiome data to improve predictive accuracy and intervention efficacy. Phenotype-based personalized nutrition remains a cornerstone of personalized dietary practice due to its clinical relevance, affordability, and strong evidence base (Verma et al., 2018). While it may not fully capture the molecular complexity underlying individual dietary responses, it offers a robust and pragmatic framework for tailoring nutrition interventions across clinical, public health, and community settings. Its continued integration with emerging omics-based approaches is expected to further strengthen its role in precision nutrition (Verma et al., 2018).

7.2.2 Genotype-Based Personalized Nutrition (Nutrigenetics)

Beyond single gene nutrient interactions, genotype-based personalized nutrition increasingly recognizes the polygenic nature of most nutrition-related traits and chronic diseases (Wang et al., 2022). Rather than relying on isolated SNPs, contemporary nutrigenetic research often employs polygenic risk scores that aggregate the effects of multiple genetic variants to better predict individual susceptibility to obesity, type 2 diabetes, cardiovascular disease, and micronutrient deficiencies (Wang et al., 2022). These composite genetic profiles enhance the predictive accuracy of nutrigenetic interventions and support more nuanced dietary recommendations tailored to an individual's inherited risk (Görman et al., 2013). Genetic variation also influences taste perception, appetite regulation, and food preferences, thereby indirectly shaping dietary behavior. Polymorphisms in taste receptor genes, such as TAS2R38, affect bitter taste sensitivity and may influence vegetable consumption, while variations in dopamine-related genes impact reward responses to high-fat or high-sugar foods (Mathers, 2017). Understanding these genetic influences enables the design of dietary strategies that account not only for metabolic differences but also for behavioral tendencies, improving adherence and long-term dietary success (Mullins et al., 2020).

Micronutrient metabolism represents a well-established application area of nutrigenetics. Genetic variants affecting vitamin D metabolism, iron absorption, and antioxidant pathways can alter individual nutrient requirements (Mullins et al., 2020). For example, polymorphisms in the GC gene influence vitamin D binding and bioavailability, potentially necessitating personalized vitamin D intake recommendations (Mullins et al., 2020). Similarly, variations in genes involved in iron metabolism, such as HFE, affect iron absorption efficiency and risk of iron overload or deficiency, highlighting the need for genotype-informed micronutrient guidance. Genotype-based personalized nutrition is particularly relevant in disease prevention and risk stratification. Individuals with genetic predispositions to insulin resistance or dyslipidaemia may benefit from early dietary modifications that reduce disease progression (Singar et al., 2024). Evidence suggests that carriers of certain APOA2 and APOE genotypes exhibit differential lipid responses to saturated fat intake, supporting tailored fat quality recommendations. Such insights strengthen the preventive potential of nutrigenetics by enabling early intervention before clinical symptoms emerge (Singar et al., 2024).

Despite its scientific promise, the translation of nutrigenetics into routine dietary practice remains challenging. Many genetic variants exert modest effects, and dietary responses are often influenced by complex interactions between genes, environment, and lifestyle factors (Verma et al., 2018). Consequently, genotype-based advice should not be interpreted in isolation but rather integrated with phenotypic, metabolic, and behavioral data to ensure accuracy and relevance (Singar et al., 2024). Overreliance on genetic information without adequate clinical context may lead to oversimplified or misleading dietary recommendations (Singar et al., 2024).

Ethical, legal, and social considerations are central to the implementation of nutrigenetics. Issues related to informed consent, genetic data ownership, confidentiality, and potential misuse of genetic information must be carefully addressed. Consumer-directed genetic testing services have increased public access to nutrigenetic data; however, variability in scientific rigor and interpretation quality raises concerns regarding misinformation and unrealistic health expectations (Mathers, 2017). Robust regulatory frameworks and professional guidance are therefore essential to ensure the responsible use of nutrigenetic information. Technological advancements are gradually enhancing the feasibility of genotype-based personalized nutrition (Wang et al., 2022). The decreasing cost of genotyping, coupled with improved bioinformatics tools, has expanded opportunities for integrating genetic data into clinical and research settings. Furthermore, artificial intelligence and machine learning approaches are increasingly used to analyze complex gene-diet interactions and generate personalized dietary recommendations with greater precision (Wang et al., 2022).

Genotype-based personalized nutrition represents a critical component of precision nutrition by offering insight into inherited dietary responses and disease susceptibility (Wang et al., 2022). While its standalone application is constrained by biological complexity and ethical considerations, nutrigenetics provides valuable predictive information that complements phenotype-based and omics-driven approaches. As scientific evidence continues to evolve and integration strategies improve, genotype-based personalization is expected to play an increasingly important role within comprehensive, multiomics personalized nutrition frameworks (Wang et al., 2022).

7.2.3 Nutrigenomics-Based Personalized Nutrition

While nutrigenetics examines how genes influence dietary responses, nutrigenomics explores how nutrients and bioactive food components influence gene expression and cellular function (Kussmann & Fay, 2008). This approach focuses on the dynamic interaction between diet and the genome, including transcriptional regulation, epigenetic modifications, and signaling pathways involved in metabolism and inflammation (Kussmann & Fay, 2008). Dietary components such as fatty acids, polyphenols, vitamins, and phytochemicals can act as molecular signals that modulate gene expression through transcription factors like peroxisome proliferator activated receptors (PPARs) and nuclear factor kappa B (NF–κB). Nutrigenomics-based personalized nutrition aims to design dietary strategies that favorably regulate gene expression profiles associated with health and disease prevention (Yang, 2023).

Although nutrigenomics provides valuable mechanistic insights, its translation into routine dietary practice remains limited due to technical complexity, high costs, and challenges in interpreting large-scale genomic data (Lagoumintzis & Patrinos, 2023). Nonetheless, it forms a critical scientific foundation for precision nutrition and future therapeutic dietary interventions (Yang, 2023).

7.2.4 Microbiome-Based Personalized Nutrition

The gut microbiome has emerged as a key determinant of individual variability in dietary responses. Microbial communities in the gastrointestinal tract influence nutrient digestion, energy harvest, immune function, and metabolic regulation. Inter-individual differences in microbial composition and functionality can significantly alter responses to dietary fiber, fats, and bioactive compounds (Kviatcovsky et al., 2021).

Microbiome-based personalized nutrition tailors dietary recommendations to promote beneficial microbial profiles and metabolic outcomes. For instance, individuals with low-fiber degrading bacterial populations may benefit from gradual fiber introduction or specific prebiotics, while those with dysbiotic microbial patterns may require targeted dietary interventions (Yeşilyurt et al., 2022).

Seminal studies have demonstrated that incorporating microbiome data improves the prediction of postprandial glycemic responses beyond traditional dietary models (Simon et al., 2023). Despite its promise, microbiome-based personalization faces challenges related to microbial variability, temporal instability, and standardization of analytical methods. Nevertheless, it represents one of the most rapidly advancing areas of personalized nutrition research (Yeşilyurt et al., 2022).

7.2.5 Metabolomics-Driven Personalized Nutrition

Metabolomics involves the comprehensive analysis of small molecule metabolites in biological samples, providing a snapshot of an individual's metabolic phenotype. Unlike genetic data, which remain relatively static, metabolomic profiles are highly dynamic and reflect recent dietary intake, metabolic health, and environmental exposures (Gibney et al., 2020). Metabolomics-driven personalized nutrition uses metabolic biomarkers to assess nutrient status, identify metabolic inefficiencies, and monitor dietary responses in real time. For example, plasma lipidomics can guide fat intake recommendations, while amino acid profiles may indicate insulin resistance or muscle protein turnover. This approach enhances the precision of dietary interventions by capturing functional metabolic outcomes rather than theoretical risk (Shawky et al., 2025). While metabolomics offers high-resolution insights, its widespread application is constrained by cost, technical expertise requirements, and data interpretation challenges. However, continued advancements in analytical platforms and bioinformatics are likely to improve its feasibility in clinical and research settings (Shawky et al., 2025).

7.2.6 Lifestyle and Behavioral Personalized Nutrition

Lifestyle and behavioral factors play a critical role in determining dietary needs and responses. Lifestyle-based personalized nutrition considers physical activity levels, sleep patterns, stress, cultural preferences, socioeconomic status, and eating behaviors when designing dietary recommendations (Ronteltap et al., 2013). Behavioral personalization aims to enhance dietary adherence by aligning recommendations with individual habits, preferences, and motivational drivers (Kussmann & Fay, 2008). For instance, dietary strategies may differ for sedentary versus physically active individuals, or for shift workers with disrupted circadian rhythms. Incorporating behavioral insights improves the sustainability and effectiveness of personalized nutrition interventions (Kussmann & Fay, 2008).

This approach is particularly relevant in public health and community nutrition settings, as it emphasizes feasibility, cultural appropriateness, and long-term behavior change. While it may lack molecular precision, lifestyle-based personalization is essential for translating scientific recommendations into real-world practice (Bush et al., 2020).

7.2.7 Digital and AI-Enabled Personalized Nutrition

The integration of digital health technologies has revolutionized personalized nutrition delivery. Mobile applications, wearable devices, continuous glucose monitors, and artificial intelligence (AI) algorithms enable real-time data collection and adaptive dietary guidance (Ordovas et al., 2018). AI-enabled personalized nutrition systems can integrate data from genetics, microbiome, metabolomics, lifestyle, and clinical parameters to generate dynamic and context-specific recommendations. These platforms improve scalability and user engagement while supporting continuous monitoring and feedback (Ronteltap et al., 2013).

7.2.8 Integrated Multiomics Personalized Nutrition

The most advanced form of personalized nutrition involves the integration of multiple data layers, including genomics, transcriptomics, metabolomics, microbiomics, and phenotypic information. Multiomics personalized nutrition aims to provide a holistic understanding of individual nutritional needs and responses (Mathers, 2017). This integrative approach aligns closely with the principles of precision health and systems biology. While currently limited to research and specialized clinical settings, multiomics strategies represent the future direction of personalized nutrition as analytical tools become more accessible and cost-effective (Boland et al., 2019). Despite their potential, digital personalized nutrition solutions raise concerns regarding data security, algorithm transparency, and equitable access. Addressing these challenges is critical for ensuring ethical and effective implementation (Boland et al., 2019).

7.3 Mechanisms of Personalized Nutrition

Personalized nutrition is grounded in the recognition that individual responses to diet are shaped by complex and interconnected biological mechanisms. Unlike conventional nutrition approaches that rely on generalized dietary guidelines, personalized nutrition seeks to understand and leverage the mechanistic pathways through which nutrients interact with the human body at genetic, molecular, microbial, metabolic, and physiological levels (Barrow et al., 2020). These mechanisms collectively determine nutrient requirements, metabolic efficiency, disease susceptibility, and dietary responses. Elucidating the underlying mechanisms of personalized nutrition is essential for translating scientific insights into effective and evidence-based dietary interventions (Barrow et al., 2020). The mechanisms of personalized nutrition operate across multiple biological scales, ranging from the genome to the whole-body phenotype, and are influenced by environmental and behavioral factors. A synthesis of the key biological mechanisms underpinning personalized nutrition and their functional relevance is presented in Table 7.2, highlighting how advances in omics technologies, systems biology, and computational analytics have enhanced the ability to characterize individual variability and integrate these mechanisms into tailored nutritional strategies (Hussain et al., 2026).

Table 7.2 Mechanisms underpinning personalized nutrition

Mechanism	Biological basis	Biomarkers	Role in personalized nutrition	References
Genetic mechanisms (Nutrigenetics)	Inherited genetic variations influence nutrient metabolism and physiological responses	SNPs (e.g., APOE, FTO, MTHFR, LCT), copy number variations	Explains inter-individual differences in response to nutrients; enables genotype-informed dietary recommendations	Barrow et al. (2020), Verma et al. (2018), and Hussain et al. (2026)
Gene expression regulation (nutrigenomics)	Dietary components modulate gene expression without altering DNA sequence	Transcription factors, epigenetic markers, mRNA expression	Identifies how nutrients activate or suppress metabolic and inflammatory pathways	Hussain et al. (2026) and Yang (2023)
Epigenetic mechanisms	Diet-induced modifications affecting gene activity	DNA methylation, histone modifications, non-coding RNAs	Explains long-term and transgenerational dietary effects; links early nutrition with chronic disease risk	Ramos-Lopez et al. (2021) and Mathers (2017)
Phenotypic and metabolic responses	Observable physiological and biochemical characteristics	BMI, body composition, lipid profile, glycemic markers, blood pressure	Provides clinically accessible data for tailoring diets based on current health status	Laddu and Hauser (2019) and Verma et al. (2018)
Gut–microbiome interactions	Diet alters gut microbial composition and activity	Microbial diversity, short-chain fatty acids, dysbiosis markers	Modulates nutrient bioavailability, immune responses, and metabolic health	Yeşilyurt et al. (2022) and Simon et al. (2023)
Metabolomic pathways	Metabolic fingerprints reflect nutrient processing and metabolic status	Blood and urine metabolites, lipidomics, amino acid profiles	Enables real-time assessment of dietary effects and metabolic flexibility	Barrow et al. (2020) and Verma et al. (2018)
Hormonal and endocrine regulation	Nutrients influence hormone secretion and sensitivity	Insulin, leptin, ghrelin, cortisol	Links diet to appetite regulation, energy balance, and metabolic disorders	Shawky et al. (2025)
Behavioral and lifestyle mechanisms	Eating behavior and lifestyle modify dietary effectiveness	Physical activity, sleep patterns, stress, food preferences	Enhances adherence and effectiveness of personalized dietary interventions	Ronteltap et al. (2013) and Bush et al. (2020)
Digital and AI-driven integration	Computational analysis of multi-source data	Wearable data, dietary logs, algorithms	Integrates genetic, phenotypic, and behavioral data into dynamic dietary guidance	Ordovas et al. (2018) and Zeevi et al. (2015)

(continued)

Table 7.2 (continued)

Mechanism	Biological basis	Biomarkers	Role in personalized nutrition	References
Multiomics integration	Combined analysis of genomics, transcriptomics, proteomics, and metabolomics	Multi-layer biological datasets	Provides a holistic and systems-level understanding of individual nutrition responses	Bush et al. (2020)

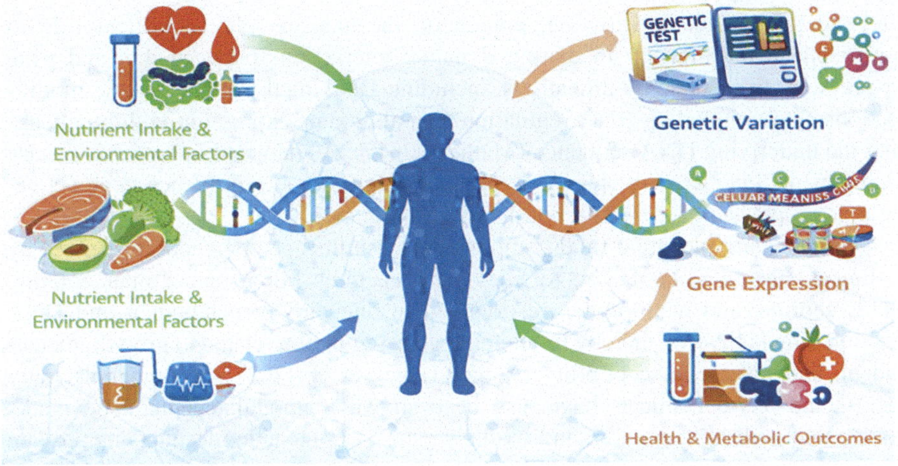

Fig. 7.1 Genetic mechanisms and nutrient–gene interactions in personalized nutrition

7.3.1 Genetic Mechanisms and Nutrient Gene Interactions

Genetic variation is a fundamental mechanism underlying inter-individual differences in dietary responses. Single-nucleotide polymorphisms (SNPs), copy number variations, and other genetic alterations can influence nutrient digestion, absorption, transport, metabolism, and excretion (Verma et al., 2018). These genetic factors form the basis of nutrigenetics, which investigates how inherited genetic differences modulate the physiological effects of nutrients and dietary patterns (Hussain et al., 2026). As illustrated in Fig. 7.1, nutrient–gene interactions occur through multiple pathways linking dietary components with genetic regulation and metabolic outcomes. For example, polymorphisms in genes involved in lipid metabolism, such as APOE, PPARα, and LDLR, can alter plasma lipid responses to dietary fat intake and influence cardiovascular disease risk (Hussain et al., 2026). Similarly, variants in the MTHFR gene affect folate metabolism and homocysteine levels, thereby modifying requirements for folate and related B vitamins. Genetic differences in lactose tolerance, mediated by polymorphisms in the LCT gene, further illustrate how inherited traits shape dietary suitability and nutrient utilization (Barrow et al., 2020).

These genetic mechanisms operate by altering enzyme activity, receptor sensitivity, or transporter efficiency, ultimately influencing metabolic pathways and health outcomes. By identifying genetic predispositions, personalized nutrition can proactively tailor dietary recommendations to mitigate disease risk and optimize nutrient adequacy (Barrow et al., 2020).

7.3.2 *Epigenetic Regulation and Dietary Modulation*

Beyond static genetic variation, epigenetic mechanisms play a critical role in personalized nutrition by mediating dynamic interactions between diet and gene expression. Epigenetic modifications, including DNA methylation, histone modifications, and non-coding RNA regulation, can alter gene expression without changing the underlying DNA sequence (Mathers, 2017). As depicted in Fig. 7.2, dietary components interact with epigenetic regulatory pathways to influence gene activation and repression across different biological contexts.

Nutrients and bioactive food components can influence epigenetic processes by acting as substrates or cofactors for enzymatic reactions. For instance, folate, vitamin B12, choline, and methionine are involved in one-carbon metabolism and DNA methylation pathways, thereby influencing gene regulation related to growth, metabolism, and disease susceptibility (Ramos-Lopez et al., 2021). Polyphenols, fatty acids, and phytochemicals have also been shown to modulate epigenetic marks associated with inflammation and oxidative stress. Epigenetic mechanisms provide

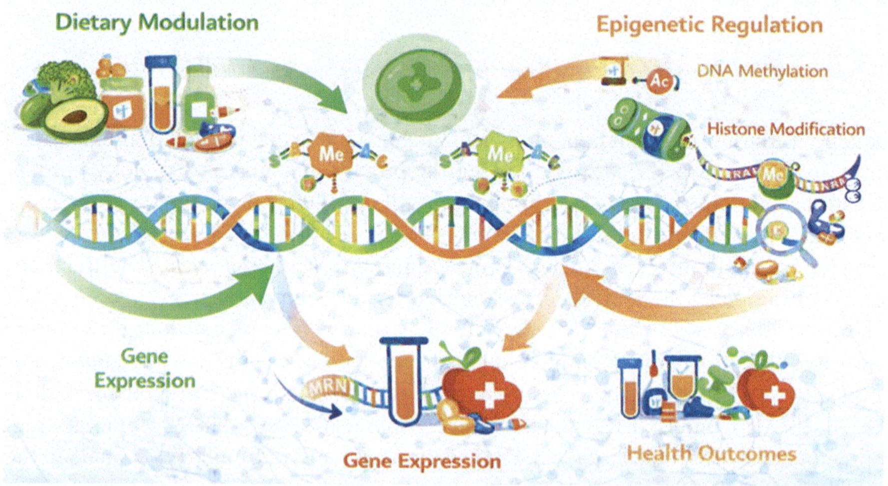

Fig. 7.2 Conceptual illustration of epigenetic regulation influenced by diet

a biological explanation for how early life nutrition, lifestyle factors, and environmental exposures can exert long-term effects on health. Personalized nutrition strategies that consider epigenetic sensitivity may be particularly relevant for disease prevention across the life course (Ramos-Lopez et al., 2021).

7.3.3 Regulation of Gene Expression by Nutrients (Nutrigenomics)

Nutrigenomics focuses on the mechanisms by which nutrients influence gene expression and cellular signaling pathways (Marcum, 2020). Nutrients act as molecular signals that interact with transcription factors, nuclear receptors, and signaling cascades to regulate metabolic processes such as lipid oxidation, glucose homeostasis, and inflammatory responses (Marcum, 2020). Fatty acids, for example, regulate the activity of peroxisome proliferator-activated receptors (PPARs), which control genes involved in lipid metabolism and energy balance. Carbohydrates influence insulin signaling and the expression of genes related to glucose transport and glycolysis. Bioactive compounds such as flavonoids and carotenoids can modulate transcription factors like nuclear factor kappa B (NF–κB), thereby influencing inflammatory pathways (Yang, 2023). These gene regulatory mechanisms enable personalized nutrition to design dietary patterns that target specific metabolic pathways and physiological outcomes. Understanding nutrigenomic responses is essential for developing precision dietary interventions tailored to individual metabolic profiles (Lagoumintzis & Patrinos, 2023).

7.3.4 Gut-Microbiome–Diet Interactions

The gut microbiome represents a central mechanistic interface between diet and host physiology. The composition and functional capacity of gut microbial communities vary widely among individuals and are shaped by genetics, diet, lifestyle, and environmental factors. These microorganisms play a crucial role in nutrient digestion, energy harvest, immune modulation, and the production of bioactive metabolites such as short-chain fatty acids (SCFAs) (Yeşilyurt et al., 2022).

Dietary components such as fiber, resistant starch, and polyphenols are metabolized by gut microbes into SCFAs, which influence glucose metabolism, lipid homeostasis, and inflammatory responses (Yeşilyurt et al., 2022). As illustrated in Fig. 7.3, diet–microbiome interactions influence key metabolic and inflammatory pathways that contribute to inter-individual variability in nutritional responses. Inter-individual differences in microbial composition can therefore lead to variability in metabolic responses to the same diet. Studies have demonstrated that incorporating microbiome data improves the prediction of postprandial glycemic responses and

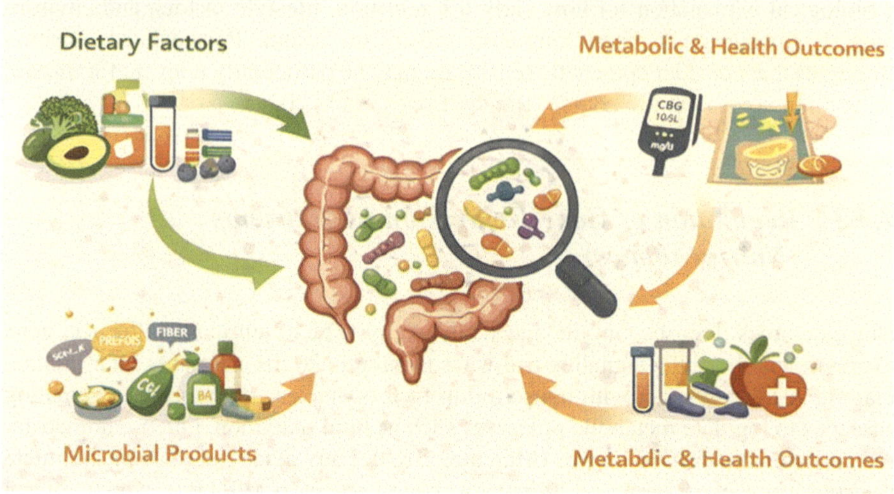

Fig. 7.3 Gut-microbiome–diet interactions in personalized nutrition

metabolic outcomes (Simon et al., 2023). Microbiome-mediated mechanisms underscore the importance of tailoring dietary recommendations to support beneficial microbial profiles. Personalized nutrition interventions may include targeted fiber intake, prebiotics, probiotics, or fermented foods to modulate gut microbial function and enhance host health (Simon et al., 2023).

7.3.5 Metabolic Pathways and Nutrient Utilization

Metabolic mechanisms play a pivotal role in determining individual responses to diet. Variations in enzymatic activity, hormonal regulation, and metabolic flexibility influence how nutrients are processed and utilized (Barrow et al., 2020). Metabolic phenotyping, including assessments of insulin sensitivity, lipid oxidation, and energy expenditure, provides insights into individual metabolic efficiency and dietary needs (Barrow et al., 2020). Metabolomics technologies enable the comprehensive analysis of metabolites involved in amino acid metabolism, lipid metabolism, and carbohydrate utilization (Hussain et al., 2026). These metabolic signatures reflect both genetic predisposition and environmental influences, offering a functional readout of dietary responses. Personalized nutrition leverages these metabolic mechanisms to optimize macronutrient distribution, meal timing, and nutrient density. By targeting specific metabolic inefficiencies, personalized dietary interventions can improve glycemic control, lipid profiles, and body composition outcomes (Hussain et al., 2026).

7.3.6 Hormonal and Endocrine Responses to Diet

Hormonal regulation represents another important mechanism in personalized nutrition. Hormones such as insulin, glucagon, leptin, ghrelin, and cortisol mediate appetite regulation, energy balance, and nutrient partitioning. Inter-individual differences in hormonal sensitivity and secretion patterns influence satiety, hunger, and metabolic responses to food (Shawky et al., 2025). For example, individuals with insulin resistance exhibit altered glucose and lipid metabolism in response to carbohydrate intake, necessitating personalized carbohydrate management strategies. Similarly, variations in leptin sensitivity affect appetite regulation and weight management outcomes. Understanding endocrine mechanisms allows personalized nutrition to tailor dietary composition and meal timing to optimize hormonal balance (Shawky et al., 2025).

7.3.7 Circadian Rhythms and Chrononutrition

Emerging evidence highlights the role of circadian rhythms in modulating metabolic responses to diet (Verma et al., 2018). The timing of food intake influences hormonal secretion, gene expression, and metabolic pathways. Disruptions to circadian rhythms, such as those experienced by shift workers, can impair glucose metabolism and increase disease risk (Franzago et al., 2023). Chrononutrition, a component of personalized nutrition, aligns dietary timing with individual circadian patterns to optimize metabolic health. Mechanistically, this involves the synchronization of nutrient intake with circadian regulation of metabolic enzymes and hormones (Franzago et al., 2023).

7.3.8 Behavioral and Environmental Modulation of Nutritional Responses

Behavioral and environmental factors interact with biological mechanisms to shape dietary responses. Physical activity, stress, sleep quality, and psychosocial factors influence metabolic pathways, hormonal regulation, and gene expression (Ronteltap et al., 2013). As illustrated in Fig. 7.4, these behavioral and environmental determinants operate alongside genetic, epigenetic, and microbiome-related mechanisms to influence personalized nutrition responses. By incorporating lifestyle patterns, cultural context, and environmental constraints, personalized nutrition strategies enhance dietary relevance, adherence, and long-term sustainability. Behavioral modulation acts synergistically with biological mechanisms, reinforcing the effectiveness of personalized dietary interventions and supporting durable improvements in health outcomes (Bush et al., 2020).

Fig. 7.4 Influence of behavioral and environmental factors on personalized nutrition

7.3.9 Integration of Multiomics Mechanisms

The most advanced mechanistic framework for personalized nutrition involves the integration of genomic, epigenomic, transcriptomic, metabolomic, and microbiomic data. Systems biology approaches enable the modeling of complex interactions between dietary components and biological networks (Ordovas et al., 2018). Multiomics integration enhances predictive accuracy and supports the development of comprehensive, individualized dietary strategies. Although currently limited by cost and analytical complexity, integrated mechanistic approaches represent the future direction of personalized nutrition research and practice (Bush et al., 2020).

7.4 Advantages and Application of Personalized Nutrition

Personalized nutrition has emerged as a transformative approach in modern nutrition science, offering tailored dietary strategies based on individual biological, physiological, behavioral, and environmental characteristics (Verma et al., 2018). Unlike conventional population-based dietary guidelines, personalized nutrition recognizes inter-individual variability in nutrient requirements and metabolic responses. By leveraging advances in genomics, metabolomics, microbiome research, and digital health technologies, personalized nutrition provides targeted dietary recommendations aimed at optimizing health outcomes and preventing diet-related diseases (Singar et al., 2024). As illustrated in Fig. 7.5, the advantages of personalized

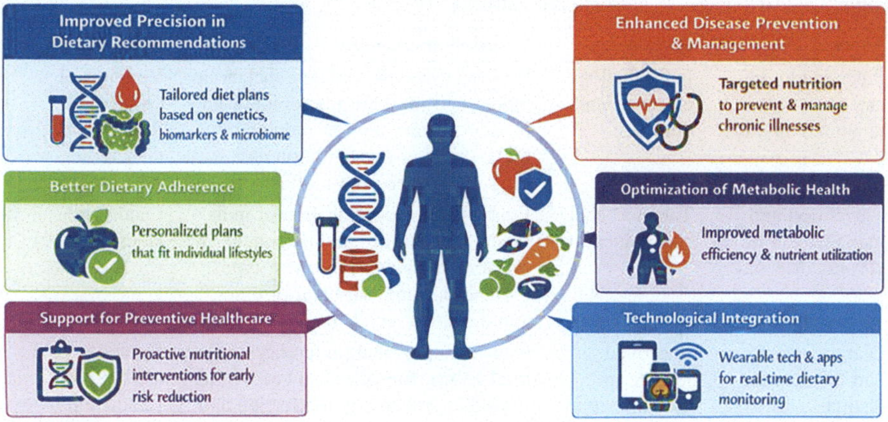

Fig. 7.5 Conceptual framework illustrating the advantages and applications of personalized nutrition

nutrition are closely linked to its practical applications across multiple domains. Accordingly, this section highlights the significant benefits of personalized nutrition and examines its diverse applications in clinical practice, public health, sports nutrition, disease prevention, and food system innovation (Singar et al., 2024).

7.4.1 Advantages of Personalized Nutrition

Personalized nutrition offers a transformative approach to dietary guidance by tailoring nutritional recommendations to individual genetic, metabolic, microbiome, and lifestyle profiles (Verma et al., 2018). Unlike conventional population-based guidelines, it enhances precision in nutrient intake, reducing the risk of deficiencies or excesses. Personalized nutrition supports effective prevention and management of chronic diseases such as obesity, diabetes, and cardiovascular disorders by addressing individual variability in dietary response (Verma et al., 2018). By incorporating behavioral and cultural factors, it improves dietary adherence and promotes sustainable lifestyle changes. Furthermore, the integration of digital health technologies enables real-time monitoring and adaptive interventions, aligning personalized nutrition with precision and preventive healthcare models (Singar et al., 2024). The main advantages of personalized nutrition and their implications for health outcomes are summarized in Table 7.3.

Table 7.3 Advantages of personalized nutrition

Aspect	Advantage	Description/relevance	References
Precision in dietary recommendations	Individual-specific nutrient guidance	Integrates genetic variants, metabolic biomarkers, and gut microbiota data to tailor dietary advice, reducing under- or over-consumption and improving intervention effectiveness	Verma et al. (2018) and Singar et al. (2024)
Prevention and management of chronic diseases	Targeted NCD risk reduction	Enables early identification of individual risk profiles and personalized dietary strategies for obesity, type 2 diabetes, cardiovascular diseases, and metabolic syndrome	Laddu and Hauser (2019)
Dietary adherence and behavioral change	Improved long-term compliance	Considers personal preferences, cultural habits, lifestyle, and behavioral readiness, enhancing motivation and sustainability of dietary interventions	Ronteltap et al. (2013) and Laddu and Hauser (2019)
Metabolic health and nutrient utilization	Optimized metabolic efficiency	Aligns dietary intake with individual metabolic phenotypes to improve energy balance, nutrient bioavailability, and metabolic outcomes	Gibney et al. (2020) and Shawky et al. (2025)
Precision and preventive healthcare	Proactive health management	Supports early dietary interventions based on genetic and metabolic predispositions, reducing disease onset and reliance on pharmacological treatments	Mathers (2017) and Shawky et al. (2025)
Technological integration	Scalable and dynamic personalization	Utilizes mobile apps, wearables, continuous glucose monitoring, and AI platforms for real-time data collection and adaptive dietary feedback	Ordovas et al. (2018) and Zeevi et al. (2015)

7.4.1.1 Improved Precision in Dietary Recommendations

One of the most significant advantages of personalized nutrition is its ability to enhance the precision of dietary advice. Traditional dietary guidelines are designed to meet the needs of the average individual, often overlooking variability in genetics, metabolism, and lifestyle (Verma et al., 2018). Personalized nutrition integrates individual-specific data, such as genetic variants, metabolic biomarkers, and gut microbiota profiles, to provide dietary recommendations that are more accurately aligned with personal needs (Singar et al., 2024). For example, individuals with genetic predispositions affecting lipid metabolism may benefit from tailored fat intake recommendations, while those with impaired glucose tolerance may require personalized carbohydrate management strategies. This precision reduces the likelihood of under- or overconsumption of nutrients and enhances the effectiveness of dietary interventions (Singar et al., 2024).

7.4.1.2 Enhanced Prevention and Management of Chronic Diseases

Personalized nutrition offers substantial advantages in the prevention and management of non-communicable diseases (NCDs), including obesity, type 2 diabetes mellitus, cardiovascular diseases, and metabolic syndrome. These conditions are strongly influenced by dietary patterns, yet individual responses to dietary interventions vary widely (Laddu & Hauser, 2019). By identifying individual risk profiles through genetic, metabolic, and microbiome assessments, personalized nutrition enables early intervention and targeted dietary modification. For instance, personalized dietary strategies have been shown to improve postprandial glycemic control and insulin sensitivity more effectively than standardized dietary recommendations (Laddu & Hauser, 2019). Such targeted approaches improve clinical outcomes and reduce healthcare burden.

7.4.1.3 Improved Dietary Adherence and Behavioral Change

Dietary adherence remains a major challenge in nutrition interventions. Personalized nutrition addresses this issue by incorporating individual preferences, cultural practices, lifestyle factors, and behavioral characteristics into dietary planning (Ronteltap et al., 2013). When dietary advice is perceived as personally relevant and achievable, individuals are more likely to adhere to recommendations. Behaviorally informed personalized nutrition strategies consider factors such as meal timing, food accessibility, motivation, and readiness for change (Laddu & Hauser, 2019). This individualized approach fosters long-term dietary compliance and sustainable behavior modification, enhancing the overall success of nutrition interventions (Laddu & Hauser, 2019).

7.4.1.4 Optimization of Metabolic Health and Nutrient Utilization

Personalized nutrition allows for the optimization of metabolic efficiency and nutrient utilization by aligning dietary intake with individual metabolic profiles (Shawky et al., 2025). Variations in enzyme activity, hormonal regulation, and metabolic flexibility influence how nutrients are processed and stored (Gibney et al., 2020). Metabolomics-driven personalized nutrition identifies metabolic inefficiencies and nutrient imbalances, enabling targeted dietary adjustments. This approach supports improved energy balance, enhanced nutrient bioavailability, and better management of metabolic disorders. As a result, personalized nutrition contributes to improved body composition, lipid profiles, and glycemic regulation (Gibney et al., 2020).

7.4.1.5 Support for Precision and Preventive Healthcare

Personalized nutrition aligns closely with the principles of precision and preventive healthcare (Shawky et al., 2025). By identifying genetic predispositions and early metabolic alterations, personalized nutrition facilitates proactive dietary interventions before the onset of disease (Mathers, 2017). This preventive focus reduces reliance on pharmacological treatments and supports holistic health management. Personalized nutrition also complements medical nutrition therapy by providing individualized dietary support alongside clinical care, enhancing treatment efficacy and patient outcomes (Shawky et al., 2025).

7.4.1.6 Integration of Technological Advancements

The integration of digital health technologies represents another key advantage of personalized nutrition. Mobile applications, wearable devices, continuous glucose monitoring systems, and artificial intelligence (AI) based platforms enable real-time data collection and personalized feedback (Ordovas et al., 2018). These technologies enhance scalability, accessibility, and user engagement, allowing personalized nutrition to be delivered beyond clinical settings. Continuous monitoring also enables dynamic dietary adjustments in response to physiological and behavioral changes (Zeevi et al., 2015).

7.4.2 Applications of Personalized Nutrition

Personalized nutrition has wide-ranging applications across clinical, public health, and lifestyle contexts. In clinical nutrition and medical nutrition therapy, it supports individualized dietary management of chronic diseases such as diabetes, cardiovascular disorders, obesity, and gastrointestinal conditions. In public health, stratified personalized approaches enhance the effectiveness of population-based dietary guidelines. Personalized nutrition is widely applied in weight management, sports and performance nutrition, and life-stage nutrition, including maternal, pediatric, and geriatric care. Emerging applications also include gut microbiome modulation, functional food and nutraceutical development, and AI-driven digital health platforms that deliver adaptive, real-time dietary recommendations to improve long-term health outcomes. The major application areas of personalized nutrition and their corresponding objectives are summarized in Table 7.4, highlighting its translational potential across healthcare and wellness domains.

Table 7.4 Applications of personalized nutrition

Application	Scope of application	Outcomes	References
Clinical nutrition and medical nutrition therapy	Disease-specific dietary management	Improved management of diabetes, cardiovascular disease, obesity, gastrointestinal disorders, and food intolerances through genotype- and microbiome-informed diets	Boland et al. (2019) and Adams et al. (2020)
Public health and population nutrition	Stratified dietary interventions	Targeted nutrition strategies for population subgroups, enhancing the effectiveness of dietary guidelines and disease prevention programs	Adams et al. (2020)
Weight management and obesity prevention	Personalized weight-loss strategies	Customized macronutrient distribution, meal timing, and behavioral support leading to improved weight loss and maintenance	Kviatcovsky et al. (2021) and Bashiardes et al. (2019)
Sports and performance nutrition	Athlete-specific dietary optimization	Enhanced performance, recovery, and injury prevention through individualized energy, hydration, and nutrient strategies	Gibney et al. (2020) and Adams et al. (2020)
Maternal, pediatric, and geriatric nutrition	Life-stage specific nutrition	Supports fetal development, childhood growth, and healthy aging by addressing changing physiological and nutritional needs	Mathers (2017) and Boland et al. (2019)
Gut health and microbiome modulation	Microbiota-targeted dietary interventions	Improved glycemic control, lipid metabolism, and inflammatory status through personalized fiber, prebiotic, and probiotic strategies	Simon et al. (2023)
Functional foods and nutraceutical development	Customized health products	Development of foods and supplements tailored to genetic and metabolic profiles, supporting consumer-driven wellness	Mohanty and Singhal (2018) and Ronteltap et al. (2013)
Digital health and nutrition platforms	Technology-enabled personalization	AI-driven, real-time dietary recommendations integrated into telehealth, corporate wellness, and consumer nutrition systems	Ordovas et al. (2018)

7.4.2.1 Clinical Nutrition and Medical Nutrition Therapy

Personalized nutrition has wide-ranging applications in clinical nutrition and medical nutrition therapy. In healthcare settings, personalized dietary interventions are increasingly used to manage conditions such as diabetes, cardiovascular disease, obesity, gastrointestinal disorders, and food intolerances (Boland et al., 2019). For example, genotype-informed dietary strategies can guide lipid and carbohydrate intake in individuals with metabolic disorders, while microbiome-based interventions may support gut health in patients with inflammatory bowel disease. Personalized nutrition enhances the effectiveness of clinical interventions by addressing individual variability in disease etiology and dietary response (Adams et al., 2020).

7.4.2.2 Public Health and Population Nutrition

Although personalized nutrition is often associated with individualized care, it also has important applications in public health. Stratified nutrition approaches use population subgroups defined by age, sex, genetic risk, or metabolic status to deliver more targeted dietary recommendations (Adams et al., 2020). Personalized nutrition can support public health initiatives aimed at reducing the prevalence of diet-related diseases by improving the relevance and impact of dietary guidelines. Digital platforms enable the scalable delivery of personalized advice, making it feasible to implement personalized nutrition strategies at the population level (Adams et al., 2020).

7.4.2.3 Weight Management and Obesity Prevention

Weight management represents one of the most prominent applications of personalized nutrition (Kviatcovsky et al., 2021). Traditional weight loss programs often yield variable results due to differences in metabolism, appetite regulation, and behavioral factors. Personalized nutrition addresses these challenges by tailoring dietary strategies to individual energy requirements, metabolic profiles, and lifestyle patterns (Kviatcovsky et al., 2021). Personalized approaches may include customized macronutrient distribution, meal timing, and behavioral support, resulting in improved weight loss outcomes and long-term weight maintenance. This individualized approach is particularly effective in addressing obesity, a multifactorial condition influenced by genetic, metabolic, and environmental factors (Bashiardes et al., 2019).

7.4.2.4 Sports and Performance Nutrition

Personalized nutrition plays a critical role in sports and performance nutrition by optimizing dietary strategies to support athletic performance, recovery, and injury prevention. Athletes exhibit significant variability in energy expenditure, nutrient requirements, and metabolic responses to training (Gibney et al., 2020). Personalized sports nutrition considers factors such as training load, body composition, metabolic phenotype, and genetic predisposition to tailor macronutrient intake, hydration strategies, and micronutrient supplementation. This approach enhances performance outcomes and supports long-term athletic health (Adams et al., 2020).

7.4.2.5 Maternal, Pediatric, and Geriatric Nutrition

Personalized nutrition has important applications across the life course, including maternal, pediatric, and geriatric nutrition (Adams et al., 2020). Nutritional requirements vary significantly across different life stages, influenced by growth, development, aging, and physiological changes (Mathers, 2017). In maternal nutrition, personalized dietary strategies can support optimal fetal development and reduce the risk of pregnancy-related complications. Pediatric personalized nutrition addresses growth patterns, food allergies, and metabolic health, while geriatric applications focus on maintaining muscle mass, bone health, and cognitive function. These life-stage-specific applications highlight the adaptability of personalized nutrition (Boland et al., 2019).

7.4.2.6 Gut Health and Microbiome Modulation

The growing understanding of the gut microbiome has expanded the application of personalized nutrition in promoting digestive and metabolic health. Microbiome-based dietary interventions aim to modulate microbial composition and function through targeted fiber intake, prebiotics, probiotics, and fermented foods (Simon et al., 2023). Personalized microbiome interventions have been shown to improve glycemic control, lipid metabolism, and inflammatory status. These applications underscore the importance of individualized dietary strategies in supporting gut-mediated health outcomes (Simon et al., 2023).

7.4.2.7 Functional Foods and Nutraceutical Development

Personalized nutrition has influenced the development of functional foods and nutraceuticals designed to target specific health outcomes or genetic profiles (Mohanty & Singhal, 2018). Advances in food technology enable the formulation of products tailored to individual nutrient needs and metabolic characteristics (Ronteltap et al., 2013). These innovations support consumer-driven health solutions and align with the growing demand for personalized wellness products. Personalized functional foods represent an intersection between nutrition science, food industry innovation, and consumer health (Mohanty & Singhal, 2018).

7.4.2.8 Digital Health and Personalized Nutrition Platforms

Digital health platforms have expanded the reach and application of personalized nutrition. AI-driven systems integrate data from genetics, microbiome, metabolomics, and lifestyle factors to deliver adaptive dietary recommendations in real time (Ordovas et al., 2018). These platforms are increasingly used in corporate wellness programs, telehealth services, and consumer nutrition applications. Digital personalized nutrition enhances accessibility and supports continuous engagement, although ethical considerations related to data privacy and equity must be addressed (Ordovas et al., 2018).

7.5 Conclusion

Personalized nutrition represents a significant evolution in nutrition science, shifting the focus from generalized dietary recommendations toward individualized, evidence-based nutritional strategies. By acknowledging inter-individual variability in genetic makeup, metabolic responses, gut microbiota composition, lifestyle behaviors, and environmental exposures, personalized nutrition offers a more precise and effective framework for optimizing health and preventing disease. The growing body of scientific evidence discussed throughout this chapter underscores the limitations of traditional "one size fits all" dietary guidelines and highlights the need for nutrition approaches that account for biological diversity and personal context. Advances in genomics, nutrigenetics, nutrigenomics, metabolomics, and microbiome research have provided critical insights into the mechanisms underlying individual dietary responses. These scientific developments have enabled the identification of genetic polymorphisms, metabolic signatures, and microbial profiles that influence nutrient metabolism, disease susceptibility, and overall physiological function. By integrating these data streams, personalized nutrition moves beyond symptom-based dietary advice toward a mechanistic understanding of nutrition-health relationships. This shift enhances the accuracy of dietary recommendations and supports targeted interventions aimed at correcting metabolic imbalances and reducing disease risk.

The applications of personalized nutrition extend across clinical practice, public health, sports and performance nutrition, weight management, and life-stage-specific nutrition. In clinical settings, personalized dietary interventions complement medical nutrition therapy by improving disease management outcomes for conditions such as diabetes, cardiovascular diseases, and gastrointestinal disorders. In public health contexts, stratified and digitally delivered personalized nutrition approaches have the potential to improve the effectiveness of dietary guidelines and reduce the burden of diet-related non-communicable diseases. Similarly, in sports nutrition and performance optimization, personalized dietary strategies support enhanced physical performance, recovery, and long-term health.

A major strength of personalized nutrition lies in its ability to improve dietary adherence and promote sustainable behavior change. By aligning nutritional advice with individual preferences, cultural practices, and lifestyle constraints, personalized nutrition enhances relevance and practicality. The incorporation of digital health technologies, including mobile applications, wearable devices, and artificial intelligence-driven platforms, further supports real-time monitoring, adaptive guidance, and continuous engagement. These technological innovations increase accessibility and scalability, enabling personalized nutrition to reach broader populations beyond traditional clinical settings.

Despite its promise, personalized nutrition also presents several challenges that must be addressed to ensure responsible and effective implementation. Scientific challenges include the integration and interpretation of complex multiomics data, the need for robust clinical validation, and the standardization of analytical methods.

Ethical and social considerations, such as data privacy, informed consent, and equitable access to personalized nutrition services, are equally critical. Without careful regulation and evidence-based practice, there is a risk of over-commercialization and misinterpretation of personalized nutrition information, particularly in consumer-driven digital platforms.

Future advancements in personalized nutrition will likely focus on the integration of multiomics data through systems biology and machine learning approaches, enabling more accurate prediction of dietary responses and health outcomes. Continued research is needed to strengthen the evidence base, establish clear clinical guidelines, and evaluate long-term health benefits. Additionally, efforts to improve affordability and accessibility will be essential for ensuring that personalized nutrition contributes to health equity rather than widening existing disparities.

In conclusion, personalized nutrition holds substantial potential to transform nutrition science, healthcare delivery, and public health strategies. By embracing individual variability and leveraging technological and scientific innovations, personalized nutrition aligns closely with the principles of precision health and preventive medicine. As research continues to advance and practical challenges are addressed, personalized nutrition is poised to play a pivotal role in shaping the future of dietary guidance and health promotion, ultimately supporting improved health outcomes and quality of life across diverse populations.

References

Adams, S. H., Anthony, J. C., Carvajal, R., Chae, L., Khoo, C. S. H., Latulippe, M. E., et al. (2020). Perspective: Guiding principles for the implementation of personalized nutrition approaches that benefit health and function. *Advances in Nutrition, 11*(1), 25–34.

Barrow, M., Bell, L., & Bell, C. (2020). Transforming personalized nutrition practice. *Nutrition Reviews, 78*(12), 1046–1051.

Bashiardes, S., Abdeen, S. K., & Elinav, E. (2019). Personalized nutrition: Are we there yet? *Journal of Pediatric Gastroenterology and Nutrition, 69*(6), 633–638.

Boland, M., Alam, F., & Bronlund, J. (2019). Modern technologies for personalized nutrition. In *Trends in personalized nutrition* (pp. 195–222). Academic.

Bush, C. L., Blumberg, J. B., El-Sohemy, A., Minich, D. M., Ordovás, J. M., Reed, D. G., & Behm, V. A. Y. (2020). Toward the definition of personalized nutrition: A proposal by the American Nutrition Association. *Journal of the American College of Nutrition, 39*(1), 5–15.

Fenech, M., El-Sohemy, A., Cahill, L., Ferguson, L. R., French, T. A. C., Tai, E. S., Milner, J., Koh, W. P., Xie, L., Zucker, M., & Buckley, M. (2011). Nutrigenetics and nutrigenomics: Viewpoints on the current status and applications in nutrition research and practice. *Journal of Nutrigenetics and Nutrigenomics, 4*(2), 69–89.

Franzago, M., Alessandrelli, E., Notarangelo, S., Stuppia, L., & Vitacolonna, E. (2023). Chrono-nutrition: Circadian rhythm and personalized nutrition. *International Journal of Molecular Sciences, 24*(3), 2571.

Gibney, M. J., Walsh, M., Brennan, L., Roche, H. M., German, B., & van Ommen, B. (2020). Metabolomics in human nutrition: Opportunities and challenges. *American Journal of Clinical Nutrition, 82*(3), 497–503.

Görman, U., Mathers, J. C., Grimaldi, K. A., Ahlgren, J., & Nordström, K. (2013). Do we know enough? A scientific and ethical analysis of the basis for genetic-based personalized nutrition. *Genes & Nutrition, 8*(4), 373–381.

Hussain, S. A., Sarker, M. I., Liu, Y., & Jin, T. Z. (2026). DNA methylation and its role in personalized nutrition: Mechanisms, clinical insights, and future perspectives. *International Journal of Molecular Sciences, 27*(2), 566.

Kussmann, M., & Fay, L. B. (2008). Nutrigenomics and personalized nutrition: Science and concept. *Personalized Medicine, 5*(5), 447–455.

Kviatcovsky, D., Zheng, D., & Elinav, E. (2021). Gut microbiome and its potential link to personalized nutrition. *Current Opinion in Physiology, 22*, 100439.

Laddu, D., & Hauser, M. (2019). Addressing the nutritional phenotype through personalized nutrition for chronic disease prevention and management. *Progress in Cardiovascular Diseases, 62*(1), 9–14.

Lagoumintzis, G., & Patrinos, G. P. (2023). Triangulating nutrigenomics, metabolomics and microbiomics toward personalized nutrition and healthy living. *Human Genomics, 17*(1), 109.

Marcum, J. A. (2020). Nutrigenetics/nutrigenomics, personalized nutrition, and precision healthcare. *Current Nutrition Reports, 9*(4), 338–345.

Mathers, J. C. (2017). Nutrigenomics in the modern era. *Proceedings of the Nutrition Society, 76*(3), 265–275.

Mohanty, S., & Singhal, K. (2018). Functional foods as personalised nutrition: Definitions and genomic insights. In *Functional food and human health* (pp. 513–535). Springer Singapore.

Mullins, V. A., Bresette, W., Johnstone, L., Hallmark, B., & Chilton, F. H. (2020). Genomics in personalized nutrition: Can you "eat for your genes"? *Nutrients, 12*(10), 3118.

Ordovas, J. M., Ferguson, L. R., Tai, E. S., & Mathers, J. C. (2018). Personalized nutrition and health. *BMJ, 361*, bmj.k2173.

Ramos-Lopez, O., Milagro, F. I., Riezu-Boj, J. I., & Martinez, J. A. (2021). Epigenetic signatures underlying inflammation: An interplay of nutrition, physical activity, metabolic diseases, and environmental factors for personalized nutrition. *Inflammation Research, 70*(1), 29–49.

Ronteltap, A., van Trijp, J. C. M., Berezowska, A., & Goossens, J. (2013). Nutrigenomics-based personalised nutritional advice: In search of a business model? *Genes & Nutrition, 8*, 153–163.

Shawky, E., Zhu, W., Tian, J., Abu El-Khair, R. A., & Selim, D. A. (2025). Metabolomics-driven prediction of vegetable food metabolite patterns: Advances in machine learning approaches. *Food Reviews International, 41*(4), 1051–1080.

Simon, M. C., Sina, C., Ferrario, P. G., Daniel, H., & Working Group "Personalized Nutrition" of the German Nutrition Society. (2023). Gut microbiome analysis for personalized nutrition: The state of science. *Molecular Nutrition & Food Research, 67*(1), 2200476.

Singar, S., Nagpal, R., Arjmandi, B. H., & Akhavan, N. S. (2024). Personalized nutrition: Tailoring dietary recommendations through genetic insights. *Nutrients, 16*(16), 2673.

Verma, M., Hontecillas, R., Tubau-Juni, N., Abedi, V., & Bassaganya-Riera, J. (2018). Challenges in personalized nutrition and health. *Frontiers in Nutrition, 5*, 117.

Wang, F., Zheng, J., Cheng, J., Zou, H., Li, M., Deng, B., et al. (2022). Personalized nutrition: A review of genotype-based nutritional supplementation. *Frontiers in Nutrition, 9*, 992986.

Yang, J. (2023). Using nutrigenomics to guide personalized nutrition supplementation for bolstering immune system. *Health Information Science and Systems, 11*(1), 4.

Yeşilyurt, N., Yılmaz, B., Ağagündüz, D., & Capasso, R. (2022). Microbiome-based personalized nutrition as a result of the 4.0 technological revolution: A mini literature review. *Process Biochemistry, 121*, 257–262.

Zeevi, D., Korem, T., Zmora, N., Israeli, D., Rothschild, D., Weinberger, A., Ben-Yacov, O., Lador, D., Avnit-Sagi, T., Lotan-Pompan, M., Suez, J., Mahdi, J. A., Matot, E., Malka, G., Kosower, N., Rein, M., Zilberman-Schapira, G., Dohnalová, L., Pevsner-Fischer, M., et al. (2015). Personalized nutrition by prediction of glycemic responses. *Cell, 163*(5), 1079–1094.

Zmora, N., Suez, J., & Elinav, E. (2019). You are what you eat: Diet, health and the gut microbiota. *Nature Reviews Gastroenterology & Hepatology, 16*, 35–56.

Chapter 8
Nanotechnology in Food Science

Abbreviations

Ag	Silver
AgNPs	Silver nanoparticles
Al_2O_3	Aluminum oxide
ALD	Atomic layer deposition
Au	Gold
AuNPs	Gold nanoparticles
CNTs	Carbon nanotubes
CO_2	Carbon dioxide
CVD	Chemical vapor deposition
E. coli	*Escherichia coli*
LbL	Layer-by-Layer (deposition)
MBE	Molecular beam epitaxy
MMT	Montmorillonite
MOCVD	Metal organic chemical vapor deposition
NPs	Nanoparticles
O_2	Oxygen
PECVD	Plasma-enhanced chemical vapor deposition
PLA	Polylactic acid
PP	Polypropylene
PVD	Physical vapor deposition
RFID	Radio-frequency identification
ROS	Reactive oxygen species
SiO_2	Silicon dioxide
SiO_x	Silicon oxide (non-stoichiometric)
SLNs	Solid lipid nanoparticles
SnO_2	Tin(IV) oxide
TiO_2	Titanium dioxide

Y. Kumar, S. Dey, *Emerging Technology in Food Processing*,
https://doi.org/10.1007/978-3-032-24505-2_8

VOCs	Volatile organic compounds
Zn^{2+}	Zinc ion
ZnO	Zinc oxide

8.1 Introduction

Nanotechnology, the science, engineering, and application of structures, devices, and systems at the nanoscale (typically 1–100 nm), is rapidly redefining many sectors, from electronics and medicine to agriculture and food systems. At the nanoscale, materials often exhibit novel physical, chemical, optical, and biological properties that differ significantly from their bulk counterparts, including enhanced reactivity, tunable surface characteristics, improved mechanical attributes, and distinctive electronic behavior. These unique features create unparalleled opportunities to devise new tools, materials, and methods with enhanced performance and precision, enabling scientists and engineers to overcome longstanding challenges and pursue new frontiers of innovation. In the domain of food science, nanotechnology is increasingly viewed as a frontier that can bolster food safety, quality, traceability, and sustainability across the food value chain (Singh et al. 2017).

Food systems face persistent and multifaceted challenges. Food safety remains a global concern, impacted by contamination from pathogens (bacteria, viruses, parasites), chemical residues (pesticides, heavy metals, mycotoxins), spoilage processes, and unexpected adulterants. Such threats compromise consumer health, shorten shelf life, strain supply chains, and impose sizable economic burdens. Traditional detection methods, decontamination techniques, and packaging solutions often struggle to keep pace with growing demands for scale, sensitivity, speed, and cost-effectiveness. Microbiological assays such as culture-based methods or polymerase chain reaction (PCR) require specialized laboratories and extended processing times, while conventional packaging materials may offer only passive protection against environmental stressors. Nanotechnology's fine-scale manipulability and high surface-area-to-volume ratios offer promising solutions to many of these obstacles, enabling advancements that extend far beyond what was previously possible (Salihah et al. 2016).

One of the most mature and impactful applications of nanotechnology in food science lies in food packaging, where nanoscale engineering transforms packaging from a passive barrier into an active, intelligent interface between food and its environment. Packaging performs several critical functions: protecting food from physical damage, preventing microbial contamination, controlling the exchange of gases and moisture, and facilitating distribution and storage. Nanocomposite films and coatings formed by embedding nanoscale fillers such as clay nanoplatelets, carbon nanotubes, nanocellulose, or graphene oxide into polymer matrices dramatically improve barrier properties by creating tortuous pathways that hinder the permeation of oxygen, carbon dioxide, water vapor, and volatile organic compounds. These enhanced barrier properties slow oxidation, moisture exchange, and microbial growth, thereby extending shelf life and preserving food quality. Beyond passive protection, nanotechnology enables active packaging systems in which functional agents (such as antimicrobials, antioxidants, or oxygen scavengers) are incorporated into carrier matrices that respond adaptively to environmental triggers. For example, silver or zinc oxide nanoparticles with intrinsic antimicrobial activity can be integrated into packaging surfaces or coatings, providing controlled release of bioactive ions that suppress microbial proliferation. Similarly, enzymes or plant extracts immobilized within nanoscale matrices may be deployed for targeted, sustained functional activity (Bratovčić et al. 2015; Kumar et al. 2017; Primožič et al. 2021).

In addition to barrier and active functions, nanotechnology enables smart or intelligent packaging capable of real-time monitoring of food conditions. Embedded nanosensors, nanoscale electrodes, or nanobiosensors can detect changes in gas composition (e.g., volatile amines from spoilage), shifts in pH, or the presence of specific microbes. Upon detecting target signals, these intelligent systems may trigger visible indicators (e.g., color changes) or electronic transmissions that communicate food status to consumers, retailers, or supply chain managers. Such real-time responsiveness enhances transparency, reduces food waste, and improves risk management (Patel et al. 2020; Shawon et al. 2020).

Nanotechnology also significantly advances food safety monitoring and quality control through the development of nanoscale sensors and assays that surpass traditional analytical methods in sensitivity, specificity, and speed. Conventional microbiological and chemical analyses, such as culture methods or chromatographic separation techniques, are often time-consuming and labor-intensive. In contrast, nanoscale biosensors integrate biological recognition elements (e.g., antibodies, aptamers, DNA probes) with nanomaterial transducers (e.g., gold nanoparticles, carbon nanotubes, graphene, quantum dots) to convert binding events with target pathogens or contaminants into measurable optical, electrochemical, or piezoelectric signals. Because of their high surface area and tunable surface chemistry, these nanomaterials facilitate detection at extremely low analyte concentrations, enabling more rapid and sensitive identification of foodborne pathogens and chemical contaminants. Some advanced platforms even permit label-free detection, in which changes in electrical impedance or surface plasmon resonance signal analyte binding occur without external labels, making them highly suitable for portable, field-deployable detection systems. Furthermore, multiplexed nanosensors allow simultaneous detection of multiple targets, a capability that greatly enhances throughput and diagnostic power (Srivastava et al. 2017; Patel et al. 2020; Anh et al. 2022).

Beyond detection and monitoring, nanotechnology can support the removal or remediation of hazardous substances from food and food-related environments. Nanomaterials such as carbon nanotubes, metal oxides, graphene oxide, and functionalized magnetic nanoparticles can be engineered as highly selective adsorbents or "nanosorbents" that bind specific chemical contaminants through adsorption, ion exchange, or chelation. Once bound, contaminants can be physically separated, for example, using magnetic fields and removed from food matrices or process water. Nanocatalysts, such as titanium dioxide under ultraviolet activation, can facilitate advanced oxidation processes to degrade organic pollutants, including pesticide residues and mycotoxins. Such remediation capabilities extend to water decontamination in food processing, enabling reuse of process water with reduced ecological impact (Bagheri et al. 2016; Navia-Mendoza et al. 2021).

Another burgeoning application of nanotechnology in food science lies in enhancing nutrient delivery, functional foods, and food processing performance. Nano-encapsulation technologies protect sensitive bioactive compounds such as vitamins, antioxidants, and probiotics by encasing them in nanoscale carriers (e.g., liposomes, polymeric nanoparticles, solid lipid nanoparticles, nanomicelles) that shield payloads from degradation and improve their stability. These nano-carriers also support controlled or targeted release in response to gastrointestinal or environmental cues, enhancing bioavailability and functional efficacy. Nanoemulsions and nanodispersions improve the solubility and dispersibility of poorly soluble compounds, contributing to enhanced sensory attributes, texture, and homogeneity in processed foods. Nanotechnology can further improve food processing operations through nanoscaled filters for separation and purification, magnetic nanoparticles for efficient component separation, and nanostructured catalysts for reaction acceleration (Silva et al. 2012; Salihah et al. 2016).

While the benefits of nanotechnology in food science are substantial, they are accompanied by important considerations related to safety, regulatory approval, environmental impact, consumer perception, and ethical use. As nanoscale materials interact with biological systems and environmental pathways in complex ways, robust safety assessment frameworks and transparent regulatory oversight are essential to ensure their responsible application. Balancing innovation with precaution will be vital as nanotechnology continues to evolve within food systems (Kumar et al. 2017).

Nanotechnology offers transformative potential across multiple facets of food science. By enabling advanced packaging systems, rapid and sensitive detection platforms, targeted remediation strategies, and enhanced nutritional delivery, nanotechnology has emerged as a important driver for future food quality, safety, and sustainability.

8.2 Nanomaterials Used in Food Systems

The application of nanotechnology in food systems is primarily based on the use of nanomaterials that exhibit unique physicochemical properties due to their extremely small size, typically ranging from 1 to 100 nanometers. At this scale, materials demonstrate altered surface area, reactivity, solubility, and biological interactions compared to their bulk counterparts. These distinctive characteristics make nanomaterials highly suitable for improving food quality, safety, processing efficiency, and nutritional value. In food systems, nanomaterials are utilized in areas such as encapsulation of bioactive compounds, enhancement of shelf life, antimicrobial protection, sensing of contaminants, and development of advanced packaging materials (Chaudhry et al. 2008; Weiss et al. 2006).

Nanomaterials used in food systems are generally designed to be food-grade, biocompatible, and safe for human consumption. They may be derived from natural food components, synthesized chemically, or produced using biological methods. Understanding their classification and properties is essential for food technologists to ensure both functionality and safety in food applications.

8.2.1 Classification of Nanomaterials

Nanomaterials used in food systems can be classified based on their origin, composition, structure, and functionality. This classification helps in selecting suitable nanomaterials for specific food applications and assessing their potential risks.

1. *Classification Based on Origin*

   (a) *Natural Nanomaterials*

   Natural nanomaterials occur inherently in many food systems without intentional engineering. Examples include casein micelles in milk, starch nanocrystals, and protein aggregates. These structures exist at the nanoscale and play a crucial role in food texture, stability, and nutrient delivery. Since they are naturally present, they are generally considered safe and well accepted by consumers (McClements, 2012).

   (b) *Engineered Nanomaterials*

   Engineered nanomaterials are intentionally designed and manufactured to achieve specific functional properties. These include nanoparticles such as silver, zinc oxide, titanium dioxide, and nanoencapsulated vitamins or antioxidants. Engineered nanomaterials offer controlled size, shape, and surface properties, making them highly effective in food preservation, packaging, and functional food development (Chaudhry et al. 2008; Morris, 2011).

2. *Classification Based on Composition*

   (a) *Organic Nanomaterials*

   Organic nanomaterials are composed of carbon-based or biological substances such as lipids, proteins, polysaccharides, and polymers. Common examples include liposomes, nanoemulsions, polymeric nanoparticles, and protein-based nanocarriers. These materials are widely used for encapsulation and delivery of bioactive compounds due to their biodegradability and compatibility with food systems (Weiss et al. 2006).

(b) *Inorganic Nanomaterials*

Inorganic nanomaterials consist of metals and metal oxides such as silver nanoparticles, zinc oxide nanoparticles, silica nanoparticles, and titanium dioxide nanoparticles. In food systems, these materials are mainly used for antimicrobial activity, food packaging, and sensing applications. For instance, silver nanoparticles are well known for their antimicrobial properties, while nano-silica is used as an anti-caking agent (EFSA, 2018).

3. *Classification Based on Structure and Dimensions*

(a) *Zero-Dimensional Nanomaterials*

These nanomaterials have all three dimensions at the nanoscale, such as nanoparticles and nanospheres. They are widely used in food applications due to their high surface-to-volume ratio, which enhances reactivity and functional performance.

(b) *One-Dimensional Nanomaterials*

These include nanofibers and nanotubes, where one dimension exceeds the nanoscale. In food packaging, nanofibers are incorporated to improve mechanical strength and barrier properties (Kuchibhatla et al. 2007).

(c) *Two-Dimensional Nanomaterials*

Nanoplates and nanolayers fall under this category. Clay nanoplatelets are commonly used in nanocomposite food packaging to enhance gas and moisture barrier properties (Duncan, 2011).

4. *Classification Based on Functionality*

(a) *Antimicrobial Nanomaterials*

These nanomaterials inhibit microbial growth and are used in food preservation and packaging. Examples include silver nanoparticles and zinc oxide nanoparticles (Weiss et al. 2006).

(b) *Nano-Delivery Systems*

These systems are designed to encapsulate and deliver nutrients, flavors, antioxidants, and probiotics with enhanced stability and bioavailability (Weiss et al. 2006).

(c) *Sensing Nanomaterials*

Nanosensors are employed for detecting pathogens, toxins, and chemical contaminants in food systems, enabling rapid and sensitive food safety monitoring (Weiss et al. 2006).

8.2.2 Properties of Food-Grade Nanomaterials

Food-grade nanomaterials exhibit a range of unique physicochemical, mechanical, optical, and biological properties that arise primarily from their nanoscale dimensions (1–100 nm). These properties differentiate them fundamentally from conventional bulk materials and underpin their growing application in food processing, preservation, packaging, safety monitoring, and delivery of bioactive compounds. Understanding these properties is essential for designing safe, effective, and regulatory-compliant nano-enabled food systems (Weiss et al. 2006; McClements, 2012).

1. *High Surface Area-to-Volume Ratio*

   One of the most defining characteristics of nanomaterials is their extremely high surface area-to-volume ratio. As particle size decreases to the nanoscale, a significantly larger fraction of atoms or molecules resides at or near the surface compared to bulk materials. This increased surface availability profoundly enhances functional performance in food systems (Chaudhry et al. 2008).

Enhanced Reactivity and Catalytic Potential
The high surface area enables greater interaction with surrounding molecules such as oxygen, moisture, enzymes, and microorganisms. Consequently, nanomaterials exhibit enhanced reactivity and catalytic activity, which is particularly beneficial in antimicrobial surfaces and antioxidant systems. For example, metal and metal oxide nanoparticles demonstrate improved antimicrobial efficiency due to greater contact with microbial cells (Duncan, 2011).

Adsorption and Binding Capacity
Nanomaterials show superior adsorption capacity for toxins, heavy metals, pesticides, and other contaminants. Nanosorbents exploit this property to remove undesirable substances from food matrices or processing water, owing to the availability of numerous active binding sites on their surfaces (Chaudhry et al. 2008).

Enhanced Biological Interactions
The high surface area also allows immobilization of a larger number of biomolecules, such as enzymes, antibodies, and DNA aptamers. This property enhances the sensitivity and specificity of nanosensors and biosensors used in food safety monitoring and smart packaging applications (Weiss et al. 2006).

Improved Barrier Performance in Packaging
In polymer-based packaging films, nanofillers such as nanoclays possess high surface area and aspect ratio, creating tortuous diffusion pathways for gases and water vapor. Even small additions (2–5%) can drastically reduce oxygen and moisture permeability, significantly extending shelf life (Duncan, 2011).

2. *Quantum Effects and Size-Dependent Behavior*

   At the nanoscale, materials exhibit size-dependent optical, electronic, and magnetic properties due to quantum mechanical effects that are negligible in bulk materials.

Quantum Confinement
When nanoparticle dimensions approach or fall below the exciton Bohr radius, quantum confinement occurs, leading to changes in the electronic band structure. This results in size-dependent absorption and emission spectra, which are exploited in sensors, color indicators, and UV-blocking applications in food packaging (McClements, 2012).

Surface Plasmon Resonance (SPR)
Metal nanoparticles such as gold and silver display surface plasmon resonance, where collective oscillations of surface electrons produce intense optical absorption peaks. Changes in SPR behavior can be used in colorimetric sensors to indicate contamination or pathogen binding in food systems (Duncan, 2011).

Optical Transparency and UV Protection
Nanomaterials enable the selective blocking of harmful ultraviolet radiation while maintaining transparency in the visible region. This property is particularly valuable in packaging, where UV protection prevents photo-oxidation of vitamins, pigments, and lipids without compromising product visibility (EFSA, 2018).

3. *Mechanical, Thermal, and Barrier Properties*

   Nanomaterials significantly enhance the structural and protective performance of food packaging materials, coatings, and edible films.

Mechanical Reinforcement
Nanoparticles reinforce polymer matrices by improving tensile strength, flexibility, toughness, and resistance to tearing or puncture. Nanoclays, silver nanoparticles, and nanofibers are commonly used to produce lightweight yet mechanically robust packaging films (Duncan, 2011).

Thermal Stability
Nanocomposites often exhibit improved thermal stability, including higher decomposition temperatures and delayed thermal degradation. Enhanced thermal resistance is critical for packaging exposed to heat during processing operations such as pasteurization, sterilization, or hot-fill packaging (McClements, 2012).

Improved Barrier Properties
By increasing diffusion path length for gases and vapors, nanofillers reduce permeability to oxygen, carbon dioxide, moisture, and volatile compounds. This slows oxidation, microbial growth, and moisture-induced spoilage, thereby extending shelf life (Chaudhry et al. 2008).

4. *Antimicrobial and Reactive Properties*

   Certain food-grade nanomaterials, especially metals and metal oxides, possess intrinsic antimicrobial activity.

Antimicrobial Mechanisms
Nanomaterials exert antimicrobial effects through mechanisms such as ion release (Ag^+, Zn^{2+}), generation of reactive oxygen species (ROS), disruption of microbial membranes, and interference with metabolic pathways. These properties are exploited in active packaging and antimicrobial coatings (Duncan, 2011).

Antioxidant Activity

Some nanomaterials also function as antioxidant systems by scavenging free radicals or inhibiting lipid oxidation. This helps preserve sensory attributes such as flavor, color, and nutritional quality of food products (Weiss et al. 2006).

5. *Tailorable and Tunable Properties*

   A major advantage of nanomaterials is the ability to precisely tailor their properties through controlled design.

   Size: Influences surface area, reactivity, bioavailability, and toxicity (McClements & Xiao, 2012).

   Shape and Aspect Ratio: Nanofibers, nanotubes, and platelets provide superior mechanical reinforcement and barrier effects compared to spherical particles (McClements & Xiao, 2012).

   Composition and Crystal Structure: Metal, metal oxide, carbon-based, or organic nanomaterials exhibit distinct functional behaviors (McClements & Xiao, 2012).

   Surface Chemistry and Functionalization: Surface coatings, charge, and hydrophobicity influence interaction with food components, migration behavior, and biocompatibility (McClements & Xiao, 2012).

   Dispersion State: Well-dispersed nanoparticles exhibit superior functional performance compared to agglomerated systems (McClements & Xiao, 2012).

6. *Optical, Electrical, and Electronic Properties*

   Nanomaterials contribute advanced optical and electronic functionalities to smart food systems.

   Transparency and UV Blocking: Balanced optical design protects food while maintaining aesthetic appeal (Duncan, 2011).

   Colorimetric Response: Nanomaterials that change color in response to chemical or biological stimuli are used in freshness indicators and intelligent packaging (Duncan, 2011).

   Electrical Conductivity: Carbon nanotubes, graphene, and metal nanoparticles enable the development of biosensors, RFID tags, and electronic detection systems in food packaging (Duncan, 2011).

7. *Biocompatibility, Biodegradability, and Safety Properties*

   Food-grade nanomaterials must meet strict safety requirements.

   Biocompatibility: Organic nanomaterials derived from proteins, polysaccharides, and lipids are generally more biocompatible and biodegradable (EFSA, 2018).

   Migration and Leaching: Migration into food depends on nanoparticle size, surface chemistry, matrix properties, food type, temperature, and contact time (EFSA, 2018).

   Toxicological Considerations: High reactivity and small size may lead to oxidative stress, cytotoxicity, or genotoxicity, emphasizing the need for rigorous characterization and risk assessment (EFSA, 2018).

8. *Functional Properties: Encapsulation, Delivery, and Emulsion Stabilization*

   Nanomaterials enable advanced functional food applications.

Nanoencapsulation and Delivery
Nanoencapsulation protects bioactive compounds from degradation and allows controlled or targeted release in the gastrointestinal tract, improving bioavailability and bioaccessibility (McClements & Xiao, 2012).

Emulsion Stabilization
Nanoparticles can stabilize Pickering emulsions by irreversibly adsorbing at oil–water interfaces. Such emulsions exhibit superior stability and are useful for the delivery of fat-soluble nutrients, flavors, and functional ingredients (McClements, 2012).

8.3 Processing of Nanomaterial Synthesis

The synthesis route used to produce nanomaterials plays a decisive role in determining their physicochemical properties, functional performance, safety, and suitability for food and food-packaging applications. Parameters such as particle size and size distribution, morphology, crystallinity, surface chemistry, degree of agglomeration, and purity are all strongly dependent on the processing method. In food systems, where nanomaterials may come into direct or indirect contact with edible products, synthesis methods must not only deliver functional performance but also minimize residual toxicity, migration risks, and environmental impact (Jafari & McClements, 2018).

Nanomaterial synthesis techniques are broadly classified into top-down and bottom-up approaches (Fig. 8.1). In top-down methods, bulk materials are broken down into nanoscale structures using physical or chemical means. These approaches are historically well established and are often attractive due to their scalability and relative simplicity. However, they may introduce surface defects, broad size distributions, and contamination, which are important concerns for food-grade applications (Table 8.1).

8.3.1 Top-Down Approaches

Top-down approaches involve the size reduction of bulk solids into nanoscale particles or nanostructures through mechanical, physical, or chemical processes. These methods do not rely on molecular self-assembly but instead on external energy input to fracture or reshape materials. While top-down synthesis is widely used in materials science and engineering, its application in food systems is more selective, mainly limited to packaging materials, fillers, coatings, and sensor surfaces rather than ingestible nanoparticles (Biswas et al. 2012; Jafari & McClements, 2018).

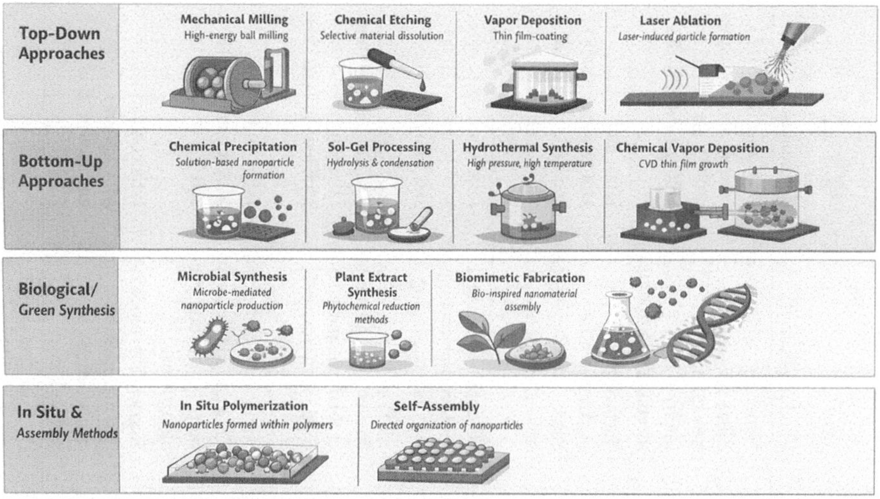

Fig. 8.1 Nanomaterial synthesis approaches in food systems

The major top-down approaches relevant to food and packaging nanotechnology include mechanical milling, chemical or reactive etching, vapor deposition techniques, and laser ablation.

8.3.1.1 Mechanical Milling/High-Energy Ball Milling

Mechanical milling, particularly high-energy ball milling, is one of the most widely used top-down techniques for producing nanostructured materials. In this process, bulk powders or solid materials are placed inside a rotating or vibrating milling chamber containing hardened balls made of steel, ceramic, or tungsten carbide. During operation, repeated impact, shear, and frictional forces between the balls and the material cause progressive particle size reduction and grain refinement, eventually leading to the formation of nanocrystalline structures. The underlying mechanism involves continuous cycles of particle fracture, cold welding, and re-fracture, accompanied by localized plastic deformation and the accumulation of microstrain within the crystal lattice. Effective control of milling parameters such as milling speed, ball-to-powder ratio, milling duration, milling atmosphere (air, inert gas, or vacuum), and temperature is essential to limit excessive agglomeration, oxidation, and contamination from the milling media. Mechanical milling offers several advantages, including the use of simple and well-established equipment, scalability for industrial production, the absence of chemical reducing agents or organic solvents, and suitability for hard materials such as metals, ceramics, and mineral fillers. As a result, it is widely employed in the preparation of nanoclays, mineral fillers, and ceramic nanoparticles. However, the method also presents important limitations, including broad particle size distributions, high densities of

Table 8.1 Processing methods for nanomaterial synthesis in food and packaging applications

Synthesis approach	Method	Principle/mechanism	Advantages	Limitations	Food applications	References
Top-down	Mechanical milling/high-energy ball milling	Physical fracture of bulk materials via impact, shear, and friction, leading to grain refinement and a nanoscale size	Scalable; simple equipment; solvent-free; suitable for hard materials	Broad size distribution; surface defects; contamination risk; limited shape control	Nanoclays and mineral fillers for polymer nanocomposites improve barrier and mechanical properties	Biswas et al. (2012), Gorrasi and Sorrentino (2015), and Jafari and McClements (2018)
Top-down	Chemical/reactive etching	Selective dissolution of material using chemical etchants to create nanoscale surface features	Precise surface modification; compatible with sensors and smart packaging	Hazardous chemicals; waste disposal concerns; not ideal for free nanoparticles	Surface-engineered packaging films; enhanced adhesion and sensor sensitivity	Jafari and McClements (2018) and Su et al. (2019)
Top-down	Vapor deposition (PVD, sputtering, MBE)	Vaporization and condensation of material onto substrates, forming thin nanocoatings	High purity; excellent thickness and composition control	High cost; limited to coatings; low throughput	Ultra-thin Al_2O_3 and SiO_2 barrier coatings on polymer packaging films	Jafari and McClements (2018) and Mandracci (2019)
Top-down	Laser ablation	Laser-induced vaporization of target material followed by nanoparticle condensation	High purity; no chemical reagents; minimal contamination	Low-yield; poor scalability; high energy consumption	Biosensors and antimicrobial nanoparticles for food safety research	Rao et al. (2014) and Jafari and McClements (2018)
Bottom-up	Chemical precipitation	Nucleation and growth of nanoparticles from supersaturated precursor solutions	Low cost; simple; scalable; applicable to metals and oxides	Agglomeration risk; residual reactants; purification required	Antimicrobial Ag and ZnO nanoparticles in biopolymer food packaging films	Jafari and McClements (2018) and Yu et al. (2018)
Bottom-up	Sol–gel processing	Hydrolysis and condensation of precursors forming nanostructured oxide networks	Molecular-level homogeneity, low processing temperature; versatile	Shrinkage/cracking during drying; organic residue removal needed	Barrier, UV-blocking, and antimicrobial nanocoatings for packaging	Jafari and McClements (2018) and Bokov et al. (2021)

Bottom-up	Chemical vapor deposition (CVD/ALD)	Gas-phase precursor decomposition on heated substrates	Atomic-scale thickness control; uniform, conformal coatings	Expensive equipment; substrate-limited; high temperatures	High-barrier inorganic nanolayers on flexible food packaging	Jafari and McClements (2018) and Mandracci (2019)
Bottom-up	Hydrothermal/ solvothermal synthesis	Nanoparticle growth under high-pressure, high-temperature solvent conditions	High crystallinity; controlled morphology; fewer defects	Specialized reactors; scale-up challenges	ZnO and TiO_2 nanoparticles for antimicrobial packaging and sensors	Qu et al. (2025)
Bottom-up	Template-assisted/self-assembly growth	Nanoparticle formation guided by physical, chemical, or biological templates	Precise shape and structural control; hierarchical architectures	Template removal complexity; contamination risk	Nanoporous membranes; structured fillers for barrier and sensing functions	Jafari and McClements (2018) and Lombardo et al. (2020)
Biological/ green	Microbial synthesis	Enzymatic reduction of metal ions by microorganisms	Eco-friendly; biocompatible; natural capping agents	Slow kinetics; polydispersity; purification challenges	Antimicrobial nanoparticles in biodegradable food packaging films	Jafari and McClements (2018) and Ghosh et al. (2021)
Biological/ green	Plant extract/ phytochemical synthesis	Reduction and stabilization by plant-derived bioactive compounds	Non-toxic; renewable; cost-effective	Batch variability; limited size control	Antimicrobial and antioxidant nanocomposites in biopolymer packaging	Jafari and McClements (2018) and Tsekhmistrenko et al. (2021)
Biological/ green	Biomimetic/ biofabrication	Nanoparticle growth guided by biological templates or mineralization pathways	High structural precision; mild conditions	Complex synthesis; low throughput; scale-up issues	Smart packagingsensors and functional coatings	Jafari and McClements (2018) and Kulkarni et al. (2023)
Hybrid	In situ polymerization/ nanofiller synthesis	Nanoparticles formed directly within the polymer matrix during processing	Excellent dispersion; strong interfacial bonding	Complex process control; reaction compatibility issues	Active and antimicrobial food packaging nanocomposites	Mittal (2011) and Jafari and McClements (2018)
Post-synthesis	Self-assembly/ directed assembly	Organization of nanoparticles via intermolecular forces into ordered structures	Tunable architecture; multifunctional surfaces	Stability and large-area uniformity challenges	Smart packaging layers and nanosensor interfaces	Mittal (2011) and Jafari and McClements (2018)

surface defects and lattice strain, potential contamination from the milling media, difficulty in achieving ultra-fine particles below 10 nm, and limited control over particle shape and surface chemistry. These drawbacks are particularly significant for food-related applications, as surface defects and contamination can influence chemical reactivity, migration behavior, and safety. Consequently, in food systems, mechanical milling is rarely used to produce nanoparticles intended for direct consumption. Instead, it is primarily applied as a preprocessing step, such as delaminating layered silicates (nanoclays) or reducing mineral fillers to the nanoscale prior to incorporation into polymer matrices for food packaging. The resulting nanocomposites exhibit enhanced mechanical strength and improved gas and moisture barrier properties, which are critical for extending food shelf life (Gorrasi & Sorrentino, 2015; Jafari & McClements, 2018).

8.3.1.2 Chemical/Reactive Etching

Chemical or reactive etching is a top-down nanofabrication technique in which bulk materials are selectively dissolved using chemical etchants such as acids, bases, or oxidizing agents to generate nanoscale features. By carefully controlling reaction conditions, this method enables the formation of nanostructures such as nanopores, nanochannels, or controlled surface roughness. The etching process proceeds through chemical reactions occurring at the solid–liquid interface, where exposed regions of the material react with the etchant and are gradually removed. To achieve spatial selectivity, masks, templates, or lithographic patterns are often applied to protect specific regions while leaving others accessible to chemical attack. The processing parameters, including etchant concentration, temperature, reaction time, and degree of agitation strongly influence the etching rate, uniformity, and final morphology of the nanostructured surface. Chemical etching offers several advantages, such as precise control over surface features, the ability to fabricate nanoporous or nanostructured surfaces, and compatibility with microfabrication, biosensing, and smart packaging technologies. However, the method also has notable limitations, including the use of aggressive and potentially hazardous chemicals, the risk of excessive surface damage or uncontrolled roughness, limited suitability for producing free nanoparticles, and concerns related to waste disposal and environmental impact. In the context of food packaging, chemical etching is primarily employed for surface modification rather than for the synthesis of particulate nanomaterials. Etched nanostructured surfaces can improve interlayer adhesion, enhance the effectiveness of antimicrobial coatings, and increase the sensitivity of sensor surfaces in smart packaging systems. Nevertheless, because of chemical safety and regulatory considerations, the application of chemical etching to food-contact materials requires strict control, validation, and compliance with food safety standards (Jafari & McClements, 2018; Su et al. 2019).

1. *Vapor Deposition*

   Vapor deposition techniques are top-down nanofabrication methods that involve the transformation of a solid or liquid precursor into a vapor phase, followed by condensation of the vapor onto a substrate to form thin films or nanostructured coatings. These methods are commonly classified under physical vapor deposition (PVD) and its related variants. In thermal evaporation, the source material is heated until it vaporizes and subsequently condenses on a cooler substrate, whereas in electron-beam evaporation, high-energy electron beams are used to locally heat and vaporize the target material. Sputtering involves the bombardment of a target with energetic ions, typically argon ions (Ar^+), which eject atoms from the target surface that then deposit onto the substrate. Molecular beam epitaxy (MBE) represents a highly controlled vapor deposition approach in which atomic or molecular beams are deposited under ultra-high vacuum conditions to form epitaxial layers with precise control over thickness and composition. Vapor deposition techniques offer several advantages, including high purity and cleanliness of deposited films, excellent control over film thickness and chemical composition, and the ability to produce uniform and conformal coatings over large surface areas. However, these methods also have limitations, such as high capital and operational costs, limited throughput, and the fact that they primarily produce surface coatings rather than free nanoparticles. In food packaging applications, vapor deposition is extensively used to apply ultra-thin inorganic barrier layers, such as aluminum oxide (Al_2O_3) or silicon oxide (SiO_2), onto polymer films. These nanometer-scale coatings significantly reduce oxygen and moisture permeability without substantially increasing film thickness or weight, thereby playing a critical role in high-barrier packaging systems designed to protect oxygen- and moisture-sensitive foods (Jafari & McClements, 2018; Mandracci, 2019).
2. *Laser Ablation*

   Laser ablation is a physical top-down nanomaterial synthesis technique in which a high-energy pulsed laser is focused onto a solid target, resulting in localized vaporization or plasma formation at the target surface. The ablated material, consisting of atoms, ions, and small clusters, subsequently cools and condenses to form nanoparticles in a surrounding medium such as a gas atmosphere, vacuum, or liquid. When the laser pulse strikes the target, extremely high localized temperatures and pressures are generated, leading to the rapid ejection of material from the surface. As these ejected species expand and cool, they undergo nucleation and growth to form nanoparticles. In liquid-phase laser ablation, the properties of the surrounding liquid, such as its polarity, viscosity, and chemical composition, play a significant role in influencing nanoparticle size, morphology, and surface chemistry. Laser ablation offers several advantages, including the production of high-purity nanoparticles without the need for chemical reducing agents or stabilizers, minimal contamination, and the ability to generate metastable or unique phases that may be difficult to obtain through conventional chemical routes. However, the technique also has important limitations, such as low yield, poor scalability for industrial production, high energy consumption, and challenges in achieving narrow particle size distributions without additional stabilization strategies. In food nanotechnology, laser ablation is mainly employed at the laboratory or research scale to synthesize clean metallic nanoparticles, such as gold or silver, for use in biosensors, antimicrobial coatings, or analytical tools for food safety and quality monitoring. Owing to its cost and scalability constraints, laser ablation is not widely adopted for large-scale food packaging applications, but it remains a valuable method for producing reference materials and high-purity nanomaterials for specialized food-related applications (Rao et al. 2014; Jafari & McClements, 2018).

8.3.2 Bottom-Up Approaches

In contrast to top-down approaches, bottom-up synthesis methods construct nanomaterials by assembling atoms, ions, or molecules into nanoscale structures through controlled chemical or physical processes. These approaches generally provide superior control over particle size, morphology, crystallinity, and surface chemistry. As a result, bottom-up methods are particularly attractive for food and food-packaging applications, where reproducibility, functionality, and safety are critical considerations (Jafari & McClements, 2018; Jagtiani, 2022).

Bottom-up approaches dominate the synthesis of nanomaterials intended for nanoemulsions, nanoencapsulation systems, antimicrobial fillers, barrier-enhancing nanocomposites, and sensing elements. They are especially well suited to producing dispersible nanoparticles with tailored surface properties that are compatible with aqueous food systems.

8.3.2.1 Chemical Synthesis

Chemical synthesis methods are among the most widely used bottom-up approaches for producing nanomaterials for food-related applications. These methods rely on controlled chemical reactions such as reduction, hydrolysis, condensation, or decomposition to generate nanoparticles from molecular precursors.

1. *Chemical Precipitation*

   Chemical precipitation is a widely used bottom-up synthesis technique in which insoluble nanoparticles are formed from a supersaturated solution of precursor ions. This process is typically initiated through chemical reduction, hydrolysis, or controlled adjustment of pH, leading first to nucleation and then to subsequent particle growth. In a typical procedure, metal salts such as silver nitrate, zinc nitrate, or iron salts are dissolved in a suitable solvent and reacted with reducing agents or precipitating agents. Once the level of supersaturation is reached, nuclei are formed, which then grow into nanoparticles through diffusion and surface reactions. To control particle size and prevent aggregation, stabilizers or capping agents such as surfactants, polymers, or biopolymers are often added during synthesis. The physicochemical properties of the resulting nanoparticles are strongly influenced by several control parameters, including precursor concentration and stoichiometry, the type and strength of the reducing or precipitating agent, solution pH and ionic strength, reaction temperature and duration, and the presence of surface-active compounds. Chemical precipitation offers multiple advantages, including simplicity, low cost, ease of scale-up, and applicability to a wide range of materials such as metals, metal oxides, and composite nanoparticles. However, the method also has limitations, including the tendency of nanoparticles to agglomerate if stabilization is inadequate, the presence of residual reactants or by-products, and the frequent need for post-synthesis purification steps to ensure food-grade quality. In food and packaging applications, chemical precipitation is extensively employed to synthesize antimicrobial nanoparticles, particularly silver and zinc oxide nanoparticles, which are subsequently incorporated into biopolymer-based films such as starch, chitosan, or gelatin. These nanocomposite films exhibit enhanced antimicrobial efficacy and improved barrier properties, thereby contributing to extended shelf life and improved safety of packaged foods (Jafari & McClements, 2018; Yu et al. 2018).

2. *Sol–Gel Processing*

   Sol–gel processing is a versatile bottom-up nanomaterial synthesis technique in which metal or metalloid precursors undergo hydrolysis and condensation reactions to form a three-dimensional nanostructured network. In this approach, metal alkoxides or inorganic salts react with water to generate hydroxylated species, which subsequently undergo condensation reactions to form M–O–M bonds, leading to the formation of a colloidal suspension or "sol." With time, continued polymerization and crosslinking cause the sol to evolve into a gel-like network. Subsequent drying and thermal treatment steps are applied to remove solvents and organic residues, resulting in the formation of nanostructured oxides or composite materials. Sol–gel processing offers several advantages, including excellent control over chemical composition and homogeneity at the molecular level, relatively low processing temperatures compared to solid-state routes, and flexibility in producing powders, coatings, or thin films. However, the method also has limitations, such as shrinkage and cracking during the drying stage, sensitivity of many precursors to moisture, and the need to carefully remove residual organic components to meet food-grade requirements. In food packaging applications, sol–gel methods are widely used to generate silica, titania, or hybrid oxide nanocoatings that enhance gas barrier performance, provide UV-blocking functionality, and impart antimicrobial activity to packaging films. Additionally, sol–gel-derived nanoparticles can be synthesized in situ within polymer matrices, which improves nanoparticle dispersion and interfacial adhesion, thereby enhancing the mechanical and functional properties of food packaging materials (Jafari & McClements, 2018; Bokov et al. 2021).

3. *Chemical Vapor Deposition*

   Chemical vapor deposition (CVD) is a bottom-up synthesis technique in which gaseous precursor molecules decompose or chemically react on a heated substrate surface, resulting in the deposition of solid material in the form of thin films or nanostructures. Several variants of CVD are commonly employed, including metal organic chemical vapor deposition (MOCVD), plasma-enhanced CVD (PECVD), and atomic layer deposition (ALD), the latter being a highly controlled, sequential process that enables atomic-scale precision in film growth. In these processes, reaction conditions such as temperature, pressure, gas composition, and exposure time govern nucleation and growth behavior, thereby determining the thickness, composition, and uniformity of the deposited layer. CVD-based techniques offer notable advantages, including excellent control over coating thickness and chemical composition, the formation of high-purity and uniform films, and conformal coverage of complex or irregular substrate geometries. However, they also present several limitations, such as the requirement for elevated temperatures or specialized chemical precursors, high capital and operational costs associated with sophisticated equipment, and substrate-dependent growth, which restricts their use primarily to surface coatings rather than free nanoparticles. In food packaging applications, CVD and ALD are extensively used to deposit ultra-thin inorganic barrier layers such as aluminum oxide (Al_2O_3) or silicon oxide (SiO_x) onto polymer films. These nanometer-scale coatings markedly reduce oxygen and moisture transmission while preserving the flexibility, transparency, and lightweight nature of the packaging materials, making them critical for high-performance packaging of oxygen- and moisture-sensitive foods (Jafari & McClements, 2018; Mandracci, 2019).

4. *Hydrothermal/Solvothermal Synthesis*

   Hydrothermal and solvothermal synthesis are bottom-up nanomaterial fabrication techniques in which chemical reactions are carried out in sealed vessels at elevated temperatures and pressures, using water or organic solvents as the reaction medium. Under these conditions, the physicochemical properties of the solvent, such as dielectric constant, viscosity, and solubility, are significantly altered, enabling controlled nucleation and growth of nanoparticles with high crystallinity and well-defined morphologies. The confined high-pressure environment promotes uniform crystal growth and reduces defect formation, while key reaction parameters, including temperature, pressure, reaction time, solvent type, pH, and the presence of mineralizers or structure-directing agents, strongly influence particle size, shape, and phase composition. Hydrothermal and solvothermal methods offer several advantages, including the production of nanoparticles with high crystallinity and purity, the ability to synthesize anisotropic nanostructures such as nanorods and nanowires, and relatively mild reaction chemistry compared to high-temperature solid-state routes. However, these techniques also have limitations, such as the need for specialized high-pressure equipment, challenges associated with large-scale production, and the requirement for post-synthesis particle recovery, washing, and drying. In food-related applications, hydrothermal synthesis is commonly employed to produce metal oxide nanoparticles, such as zinc oxide (ZnO) and titanium dioxide (TiO_2), which are subsequently used in antimicrobial packaging materials, active coatings, and sensor systems designed to enhance food safety and shelf life (Qu et al. 2025).

5. *Template/Templated Growth/Self-Assembly*

Template-assisted growth is a bottom-up nanomaterial synthesis approach in which preexisting physical, chemical, or biological structures are used to guide the formation of nanomaterials with controlled size, shape, and spatial arrangement. In this method, precursor species are deposited within or onto a template, and subsequent chemical reactions lead to the formation of nanostructures that replicate the geometry of the template. A wide variety of templates can be employed, including hard templates such as porous membranes, silica spheres, or anodic aluminum oxide; soft templates such as micelles, block copolymers, and surfactant assemblies; and biological templates, including proteins, polysaccharides, or DNA. Template-assisted synthesis offers important advantages, particularly the ability to achieve precise control over nanomaterial morphology and to fabricate ordered or hierarchical nanostructures that are difficult to obtain through conventional synthesis routes. However, the technique also has limitations, as template removal can be complex and may require harsh chemical or thermal treatments, and incomplete removal can lead to contamination from residual template materials. In food and packaging applications, template-assisted growth is used to fabricate nanoporous membranes, nanoscale coatings, and structured nanofillers that enhance gas and moisture barrier properties or enable advanced sensing and active packaging functionalities. These structured nanomaterials contribute to improved shelf life, safety, and quality monitoring in modern food packaging systems (Jafari & McClements, 2018; Lombardo et al. 2020).

8.3.3 Biological/Green Synthesis

With increasing awareness of environmental sustainability, food safety, and consumer acceptance, biological or green synthesis approaches have gained significant importance in food nanotechnology. These methods utilize biological systems or bio-derived compounds to synthesize nanomaterials under mild conditions, avoiding toxic solvents, harsh reducing agents, and extreme processing parameters. For food and packaging applications where nanomaterials may directly contact food, green synthesis offers advantages in terms of biocompatibility, reduced toxicity, and regulatory acceptance (Jafari & McClements, 2018).

Biological synthesis methods typically rely on the reducing and stabilizing capabilities of microorganisms, plant-derived phytochemicals, or biomolecules that act as natural capping agents. These biomolecules not only facilitate nanoparticle formation but also enhance stability and reduce aggregation.

8.3.3.1 Microbial Synthesis

Microbial synthesis is a biological or green nanomaterial fabrication approach that employs microorganisms such as bacteria, fungi, yeast, or algae to produce nanoparticles through enzymatic or metabolic processes. In this method, microorganisms are capable of reducing metal ions either intracellularly or extracellularly, leading to the formation of metallic or metal oxide nanoparticles. The underlying mechanism involves enzymes, such as reductases, and various metabolites present within or secreted by microbial cells, which reduce metal ions, including Ag^{+}, Au^{3+}, or Zn^{2+}, to their corresponding nanoscale forms. During this process, proteins, polysaccharides, and other biomolecules produced by the microorganisms often act as natural capping and stabilizing agents, preventing nanoparticle aggregation and enhancing colloidal stability. Microbial synthesis offers several advantages, including mild reaction conditions at ambient temperature and pressure, elimination of toxic reducing or stabilizing chemicals, improved biocompatibility due to natural surface functionalization, and the potential for scale-up using established fermentation technologies. However, the method also has limitations, such as slower reaction rates compared to conventional chemical synthesis, polydispersity, and limited control over particle size and morphology, challenges associated with nanoparticle separation and purification, and variability in nanoparticle characteristics between different microbial strains or cultures. In food and packaging applications, microbially synthesized silver and zinc oxide nanoparticles have been incorporated into biodegradable polymer films to impart antimicrobial activity, thereby extending food shelf life. Owing to their biomolecule-capped surfaces, these nanoparticles may exhibit reduced toxicity compared to chemically synthesized counterparts, making microbial synthesis an attractive strategy for producing nanomaterials intended for food-contact applications (Jafari & McClements, 2018; Ghosh et al. 2021).

8.3.3.2 Plant Extract/Phytochemical Synthesis

Plant-extract–mediated synthesis, commonly referred to as phytochemical synthesis, is a green nanomaterial fabrication approach that utilizes extracts derived from plant parts such as leaves, roots, fruits, peels, or seeds as natural reducing and stabilizing agents. These plant extracts are rich in bioactive compounds, including polyphenols, flavonoids, alkaloids, terpenoids, sugars, and organic acids, which play a crucial role in nanoparticle formation. In this method, phytochemicals present in the extract reduce metal ions into their nanoscale forms while simultaneously acting as capping agents that stabilize the nanoparticles and inhibit aggregation. Reaction conditions such as extract concentration, solution pH, temperature, and reaction time strongly influence nanoparticle size, morphology, and stability. Phytochemical synthesis offers several advantages, including high eco-friendliness, non-toxicity, simplicity, cost-effectiveness, and the use of readily available and renewable raw materials. In addition, the biomolecule-coated surfaces of the resulting nanoparticles enhance biocompatibility, making them particularly suitable for food-related applications. However, this approach also has limitations, such as batch-to-batch variability in plant extract composition, limited control over nanoparticle uniformity and size distribution, and the need for purification steps to remove excess or unbound biomolecules. In food and packaging applications, plant-mediated synthesis of silver, gold, zinc oxide, and copper oxide nanoparticles has been widely explored for use in antimicrobial packaging materials, sensors, and active coatings. These nanoparticles are often incorporated into biopolymer matrices such as starch, cellulose, or chitosan, resulting in biodegradable packaging materials with enhanced antimicrobial and antioxidant properties that contribute to improved food safety and extended shelf life (Jafari & McClements, 2018; Tsekhmistrenko et al. 2021).

8.3.3.3 Biomimetic/Biofabrication

Biomimetic or biofabrication approaches to nanomaterial synthesis draw inspiration from natural mineralization processes observed in biological systems, such as the formation of silica structures in diatoms or magnetite nanoparticles in magnetotactic bacteria. In these approaches, biological molecules or structures are deliberately employed to guide nanoparticle nucleation and growth in a controlled and hierarchical manner. Proteins, peptides, polysaccharides, DNA, or even whole biological architectures can function as scaffolds or templates that regulate nanoparticle size, shape, orientation, and spatial organization. Depending on the synthesis strategy, these biological templates may either remain incorporated within the final nanostructure or be removed after synthesis to yield purely inorganic materials. Biomimetic synthesis offers several advantages, including high precision in structural control, the ability to generate hierarchical and highly ordered nanostructures, and the use of mild synthesis conditions that are compatible with sensitive biological components. Furthermore, the integration of biological functionality such as recognition, binding, or catalytic activity into nanomaterials is a unique and valuable feature of this approach. However, biomimetic and biofabrication strategies also face limitations, including complex and multistep synthesis protocols, low production throughput, and significant challenges in scaling up for industrial applications. In the context of food nanotechnology, biomimetic approaches are primarily explored for the development of advanced food safety sensors, smart packaging systems, and surface-engineered coatings rather than for bulk nanoparticle production. Their capacity to produce highly ordered and functional nanostructures is particularly valuable for detection, signaling, and responsive applications in food systems, where precision and reliability are critical (Jafari & McClements, 2018; Kulkarni et al. 2023).

8.3.4 In Situ Polymerization/In Situ Nanofiller Synthesis

In situ polymerization or in situ nanofiller synthesis is an advanced nanocomposite fabrication strategy in which nanoparticles are generated directly within a polymer matrix during polymerization or film formation. Instead of synthesizing nanoparticles separately and subsequently dispersing them into the polymer, precursor compounds such as metal salts or metal alkoxides are first incorporated into the polymer solution or monomer system and then converted into nanoparticles during processing. The underlying mechanism involves the nucleation and growth of nanoparticles through reduction, precipitation, or sol–gel reactions that occur simultaneously with polymerization or matrix formation. Polymer chains surrounding the growing nanoparticles act as physical barriers that restrict particle growth and serve as stabilizing agents, resulting in improved dispersion and reduced aggregation. This approach offers several advantages, including enhanced nanoparticle dispersion, strong interfacial bonding between the nanofiller and polymer matrix, fewer processing steps, and improved consistency of nanocomposite properties. However, in situ synthesis also presents challenges, such as complex process control, compatibility issues between polymerization reactions and nanoparticle formation, and the potential for interference with polymer structure or functional properties. In food packaging applications, in situ nanofiller synthesis is widely employed to produce nanocomposite films based on biopolymers such as chitosan, starch, polyvinyl alcohol, or gelatin, embedded with antimicrobial or barrier-enhancing nanoparticles. These nanocomposites exhibit improved mechanical strength, reduced gas and moisture permeability, and enhanced antimicrobial activity, making them particularly suitable for active and functional food packaging systems (Mittal, 2011; Jafari & McClements, 2018).

8.3.5 Self-Assembly/Directed Assembly Post-Synthesis

Self-assembly or directed assembly post-synthesis refers to the organization of pre-synthesized nanoparticles into ordered structures such as films, multilayers, coatings, or arrays through controlled physical or chemical interactions. In this approach, individual nanoparticles are first synthesized and stabilized in suspension, after which external forces or interfacial interactions drive their assembly into higher-order architectures. Common assembly mechanisms include electrostatic attraction, hydrogen bonding, van der Waals forces, hydrophobic interactions, and specific ligand–receptor binding. Directed assembly techniques such as layer-by-layer (LbL) deposition, Langmuir–Blodgett or Langmuir–Schaefer transfer, dip-coating, spin coating, and electrophoretic deposition enable precise control over film thickness, composition, and spatial arrangement at the nanoscale. Self-assembly offers significant advantages, including tunable structural organization, high reproducibility, and the ability to fabricate multifunctional nanostructured surfaces without extensive chemical modification. However, challenges remain in achieving large-area uniformity, maintaining long-term structural stability, and preventing nanoparticle aggregation during assembly. In food and packaging applications, self-assembled nanostructures are particularly valuable for creating smart packaging layers, sensor interfaces, and functional coatings that respond to environmental changes such as temperature, pH, gas composition, or microbial activity. These ordered nanostructures enhance detection sensitivity, barrier performance, and active functionality, thereby contributing to improved food quality monitoring, safety, and shelf-life extension (Mittal, 2011; Jafari & McClements, 2018).

8.4 Types of Nanomaterials in Food Science

Nanomaterials used in food science vary widely in composition, structure, and functionality. Their selection depends on the intended application, such as food packaging, preservation, safety monitoring, or delivery of functional ingredients. Based on material composition, nanomaterials used in food systems can be broadly classified into metallic/metal oxide, carbon-based, natural/biopolymeric, silicon-based, and hybrid/composite nanomaterials (Table 8.2 and Fig. 8.2).

8.4.1 Metallic/Metal Oxide Nanomaterials

Metallic and metal oxide nanomaterials are among the most extensively studied and applied nanomaterials in food science, particularly for antimicrobial activity, UV protection, and enhancement of packaging performance. Their functionality arises from high surface reactivity, ion release, photocatalytic activity, and interaction with microbial cell membranes.

Silver nanoparticles (AgNPs) are the most widely used metallic nanomaterials in food packaging due to their strong and broad-spectrum antimicrobial activity. Their effectiveness is attributed to the release of Ag^+ ions, disruption of bacterial cell membranes, interaction with thiol groups of proteins, and induction of oxidative stress within microbial cells. AgNPs have been incorporated into polymer films, coatings, and sensors to inhibit bacterial and fungal growth and extend the shelf life of food products (Duncan, 2011; Jafari & McClements, 2018). However, concerns regarding nanoparticle migration, bioaccumulation, and long-term toxicity necessitate careful control of particle size, concentration, and immobilization within packaging matrices.

Zinc oxide (ZnO) nanoparticles are widely used because of their antimicrobial activity, UV-blocking capability, and relatively low cost. ZnO nanoparticles embedded in biodegradable polymer films such as starch, chitosan, or cellulose enhance mechanical strength, reduce gas permeability, and inhibit microbial growth. ZnO is also considered relatively safe and is approved for certain food-contact applications, although nanoscale behavior is still under regulatory evaluation (Chaudhry et al. 2008).

Titanium dioxide (TiO_2) nanoparticles are primarily used for UV protection and photocatalytic activity. When incorporated into food packaging films, TiO_2 nanoparticles protect food components from photooxidation, thereby preserving color, flavor, and nutritional quality. Under UV illumination, TiO_2 can also exhibit antimicrobial activity through reactive oxygen species generation (Saied et al. 2025).

Table 8.2 Types of nanomaterials used in food science and their applications

Nanomaterial type	Examples	Properties/ functions	Food science applications	References
Metallic/ metal oxide nanomaterials	Silver nanoparticles (AgNPs), Zinc oxide (ZnO), Titanium dioxide (TiO_2), Copper oxide, Iron oxide, Magnesium oxide	Strong antimicrobial activity, UV protection, photocatalytic activity, ion release, high surface reactivity	Antimicrobial food packaging, UV-protective films, shelf-life extension, sensing applications	Duncan (2011), Chaudhry et al. (2008), Jafari and McClements (2018), and Saied et al. (2025)
Carbon-based nanomaterials	Graphene, Graphene oxide (GO), Carbon nanotubes (CNTs), Carbon dots	High mechanical strength, excellent gas/moisture barrier, electrical conductivity, fluorescence	High-barrier packaging, smart packaging, freshness indicators, food safety sensors	Duncan (2011), Jafari and McClements (2018), and Abdelmonem (2020)
Natural/ biopolymeric nanomaterials	Nanocellulose (CNF, CNC), starch nanoparticles, chitosan nanoparticles, protein-based nanoparticles	Biodegradability, biocompatibility, film-forming ability, antimicrobial activity, encapsulation capacity	Edible coatings, biodegradable packaging, delivery of bioactive compounds, Pickering emulsions	Jafari and McClements (2018) and Baskaran et al. (2026)
Silicon/silica-based nanomaterials	Silica nanoparticles (SiO_2), Silica aerogels, porous silica nanostructures	Improved mechanical strength, moisture absorption, thermal insulation, barrier enhancement	Reinforcement of packaging films, sol–gel coatings, moisture control in packaging	Canham (2007) and Chaudhry et al. (2008)
Hybrid/composite nanomaterials	Chitosan–ZnO, starch–ZnO, cellulose–Ag, PLA–graphene, protein–inorganic nanoflowers	Synergistic enhancement of mechanical, antimicrobial, and barrier properties; multifunctionality	Active and smart packaging, controlled delivery systems, sustainable packaging materials	Jafari and McClements (2018) and Abd-Elsalam (2020)

Other metal or metal oxide nanomaterials, such as copper oxide, iron oxide, and magnesium oxide nanoparticles, are explored for antimicrobial coatings, sensing applications, and nutrient delivery systems. Overall, metallic and metal oxide nanomaterials provide powerful functional advantages but require rigorous safety assessment for food-related use.

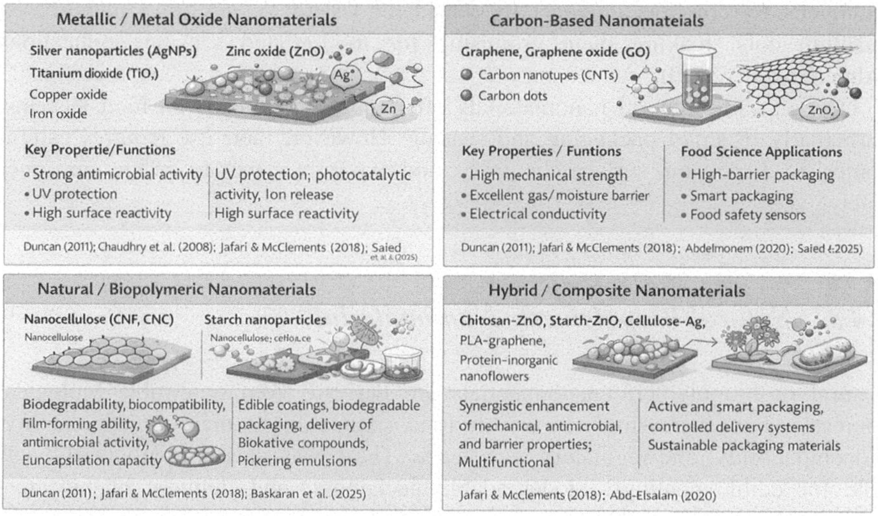

Fig. 8.2 Types of nanomaterials in food science

8.4.2 Carbon-Based Nanomaterials

Carbon-based nanomaterials include graphene, graphene oxide (GO), carbon nanotubes (CNTs), and carbon dots. These materials are valued for their exceptional mechanical strength, high aspect ratio, electrical conductivity, and unique optical properties, which make them suitable for advanced food packaging and sensing applications.

Graphene and graphene oxide (GO) are two-dimensional nanomaterials with outstanding mechanical strength and excellent barrier properties against gases and moisture. When dispersed in polymer matrices, GO sheets create highly tortuous diffusion pathways that significantly reduce oxygen and water vapor permeability. In addition, graphene-based nanocomposites improve tensile strength, thermal stability, and UV resistance of packaging films. Depending on surface functionalization, graphene oxide may also exhibit antimicrobial properties (Jafari & McClements, 2018; Abdelmonem, 2020).

Carbon nanotubes, which are cylindrical nanostructures with extremely high aspect ratios, have been investigated for enhancing mechanical strength and electrical conductivity in polymer composites. Their electrical properties are particularly useful in sensor-based packaging systems, such as RFID tags or freshness indicators. However, due to concerns related to inhalation toxicity and biopersistence, CNTs are less commonly used in direct food-contact materials and are mainly restricted to indirect applications such as sensors (Duncan, 2011; Abdelmonem, 2020).

Carbon dots and related quantum-dot-like carbon nanomaterials exhibit fluorescence and photoluminescence properties. These optical characteristics are exploited in food safety applications, including the detection of spoilage gases, toxins, or pathogens. Carbon dots are generally considered less toxic than heavy-metal-based quantum dots, making them promising for food-related sensing applications (Abdelmonem, 2020).

Overall, carbon-based nanomaterials offer multifunctionality in food systems, particularly in smart packaging and sensing. However, their use requires careful control of dispersion, surface chemistry, and exposure routes to ensure consumer safety.

8.4.3 Natural/Biopolymeric Nanomaterials

Natural or biopolymeric nanomaterials are derived from renewable biological sources and are increasingly favored in food science due to their biodegradability, biocompatibility, and regulatory acceptance. These materials are particularly suitable for edible coatings, biodegradable packaging, and delivery of bioactive compounds.

Nanocellulose, including cellulose nanofibers (CNFs) and cellulose nanocrystals (CNCs), is one of the most promising biopolymeric nanomaterials. Derived from plant biomass, nanocellulose exhibits excellent mechanical strength, high surface area, and good barrier properties. When incorporated into packaging films, nanocellulose enhances tensile strength and reduces oxygen and moisture permeability. It is also used to stabilize Pickering emulsions and form edible coatings (Jafari & McClements, 2018).

Starch nanoparticles and starch nanocrystals are produced from native starch sources such as corn or potato through acid hydrolysis, high-pressure homogenization, or ultrasonication. These nanomaterials improve film-forming ability, reduce brittleness, and enhance biodegradability. However, their inherent hydrophilicity necessitates modification or blending with other materials to improve moisture resistance.

Chitosan nanoparticles, derived from chitin, possess intrinsic antimicrobial activity due to their positive charge and ability to disrupt microbial cell membranes. At the nanoscale, chitosan exhibits enhanced surface area and antimicrobial efficiency. Chitosan-based nanomaterials are widely used in edible coatings, antimicrobial films, and composite systems with metal nanoparticles (Baskaran et al. 2026).

Protein-based nanoparticles from whey, soy, zein, or gelatin are employed as carriers for flavors, antioxidants, vitamins, and probiotics. Their ability to form films and encapsulate bioactive compounds makes them valuable for functional food and packaging applications (Baskaran et al. 2026).

8.4.4 Silicon-/Silica-Based Nanomaterials

Silicon- and silica-based nanomaterials are widely used in food science, primarily as reinforcing agents, barrier enhancers, and moisture control materials. Silica nanoparticles (SiO_2) are commonly incorporated into polymer matrices to improve mechanical strength, thermal stability, and barrier performance.

Nano-silica fillers increase the stiffness and durability of packaging films while maintaining transparency. Silica nanoparticles are also used in sol–gel-derived coatings that enhance oxygen and moisture barrier properties of polymer films. In addition, silica aerogels and porous silica nanostructures exhibit extremely high surface area and are used for moisture absorption and thermal insulation in packaging systems.

Silicon-based nanomaterials can also act as carriers for active compounds or be surface-modified to impart hydrophobicity, thereby improving the water resistance of biodegradable packaging materials. However, concerns remain regarding long-term exposure and inhalation risks, emphasizing the need for controlled incorporation and immobilization within solid matrices (Canham, 2007; Chaudhry et al. 2008).

8.4.5 Hybrid/Composite Nanomaterials

Hybrid or composite nanomaterials combine two or more types of nanomaterials to achieve synergistic improvements in functionality. In food science, hybrid systems are particularly attractive because they allow simultaneous enhancement of mechanical strength, barrier properties, antimicrobial activity, and sustainability.

Common examples include biopolymer–metal oxide composites such as chitosan–ZnO, starch–ZnO, or cellulose–silver nanocomposites, which combine biodegradability with antimicrobial performance. Similarly, graphene or graphene oxide incorporated into PLA or cellulose-based matrices significantly improves mechanical and barrier properties while maintaining eco-friendliness.

More advanced hybrid systems include protein–inorganic nanoflowers, which consist of proteins integrated with inorganic components to form hierarchical nanostructures. These materials are explored for pigment stabilization, enzyme immobilization, and controlled delivery of bioactive compounds.

Hybrid nanomaterials help mitigate the limitations of individual components, such as brittleness, water sensitivity, or toxicity, while enhancing multifunctionality. As a result, composite nanomaterials represent a key direction for future food packaging and functional food technologies (Jafari & McClements, 2018; Abd-Elsalam, 2020).

8.5 Application of Nanomaterials in Food Science

8.5.1 Nanotechnology in Food Packaging (Barrier, Mechanical, and Durability Enhancements)

Nanotechnology has revolutionized food packaging by enabling advanced materials that significantly enhance barrier properties, mechanical strength, and durability relative to conventional polymers. The integration of nanocomposite films and coatings comprising nanofillers such as montmorillonite (MMT) nanoclay, graphene oxide, silica, TiO_2, and ZnO into polymer matrices introduces a tortuous path for gas and moisture diffusion, effectively slowing the movement of oxygen, carbon dioxide, and water vapor through packaging materials (Barrier Property Enhancement). This "maze" or "brick-and-mortar" effect is especially notable in layered nanoclay structures, where overlapping platelets extend diffusion pathways and reduce permeation rates, thereby controlling oxidation and moisture exchange and ultimately prolonging shelf life. Polymer–clay nanocomposites, such as MMT dispersed in polyethylene or polylactic acid (PLA), have been reported to reduce oxygen permeability by over 50%, directly translating to improved protection against oxidative spoilage and microbial growth (Saied et al. 2025) (Fig. 8.3).

In addition to barrier performance, nanoparticles serve as reinforcing fillers that enhance the mechanical properties of packaging films. When nanoparticles like TiO_2 or SiO_2 are uniformly dispersed in polymers such as polypropylene (PP), strong interfacial interactions (e.g., hydrogen bonding, van der Waals forces) restrict polymer chain mobility, improving tensile strength, stiffness, and resistance to

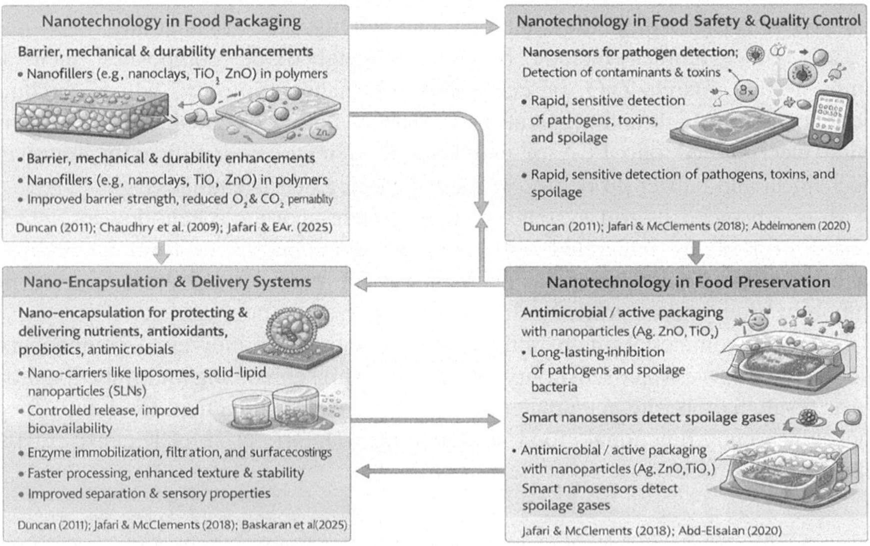

Fig. 8.3 Application of nanomaterials in food science

tearing (mechanical strength, stiffness, and durability). This reinforcement not only reduces brittleness but also enhances puncture and impact resistance, vital for handling, transport, and storage. Because many nanofillers are highly efficient at low weight percentages (1–5 wt%), they allow reduction of overall polymer content or film thickness without sacrificing performance, yielding lighter materials with lower costs and environmental impact. For example, incorporating 2 wt% MMT nanoclay into PP has been shown to reduce packaging weight by approximately 15% while maintaining mechanical integrity (lightweight, flexibility, and material efficiency). Moreover, the high aspect ratio of certain nanofillers enables considerable reinforcement at low loading, while low-density fillers such as hollow silica or nanocellulose can decrease density and improve flexibility. In sum, nanotechnology in packaging fosters multifunctional films that address fundamental needs in barrier enhancement, mechanical robustness, and sustainability, main drivers for industry adoption and food quality preservation (Saied et al. 2025).

8.5.2 Nanotechnology in Food Safety and Quality Control (Nanosensors for Pathogen Detection; Detection of Contaminants and Toxins)

Ensuring food safety and quality is a cornerstone of modern food science, and nanotechnology has introduced powerful tools, particularly nanosensors and biosensors, that significantly surpass traditional detection technologies in speed, sensitivity, and specificity. Nanomaterials such as metal oxides (SnO_2, ZnO), carbon nanotubes (CNTs), and graphene serve as the active layer in gas or volatile compound sensors integrated into smart packaging systems to detect spoilage indicators like ammonia, hydrogen sulfide, ethanol, and other volatile organic compounds (VOCs) (user text, gas/volatile compound sensors). The high surface area of these nanostructures facilitates substantial adsorption of target gases, triggering measurable changes in electrical resistance or conductivity that correlate with spoilage progression. For instance, graphene-based resistive sensors have been developed to detect ethanol and volatile amines, enabling real time, non-invasive monitoring of food condition and reducing waste by informing consumers and distributors of freshness status (Vijayakumar et al. 2023).

Beyond monitoring gases, nanotechnology has significantly impacted pathogen and contaminant detection through the development of biosensors that harness functionalized nanomaterials as signal transducers and amplifiers. In these systems, nanoparticles are functionalized with biorecognition elements such as antibodies, aptamers, or enzymes that selectively bind target analytes (Biosensors/Contaminant Detectors). When a pathogen (e.g., *E. coli*) binds to its corresponding receptor, the event induces a change in the nanomaterial's electrical, optical, or electrochemical properties, producing a measurable signal. Gold nanoparticles (AuNPs) conjugated with specific antibodies have been widely investigated for rapid detection of bacterial pathogens in meat samples, demonstrating response times of minutes rather than the days required by culture-based methods. Magnetic nanoparticles, when paired with bioluminescent assays, can concentrate pathogens from complex food matrices and enhance detection sensitivity, further illustrating the utility of nanotechnology in food safety diagnostics (Vijayakumar et al. 2023).

Moreover, integrating these nanosensors directly into food packaging or into portable detection devices enables on-site assessment of contaminants, allergens, toxins, or pathogens without specialized laboratory infrastructure. Quantum-dot-labeled nanoparticles combined with fluorescence assays have also been employed for toxin detection, offering high sensitivity and multiplexing potential. Smart packaging systems marrying sensor capability with barrier function, such as RFID tags coated with nanomaterials that relay humidity or contamination data, represent a move toward comprehensive quality control platforms that enhance food safety from production to consumption. Collectively, nano-enabled sensors and biosensors are transforming food quality and safety monitoring by delivering rapid, accurate, and cost-effective detection across diverse food systems (Vijayakumar et al. 2023).

8.5.3 Nano-Encapsulation and Delivery Systems

Nano-encapsulation has emerged as a pivotal technology in food science, offering transformative capabilities for protecting, delivering, and releasing functional ingredients such as nutrients, antioxidants, flavors, probiotics, and antimicrobials. At its core, nano-encapsulation refers to the entrapment of active substances within nanoscale carriers typically below 1000 nm, resulting in enhanced stability, controlled release, improved solubility, and heightened bioavailability. Unlike conventional microencapsulation, the nanoscale domain increases surface area and interaction with biological systems, enabling more efficient transport of actives through the gastrointestinal tract and targeted delivery at desired sites (Kaur et al. 2022).

Mechanistically, nano-encapsulation involves the formation of a protective shell around a functional core through interfacial assembly governed by hydrophobic interactions, electrostatic forces, hydrogen bonding, or van der Waals attractions. Shell materials may include proteins, polysaccharides (e.g., chitosan, alginate), or lipid matrices, which self-assemble around the encapsulated agent. Stabilization of these structures is achieved through electrostatic repulsion, steric hindrance, or cross-linking, preventing aggregation and ensuring dispersion uniformity. Subsequently, processes such as spray drying, freeze-drying, or solidification of lipid carriers lock the nanoscale capsules into stable forms suitable for incorporation into beverages, foods, coatings, or packaging systems (Kaur et al. 2022).

Nanoemulsions, oil-in-water or water-in-oil dispersions with nanoscale droplets stabilized by surfactants or biopolymers, represent a common form of nano-encapsulation. These systems are particularly effective for delivering lipophilic bioactives (e.g., vitamins, carotenoids) that typically exhibit poor water solubility. By reducing particle size to the nanoscale, solubility and dispersibility in aqueous food matrices are enhanced, while degradation due to light, oxygen, or pH variations is mitigated. Furthermore, nanoencapsulation can mask undesirable tastes or odors, improving sensory quality without compromising functional efficacy (Kaur et al. 2022).

Embedding nano-encapsulated antimicrobials or antioxidants into edible films or coatings creates active packaging solutions that not only extend shelf life but also maintain food quality. For example, nanoemulsions loaded with antimicrobial compounds can continuously release actives in response to environmental cues, retarding microbial growth on food surfaces. Similarly, biopolymer nanoparticles such as chitosan or alginate can be engineered for targeted probiotic delivery, enhancing survival through digestion and maximizing health benefits. The controlled release characteristic of nano-encapsulation allows for sustained activity over extended periods, minimizing the need for high initial doses and reducing the risk of undesirable migration into food (Kaur et al. 2022).

The nano-encapsulation and delivery systems represent a versatile suite of technologies that integrate protection, controlled release, improved bioavailability, and sensory enhancement.

8.5.4 Nanotechnology in Food Processing

Nanotechnology is increasingly integrated into food processing to improve efficiency, enhance functionality, and enable novel preservation or quality enhancement strategies. One significant application lies in the immobilization of enzymes on nanostructured carriers, which enhances enzyme stability, catalytic activity, and reusability, critical factors in processes such as hydrolysis, oxidation, and other modification reactions. Nanostructured catalysts, due to their high surface area and tunable surface chemistry, facilitate faster reaction kinetics and more uniform product outcomes, contributing to improved process consistency and reduced processing times (Gowda et al. 2024).

Additionally, nanoporous membranes and nanofiltration systems have become powerful tools for separation, concentration, and purification of food components such as proteins, phenolics, and flavor compounds. The precise pore size distribution and surface characteristics of nanofiltration membranes enable selective separation at the molecular level, enhancing yield and product quality while preserving heat-sensitive components. Nanofiltration also supports water recovery and waste reduction in processing streams, aligning with sustainability goals (Gowda et al. 2024).

Nanocoatings on food surfaces composed of edible biopolymers like nanocellulose or chitosan nanoparticles offer direct functional benefits by reducing moisture loss, controlling gas exchange, and creating physical barriers against microbial invasion. These coatings can be fortified with additional active agents such as antimicrobials or antioxidants, further enhancing food preservation in minimally processed fruits, vegetables, and meats. Because these coatings are edible and biocompatible, they do not alter the food's sensory attributes while delivering protective functionality (Gowda et al. 2024).

Moreover, nanocarriers embedded within food matrices can be engineered for triggered release, where active agents such as flavors, nutrients, or antimicrobials are released in response to specific stimuli such as pH changes, temperature shifts, or microbial enzyme activity. This smart release mechanism tailors functionality to the product's environment, for example, releasing preservatives only when spoilage conditions arise (Gowda et al. 2024).

Nanostructured ingredients also play roles in improving sensory properties and textural attributes. Nano-scale flavor encapsulation can enhance aroma release and perception during consumption, while nano-fibrillar cellulose and silica nanoparticles may serve as thickeners or anti-caking agents in powdered foods. By influencing mouthfeel and stability, these materials contribute to higher-quality end products that appeal to consumer preferences (Gowda et al. 2024).

The nanotechnology in food processing delivers multifunctional benefits spanning catalytic enhancement, separation technology, edible surface treatments, triggered release systems, and sensory improvements, all of which contribute to more efficient, sustainable, and quality-driven food production.

8.5.5 Nanotechnology in Food Preservation

Nanotechnology has significantly impacted food preservation by enabling innovations that extend shelf life, improve safety, and maintain quality through advanced material design and functional delivery systems. One of the foremost preservation applications is the development of antimicrobial packaging materials that go beyond passive barrier functions to actively inhibit microbial growth on food surfaces. Nanoparticles with inherent antimicrobial properties such as silver (Ag), zinc oxide (ZnO), and titanium dioxide (TiO_2) have been incorporated into films and coatings, where they interact with microbial cell membranes, generate reactive oxygen species (ROSs), and release metal ions that interfere with cellular processes, leading to microbial inactivation. These mechanisms collectively disrupt membrane integrity, induce oxidative stress, and inhibit key metabolic functions, thereby reducing the prevalence of spoilage organisms and pathogens (Mousavi & Karami, 2022; Singh & Kumar, 2023).

In addition to materials with intrinsic activity, nanotechnology enables controlled-release systems that provide sustained antimicrobial effects. Nanoparticles can serve as carriers for antimicrobial agents such as essential oils, enzymes, or bacteriocins, allowing gradual release in response to environmental triggers like moisture or pH changes. Controlled release prevents burst effects and maintains effective concentrations of actives over extended periods, enhancing preservation without the need for large initial doses. This approach minimizes potential migration into the food while offering stimulus-responsive protection that aligns with dynamic spoilage processes (Mousavi & Karami, 2022; Singh & Kumar, 2023).

Nano-coatings and edible films represent another preservation frontier, as they can be applied directly to food surfaces to retard moisture loss and create micro-barriers against gas exchange. When fortified with encapsulated preservatives or antioxidants, these coatings maintain food quality by shielding sensitive components from oxidation and dehydration. For example, nanoemulsions containing antimicrobial compounds can be integrated into edible films, enabling sustained protective action that delays spoilage and extends sensory quality (Mousavi & Karami, 2022).

Beyond direct antimicrobial strategies, nanotechnology also supports preservation through smart detection systems that signal spoilage before quality deterioration becomes irreversible. Nanosensors embedded in packaging can detect gases associated with spoilage such as ammonia or volatile amines providing early warning indicators that help reduce waste and ensure timely consumption. Biosensors targeting specific pathogens or toxins further enhance safety by enabling rapid on-site detection without specialized laboratory equipment (Mousavi & Karami, 2022; Singh & Kumar, 2023).

Together, these nanotechnology-enabled preservation methods deliver a multifaceted approach to extending shelf life, improving safety, and maintaining food quality, ushering in intelligent preservation systems that are active, responsive, and integrated into food packaging and processing workflows.

8.6 Conclusion

Nanotechnology has demonstrated immense potential to transform food science by offering innovative solutions across food packaging, safety monitoring, processing, preservation, and nutritional enhancement. Through the manipulation of materials at the nanoscale, it is possible to develop packaging systems with superior barrier, mechanical, and durability properties that extend shelf life and reduce food waste. The integration of active and smart packaging technologies such as antimicrobial nanocomposites and nanosensors has further shifted packaging from a passive containment role to an intelligent system capable of monitoring food quality in real time. In parallel, nano-enabled biosensors have significantly improved the speed, sensitivity, and reliability of pathogen and contaminant detection, supporting more effective food safety management and traceability throughout the supply chain.

Beyond packaging and detection, nanotechnology has expanded the horizons of food processing and nutrition. Nano-encapsulation and nanoemulsion systems have enabled the protection, controlled release, and enhanced bioavailability of nutrients, antioxidants, probiotics, and other functional ingredients. Nanomaterials also support improved processing efficiency through enzyme immobilization, selective separations, catalytic reactions, and surface coatings that enhance quality and stability. Collectively, these advances highlight nanotechnology's capacity to improve food quality, safety, functionality, and sustainability while addressing the demands of a growing global population.

Despite these promising developments, several challenges must be addressed to ensure the responsible and widespread adoption of nanotechnology in food systems. A primary concern is the safety and toxicological impact of engineered nanomaterials. Due to their small size and high reactivity, nanoparticles may interact with biological systems in unpredictable ways, raising questions about their long-term effects on human health and the environment. Comprehensive risk assessment, standardized testing protocols, and life-cycle analyses are essential to evaluate nanoparticle migration, accumulation, and biodegradation in food and ecological systems. Regulatory frameworks also remain fragmented and inconsistent across regions, creating uncertainty for industry stakeholders and slowing commercialization.

Consumer perception and acceptance represent another critical challenge. Public concerns regarding "nanofoods," labeling transparency, and perceived risks can influence market adoption. Clear communication, evidence-based risk assessment, and transparent regulatory policies will be crucial to building consumer trust. Additionally, technical challenges such as scalability, cost-effectiveness, reproducibility of nanomaterial synthesis, and integration into existing food processing infrastructure must be addressed to enable industrial implementation.

Looking forward, future research should focus on the development of safer, biodegradable, and bio-based nanomaterials, particularly those derived from natural polymers and green synthesis routes. Advances in multifunctional nanocomposites, intelligent packaging with digital connectivity, and portable nanosensor platforms are expected to further enhance food safety and traceability. The convergence of nanotechnology with artificial intelligence, the Internet of Things (IoT), and advanced data analytics may enable predictive food quality management systems. Ultimately, interdisciplinary collaboration among food scientists, material scientists, toxicologists, regulators, and policymakers will be essential to harness the full potential of nanotechnology while ensuring safety, sustainability, and societal acceptance.

References

Abdelmonem, A. M. (2020). Application of carbon-based nanomaterials in food preservation area. In *Carbon nanomaterials for agri-food and environmental applications* (pp. 583–593). Elsevier.

Abd-Elsalam, K. A. (2020). Multifunctional hybrid nanomaterials for sustainable agri-food and ecosystems: A note from the editor. In *Multifunctional hybrid nanomaterials for sustainable agri-food and ecosystems* (pp. 1–19). Elsevier.

Anh, N. H., Doan, M. Q., Dinh, N. X., Huy, T. Q., Tri, D. Q., Loan, L. T. N., et al. (2022). Gold nanoparticle-based optical nanosensors for food and health safety monitoring: Recent advances and future perspectives. *RSC Advances, 12*(18), 10950–10988.

Bagheri, H., Asgharinezhad, A. A., & Ebrahimzadeh, H. (2016). Determination of trace amounts of Cd (II), Cu (II), and Ni (II) in food samples using a novel functionalized magnetic nanosorbent. *Food Analytical Methods, 9*(4), 876–888.

Baskaran, R., & Thirugnanasambandham, T. (2026). Advances in sustainable polymeric materials derived from renewable biopolymers integrated with nanotechnology for active food preservation: A review. *Polymer-Plastics Technology and Materials, 65*, 100–139.

Biswas, A., Bayer, I. S., Biris, A. S., Wang, T., Dervishi, E., & Faupel, F. (2012). Advances in top–down and bottom–up surface nanofabrication: Techniques, applications & future prospects. *Advances in Colloid and Interface Science, 170*(1–2), 2–27.

Bokov, D., Turki Jalil, A., Chupradit, S., Suksatan, W., Javed Ansari, M., Shewael, I. H., et al. (2021). Nanomaterial by sol-gel method: Synthesis and application. *Advances in Materials Science and Engineering, 2021*(1), 5102014.

Bratovčić, A., Odobašić, A., Ćatić, S., & Šestan, I. (2015). Application of polymer nanocomposite materials in food packaging. *Croatian Journal of Food Science and Technology, 7*(2), 86–94.

Canham, L. T. (2007). Nanoscale semiconducting silicon as a nutritional food additive. *Nanotechnology, 18*(18), 185704.

Chaudhry, Q., Scotter, M., Blackburn, J., Ross, B., Boxall, A., Castle, L., et al. (2008). Applications and implications of nanotechnologies for the food sector. *Food Additives and Contaminants, 25*(3), 241–258.

Duncan, T. V. (2011). Applications of nanotechnology in food packaging and food safety. *Journal of Colloid and Interface Science, 363*(1), 1–24.

EFSA Scientific Committee, Hardy, A., Benford, D., Halldorsson, T., Jeger, M. J., Knutsen, H. K., et al. (2018). Guidance on risk assessment of the application of nanoscience and nanotechnologies in the food and feed chain: Part 1, human and animal health. *EFSA Journal, 16*(7), e05327.

Ghosh, S., Ahmad, R., Zeyaullah, M., & Khare, S. K. (2021). Microbial nano-factories: Synthesis and biomedical applications. *Frontiers in Chemistry, 9*, 626834.

Gorrasi, G., & Sorrentino, A. (2015). Mechanical milling as a technology to produce structural and functional bio-nanocomposites. *Green Chemistry, 17*(5), 2610–2625.

Gowda, B. N., Rao, B. V., & Mayegowda, S. B. (2024). Bionanomaterials in improving food quality and safety. In *Bionanomaterials for industrial applications* (pp. 149–164). CRC Press.

Jafari, S. M., & McClements, D. J. (Eds.). (2018). *Nanoemulsions: Formulation, applications, and characterization*. Academic Press.

Jagtiani, E. (2022). Advancements in nanotechnology for food science and industry. *Food Frontiers, 3*(1), 56–82.

Kaur, G., Kaur, R., Rani, N., & Kaur, S. (2022). Nanoencapsulation of probiotics in food packaging. In *Nanotechnology in intelligent food packaging* (pp. 175–202). John Wiley & Sons.

Kuchibhatla, S. V., Karakoti, A. S., Bera, D., & Seal, S. (2007). One dimensional nanostructured materials. *Progress in Materials Science, 52*(5), 699–913.

Kulkarni, D., Sherkar, R., Shirsathe, C., Sonwane, R., Varpe, N., Shelke, S., et al. (2023). Biofabrication of nanoparticles: Sources, synthesis, and biomedical applications. *Frontiers in Bioengineering and Biotechnology, 11*, 1159193.

Kumar, N., Kaur, P., & Bhatia, S. (2017). Advances in bio-nanocomposite materials for food packaging: A review. *Nutrition & Food Science, 47*(4), 591–606.

Lombardo, D., Calandra, P., Pasqua, L., & Magazù, S. (2020). Self-assembly of organic nanomaterials and biomaterials: The bottom-up approach for functional nanostructures formation and advanced applications. *Materials, 13*(5), 1048.

Mandracci, P. (Ed.). (2019). *Chemical vapor deposition for nanotechnology*. BoD–Books on Demand.

McClements, D. J. (2012). Nanoemulsions versus microemulsions: Terminology, differences, and similarities. *Soft Matter, 8*(6), 1719–1729.

McClements, D. J., & Xiao, H. (2012). Potential biological fate of ingested nanoemulsions: Influence of particle characteristics. *Food & Function, 3*(3), 202–220.

Mittal, V. (Ed.). (2011). *Nanocomposites with biodegradable polymers: Synthesis, properties, and future perspectives* (Vol. 68). Oxford University Press.

Morris, V. J. (2011). Emerging roles of engineered nanomaterials in the food industry. *Trends in Biotechnology, 29*(10), 509–516.

Mousavi, S. S., & Karami, A. (2022). Graphene oxide from natural products and its applications in the agriculture and food industry. In *Graphene from natural sources* (pp. 95–116). CRC Press.

Navia-Mendoza, J. M., Filho, O. A. E., Zambrano-Intriago, L. A., Maddela, N. R., Duarte, M. M. M. B., Quiroz-Fernández, L. S., et al. (2021). Advances in the application of nanocatalysts in photocatalytic processes for the treatment of food dyes: A review. *Sustainability, 13*(21), 11676.

Patel, G., Pillai, V., Bhatt, P., & Mohammad, S. (2020). Application of nanosensors in the food industry. In *Nanosensors for smart cities* (pp. 355–368). Elsevier.

Primožič, M., Knez, Ž., & Leitgeb, M. (2021). (Bio) nanotechnology in food science—Food packaging. *Nanomaterials, 11*(2), 292.

Qu, B., Xiao, Z., & Luo, Y. (2025). Sustainable nanotechnology for food preservation: Synthesis, mechanisms, and applications of zinc oxide nanoparticles. *Journal of Agriculture and Food Research, 19*, 101743.

Rao, S. V., Podagatlapalli, G. K., & Hamad, S. (2014). Ultrafast laser ablation in liquids for nanomaterials and applications. *Journal of Nanoscience and Nanotechnology, 14*(2), 1364–1388.

Saied, M., Bassyouni, M., El Fray, M., & Hasanin, M. S. (2025). Recent advances in food packaging technology using nanotechnology. *Packaging Technology and Science, 38*, 775.

Salihah, N. T., Hossain, M. M., Lubis, H., & Ahmed, M. U. (2016). Trends and advances in food analysis by real-time polymerase chain reaction. *Journal of Food Science and Technology, 53*(5), 2196–2209.

Shawon, Z. B. Z., Hoque, M. E., & Chowdhury, S. R. (2020). Nanosensors and nanobiosensors: Agricultural and food technology aspects. In *Nanofabrication for smart nanosensor applications* (pp. 135–161). Elsevier.

Silva, H. D., Cerqueira, M. Â., & Vicente, A. A. (2012). Nanoemulsions for food applications: Development and characterization. *Food and Bioprocess Technology, 5*(3), 854–867.

Singh, R., & Kumar, S. (2023). Regulatory and safety concerns regarding the use of active nanomaterials in food industry. In *Nanotechnology advancement in agro-food industry* (pp. 269–306). Springer Nature Singapore.

Singh, T., Shukla, S., Kumar, P., Wahla, V., Bajpai, V. K., & Rather, I. A. (2017). Application of nanotechnology in food science: Perception and overview. *Frontiers in Microbiology, 8*, 1501.

Srivastava, A. K., Dev, A., & Karmakar, S. (2017). Nanosensors for food and agriculture. In *Nanoscience in food and agriculture* (Vol. 5, pp. 41–79). Springer International Publishing.

Su, L., Song, Y., Fu, C., & Tang, D. (2019). Etching reaction-based photoelectrochemical immunoassay of aflatoxin B1 in foodstuff using cobalt oxyhydroxide nanosheets-coating cadmium sulfide nanoparticles as the signal tags. *Analytica Chimica Acta, 1052*, 49–56.

Tsekhmistrenko, S. C., Bityutskyy, V., Tsekhmistrenko, O., Merzlo, S., Tymoshok, N., Melnichenko, A., et al. (2021). Bionanotechnologies: Synthesis of metals'nanoparticles with using plants and their applications in the food industry: A review. *The Journal of Microbiology, Biotechnology and Food Sciences, 10*(6), e1513.

Vijayakumar, G., Venkatesan, S. A., Kannan, V. A., & Perumal, S. (2023). Detection of food toxins, pathogens, and microorganisms using nanotechnology-based sensors. In *Nanotechnology applications for food safety and quality monitoring* (pp. 155–170). Academic Press.

Weiss, J., Takhistov, P., & McClements, D. J. (2006). Functional materials in food nanotechnology. *Journal of Food Science, 71*(9), R107–R116.

Yu, H., Park, J. Y., Kwon, C. W., Hong, S. C., Park, K. M., & Chang, P. S. (2018). An overview of nanotechnology in food science: Preparative methods, practical applications, and safety. *Journal of Chemistry, 2018*(1), 5427978.

Chapter 9
Smart Food Packaging

Abbreviations

2D	Two-dimensional
IoT	Internet of things
MAP	Modified atmosphere packaging
NFC	Near field communication
QR	Quick response
RFID	Radio frequency identification
SFP	Smart food packaging
TFP	Traditional food packaging
TTIs	Time temperature indicators

9.1 Introduction

In the realm of food processing and distribution, ensuring appropriate packaging of products plays a critical role in preserving their integrity and appeal. A strategically designed packaging system not only makes a product look better on the shelf, but it also helps keep important nutrients in the product and keeps it stable overall (Ahari & Soufiani, 2021). Packaging systems have been organized into three hierarchical categories. The innermost layer, which forms a direct barrier around the edible product, plays a crucial role in preserving freshness and preventing contamination due to its immediate proximity to the consumables and has been termed as primary packaging (Ahari & Soufiani, 2021). The next layer, which is the secondary packaging, serves to group individual units for easier handling and branding during retail. Finally, the outermost layer, known as the tertiary packaging, has been designed for efficient logistics, enabling safe transportation and storage on a larger scale (Ahari & Soufiani, 2021). However, even with this multilayered system, complete assurance

Y. Kumar, S. Dey, *Emerging Technology in Food Processing*,
https://doi.org/10.1007/978-3-032-24505-2_9

of food hygiene and protection is not always achieved (Muncke, 2014). The main advantages of food packaging, including primary, secondary, and tertiary packaging, include ease of transportation, preservation of food integrity and quality, protection from dangerous particles and chemicals, and prevention of pests and bacterial growth. Through labelling, packaging also provides customers with the information they need about the products. The primary information provided by labelling is the product name, contents list, method of consumption, cost, and expiration date (Ahari & Soufiani, 2021).

The degradation of foods like kiwifruit, apples, tomatoes, and bananas throughout the stages of harvesting, distribution, storage, and retail is largely driven by microbial activity, metabolic processes, and oxidative reactions. These factors play a central role in compromising the quality and shelf life of perishable produce. Communication, protection, containment, and convenience represent the primary functions of traditional food packaging (TFP). Essentially, TFP operates as a conventional system that offers physical support and shields food from external factors during processing, distribution, transportation, and storage (Azeredo & Correa, 2021). Generally, well-designed TFP aids in food preservation but functions mainly as a passive barrier against chemical, physical, and microbial deterioration (Robertson, 2019). It is also recognized that TFP contributes significantly to the generation of packaging waste each year. Nevertheless, advancements in technology and changing lifestyles have increased consumer expectations for safe, high-quality foods with ease of handling and extended shelf life. Consequently, traditional packaging systems are increasingly inadequate in meeting these demands, highlighting the need for suitable alternative packaging solutions (Ahari & Soufiani, 2021).

In recent times, significant attention has been directed toward the development of advanced food packaging systems. These innovations involve the application of diverse biomaterials and emerging technologies, as well as the integration of sensors, indicators, essential oils, and plant-derived extracts. The primary objective of such approaches is to maintain food quality and nutritional integrity, prolong shelf life, and reduce the environmental burden associated with packaging waste (Muncke, 2014). Within this domain, two major categories are recognized: smart packaging, which tracks and conveys information about the condition of the food, and active packaging, which directly interacts with the product to improve its stability and preservation (Ahari & Soufiani, 2021). Smart food packaging (SFP), in particular, allows consumers to assess changes in food quality over time through the inclusion of indicator compounds in the packaging matrix that signal freshness or spoilage (Ahari & Soufiani, 2021).

9.2 Types of Smart Food Packaging

Smart food packaging represents an advanced generation of packaging systems designed not only to contain and protect food products but also to monitor, sense, record, and communicate information related to food quality, safety, and environmental conditions throughout the supply chain (Ahari & Soufiani, 2021). Unlike traditional food packaging, which primarily functions as a passive barrier against external physical, chemical, and biological hazards, smart packaging integrates functional components such as indicators, sensors, data carriers, and communication devices (Azeredo & Correa, 2021). These components provide real time or cumulative information regarding the condition of the packaged food, thereby supporting informed decision-making by manufacturers, distributors, retailers, and consumers. The increasing global demand for safe, high-quality, minimally processed foods with extended shelf life has significantly accelerated research, development, and adoption of smart food packaging technologies (Ahari & Soufiani, 2021; Kerry et al., 2006).

Smart food packaging systems can be categorized into indicator-based packaging, sensor-based packaging, data-carrier–based packaging, and integrated or interactive intelligent systems, each serving a specific function in quality monitoring, safety assurance, and supply chain management. These categories are not mutually exclusive; rather, they often overlap and complement one another in advanced packaging applications (Bhatlawande et al., 2024).

Among these, indicator-based smart packaging represents one of the most extensively studied and commercially implemented forms due to its simplicity, cost-effectiveness, and ease of interpretation (Bhatlawande et al., 2024). Indicators are designed to provide visual information about the condition of packaged food by undergoing a detectable change, typically in color, shape, or position, in response to specific environmental or chemical stimuli (Heo & Lim, 2024). Unlike sensors, indicators do not continuously measure parameters but instead provide qualitative or semi-quantitative information that can be readily understood without specialized equipment. This makes indicator-based systems particularly attractive for consumer-facing applications. Based on the monitored parameter, indicators are commonly classified into time temperature indicators (TTIs), freshness indicators, and gas indicators (Yam et al., 2005; Robertson, 2019).

Time temperature indicators (TTIs) are among the most widely researched smart packaging tools and are specifically designed to monitor the cumulative thermal history of food products. Since temperature abuse is a primary cause of food spoilage, microbial growth, and quality deterioration, TTIs play a critical role in ensuring cold chain integrity (Azeredo & Correa, 2021). These indicators undergo irreversible physical, chemical, or enzymatic changes at rates that correlate with temperature-dependent reactions occurring within the food (Salgado et al., 2021). TTIs are further categorized as critical temperature indicators, partial history indicators, or full history indicators depending on their response mechanism and sensitivity. By providing a clear visual signal when a product has been exposed to unfavourable temperature conditions, TTIs help prevent the distribution and consumption of unsafe foods, reduce food waste, and improve inventory and logistics management (Taoukis & Labuza, 2003; Kerry et al., 2006; Bhatlawande et al., 2024).

Another important subclass of indicator-based systems is freshness indicators, which directly reflect the quality status of food by responding to metabolites produced during spoilage processes (Azeredo et al., 2019). These indicators are particularly relevant for protein-rich foods such as meat, fish, and poultry, where microbial degradation leads to the formation of volatile basic nitrogen compounds, organic acids, and sulphur-containing gases. Freshness indicators are typically incorporated into labels or films that change color in response to pH variation or specific spoilage compounds (Bhatlawande et al., 2024). Recently, pH-sensitive dyes derived from natural sources such as anthocyanins, curcumin, and betalains have gained considerable attention due to their biodegradability, non-toxicity, and compatibility with sustainable packaging materials. Such indicators provide real-time information on food freshness, enabling consumers to assess product quality more accurately than conventional date labeling systems (Ahari & Soufiani, 2021; Kuswandi et al. 2011a).

Gas indicators form another subset of indicator-based smart packaging and are designed to detect the presence or concentration of gases within the package headspace. These systems are especially relevant for foods packaged under modified atmosphere packaging (MAP) or vacuum conditions (Heo & Lim, 2024). Oxygen indicators are widely used to detect oxygen ingress resulting from package leakage or seal failure, which can accelerate oxidative spoilage and microbial growth. Similarly, indicators for carbon dioxide and hydrogen sulphide are used to monitor microbial activity in meat and seafood products. By signalling deviations from the desired internal atmosphere, gas indicators enhance quality control and food safety during storage and distribution (Mills, 2005; Yam et al., 2005).

While indicators offer simplicity and affordability, sensor-based smart packaging systems provide more precise and quantitative information by continuously detecting and measuring specific physical, chemical, or biological parameters. Sensors generally consist of a receptor element that interacts with the target analyte and a transducer that converts this interaction into a measurable signal, such as an electrical, optical, or thermal output (Azeredo & Correa, 2021). Based on their detection mechanism, smart packaging sensors are broadly classified into chemical sensors, biosensors, and physical sensors (Kerry et al., 2006).

Chemical sensors are designed to detect compounds associated with spoilage, contamination, or changes in the package environment. These sensors can measure parameters such as pH, oxygen concentration, carbon dioxide levels, and volatile organic compounds with high sensitivity (Azeredo & Correa, 2021). Oxygen and carbon dioxide sensors are widely used in MAP systems to ensure that internal gas composition remains within optimal limits for food preservation. Compared to indicators, chemical sensors offer improved accuracy and reliability; however, their relatively higher cost, technical complexity, and power requirements have limited their widespread commercial adoption (Mills, 2005).

Biosensors represent a more advanced and highly specific category of sensor-based smart packaging (Sheibani et al., 2024). These systems utilize biological recognition elements such as enzymes, antibodies, nucleic acids, or whole cells to detect specific targets, including foodborne pathogens, toxins, and allergens (Salgado et al., 2021). Biosensors capable of detecting microorganisms such as *Escherichia coli*, *Salmonella*, and *Listeria monocytogenes* have demonstrated strong potential for enhancing food safety by providing early warning signals of contamination. Despite their high sensitivity and specificity, challenges related to stability, cost, large-scale manufacturability, and regulatory approval have limited their commercial deployment in food packaging applications (Realini & Marcos, 2014).

In addition to chemical and biological sensing, physical sensors integrated into smart packaging systems are used to monitor environmental conditions such as temperature, humidity, pressure, and mechanical stress (Azeredo & Correa, 2021). These sensors are particularly valuable for tracking transportation and storage conditions of perishable foods. Recent advances in miniaturization, flexible electronics, and printable sensor technologies have enabled the development of lightweight, low-cost sensors that can be seamlessly integrated into packaging materials. When combined with data logging and wireless communication technologies, physical sensors significantly improve traceability and supply chain transparency (Kuswandi et al. 2011b).

Beyond indicators and sensors, data carrier-based smart packaging systems focus primarily on information storage, identification, and communication rather than direct quality sensing (Bhatlawande et al., 2024). Technologies such as barcodes, two-dimensional (2D) codes, radio frequency identification (RFID), and near field communication (NFC) fall under this category. Among these, RFID has emerged as one of the most powerful tools for smart food packaging applications (Bhatlawande et al., 2024). RFID systems consist of tags, readers, and databases that enable automatic identification and real-time tracking of food products throughout the supply chain. RFID-enabled packaging can store and transmit information related to product origin, processing history, storage conditions, and expiration dates, thereby enhancing traceability, inventory management, and protection against counterfeiting (Jones et al., 2005; Regattieri et al., 2007).

More recently, integrated or interactive intelligent packaging systems have been developed by combining sensing technologies with data carrier and communication functionalities. These systems not only monitor food quality but also actively communicate information to stakeholders through digital platforms, cloud-based databases, or smartphones. For example, smart labels that integrate freshness indicators with QR codes or NFC technology allow consumers to access detailed information regarding product quality, safety, sustainability credentials, and storage recommendations (Azeredo & Correa, 2021). Such systems align well with the growing emphasis on digitalization, transparency, and consumer engagement in the modern food sector (Ahari & Soufiani, 2021; Realini & Marcos, 2014).

The scope of smart food packaging has further expanded with the emergence of printed, nano-enabled, and hybrid smart packaging systems. Printed smart packaging utilizes advances in printable electronics and functional inks to fabricate sensors and circuits directly onto packaging substrates, enabling flexible, lightweight, and cost-effective solutions (Motelica et al., 2020). Nano-enabled smart packaging incorporates nanomaterials to enhance sensitivity, responsiveness, and multifunctionality, though regulatory and safety considerations remain critical. Hybrid systems integrate multiple smart features such as indicators, sensors, and RFID within a single package to provide comprehensive quality and traceability

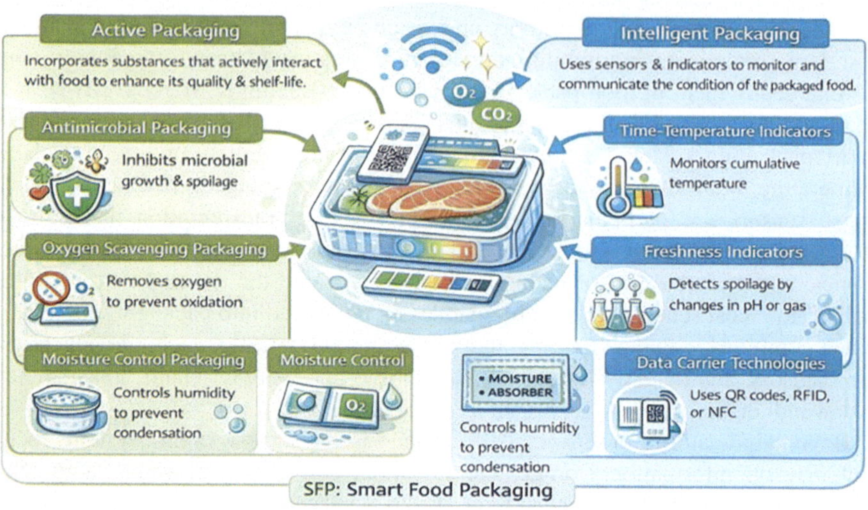

Fig. 9.1 Types of smart food packaging, illustrating active and intelligent packaging with their subcategories and functional features

information, particularly for high-value and high-risk food products (Mills, 2005; Robertson, 2019).

In recent years, increasing environmental awareness has driven the development of biodegradable and sustainable smart food packaging systems. The incorporation of natural polymers, plant-based sensing compounds, and eco-friendly substrates addresses concerns associated with plastic waste and electronic disposal (Salgado et al., 2021). Smart packaging based on biopolymers such as starch, chitosan, cellulose, and polylactic acid not only provides intelligent functionality but also supports circular economy and sustainability goals (Siracusa et al., 2008; Robertson, 2019).

Smart food packaging encompasses a diverse and continuously evolving range of technologies aimed at enhancing food quality monitoring, safety assurance, and supply chain efficiency (Salgado et al., 2021). Figure 9.1 depicts the types of smart food packaging, illustrating active and intelligent packaging along with their respective subcategories and functional features. Indicator-based systems provide simple and cost-effective solutions for monitoring freshness and temperature changes, whereas sensor-based systems offer higher accuracy, sensitivity, and specificity in detecting physicochemical and microbiological changes in food products. Data carrier technologies facilitate traceability, authentication, and efficient information management throughout the supply chain, while integrated intelligent systems bridge the gap between physical packaging and digital communication platforms (Regattieri et al., 2007). With ongoing technological advancements and an increasing emphasis on sustainability, smart food packaging is expected to play a pivotal role in meeting modern consumer expectations and addressing global challenges related to food safety, quality assurance, and food waste reduction. Table 9.1 presents an overview of the different classifications of smart food packaging systems, highlighting their functional principles and practical applications in the food industry.

9.3 Processing of Smart Food Packaging

The processing of smart food packaging involves the systematic integration of functional materials, sensing components, and information technologies into conventional packaging systems to enable monitoring, communication, and quality assessment of food products throughout the supply chain (Salgado et al., 2021). Unlike traditional packaging processes that primarily focus on forming, filling, and sealing operations, smart packaging processing encompasses additional stages such as material functionalization, incorporation of indicators or sensors, activation of intelligent elements, and compatibility assessment with food matrices (Yam et al., 2005). These processing strategies are designed to ensure that smart packaging components perform reliably without compromising food safety, product quality, or regulatory compliance (Yam et al., 2005; Robertson, 2019).

The initial stage in the processing of smart food packaging is the selection and preparation of base packaging materials. Conventional polymers such as polyethylene, polypropylene, polyethylene terephthalate, and polyvinyl chloride have been widely used due to their mechanical strength, barrier properties, and processability (Drago et al., 2020). However, growing environmental concerns and sustainability goals have driven the adoption of biodegradable and biobased polymers, including starch, cellulose, chitosan, polylactic acid, and protein-based films (Drago et al., 2020). The base material must be compatible with smart components, possess adequate thermal and chemical stability during processing, and ensure safe contact with food products (Siracusa et al., 2008; Ahari & Soufiani, 2021). Material preparation often involves extrusion, casting, or coating processes to achieve the desired thickness, flexibility, and surface characteristics for subsequent functionalization.

A critical processing step in smart food packaging is the incorporation of indicators and sensing agents into the packaging matrix or as separate labels. Indicator integration can be achieved through blending, surface coating, printing, or lamination techniques (Chen et al., 2020). For instance, time temperature indicators are commonly produced using enzymatic, polymeric, or diffusion-based systems that respond to cumulative thermal exposure. These indicators are either embedded within the packaging film or attached as self-adhesive labels during the packaging process (Chen et al., 2020). Freshness indicators, particularly those based on pH-sensitive dyes or volatile compound detection, are often immobilized within polymer matrices or encapsulated to prevent direct contact with food while allowing interaction with headspace gases (Kuswandi et al. 2011a, b; Mills, 2005).

The processing of smart packaging increasingly emphasizes the use of natural and bio-derived sensing compounds, especially plant-based pigments such as

Table 9.1 Classification of smart food packaging systems, their functions, and applications

Type of smart food packaging	Sub-category	Function	Typical components/ Examples	Major applications	References
Indicator-based packaging	Time temperature indicators (TTIs)	Monitors cumulative temperature exposure and thermal abuse	Enzymatic, chemical, or diffusion-based indicators showing irreversible color change	Cold chain monitoring of meat, dairy, seafood, frozen foods	Ahari and Soufiani (2021) and Azeredo & Correa (2021)
	Freshness indicators	Detects spoilage metabolites and pH changes associated with microbial activity	pH-sensitive dyes (anthocyanins, curcumin, betalains), color-changing labels	Meat, fish, poultry, protein-rich foods	Kuswandi et al. (2011a, b) and Ahari and Soufiani (2021)
	Gas indicators	Detects the presence or concentration of gases in the package headspace	Oxygen, CO_2, H_2S indicators	MAP foods, vacuum-packed meat and seafood	Mills (2005), Yam et al. (2005), and Rodrigues et al. (2021)
Sensor-based packaging	Chemical sensors	Quantitative detection of chemical changes related to spoilage or atmosphere	pH sensors, oxygen and CO_2 sensors, VOC sensors	MAP systems, high-value perishables	Kerry et al. (2006) and Mills (2005)
	Biosensors	Uses biological recognition elements to detect specific targets	Enzymes, antibodies, DNA probes for pathogens	Pathogen detection (*E. coli*, *Salmonella*, *Listeria*)	Salgado et al. (2021)
	Physical sensors	Monitors environmental and mechanical conditions	Temperature, humidity, pressure, shock sensors	Transportation and storage monitoring	Kuswandi et al. (2011a, b)
Data carrier–based packaging	Barcodes and 2D codes	Product identification and static information storage	Barcodes, QR codes	Retail, inventory management, consumer information	Regattieri et al. (2007)
	RFID and NFC	Real-time tracking, traceability, and wireless communication	RFID tags, NFC chips	Supply chain management, anti-counterfeiting, exports	Jones et al. (2005) and Regattieri et al. (2007)

(continued)

Table 9.1 (continued)

Type of smart food packaging	Sub-category	Function	Typical components/ Examples	Major applications	References
Integrated/ intelligent packaging	Interactive smart packaging	Combines sensing with digital communication	Smart labels with QR/NFC and indicators	Consumer engagement, transparency, traceability	Ahari and Soufiani (2021) and Realini and Marcos (2014)
	Hybrid smart systems	Multiple smart functions in one package	TTIs and freshness indicators, and RFID	High-risk and high-value foods	Yam et al. (2005) and Kerry et al. (2006)
Advanced smart packaging	Printed smart packaging	Uses printable electronics for flexible sensors	Printed conductive inks, flexible circuits	Large-scale, low-cost smart packaging	Mills (2005)
	Nano-enabled smart packaging	Enhances sensitivity using nanomaterials	Silver nanoparticles, ZnO, TiO_2	High sensitivity spoilage detection	Robertson (2019) and Motelica et al. (2020)
Sustainable smart packaging	Bio-based smart packaging	Combines smart function with biodegradability	Starch, chitosan, cellulose, PLA, and natural indicators	Eco-friendly food packaging	Siracusa et al. (2008), Robertson (2019), and Salgado et al. (2021)

anthocyanins, curcumin, and betalains (Bhatlawande et al., 2024). These compounds are extracted, purified, and stabilized prior to incorporation into packaging materials. Processing methods such as solvent casting, electrospinning, and nanoencapsulation are employed to enhance the stability, sensitivity, and controlled response of these natural indicators (Motelica et al., 2020). Encapsulation techniques using cyclodextrins, liposomes, or biopolymer matrices protect the active compounds from degradation during thermal processing and storage while ensuring a visible and reversible response to quality changes in food products (Ahari & Soufiani, 2021).

Another important aspect of smart food packaging processing is the fabrication and integration of sensors, which requires greater technical precision than indicator-based systems (Mkhari et al., 2025). Chemical and physical sensors are typically manufactured using microfabrication, thin film deposition, or printable electronics technologies. Advances in inkjet printing, screen printing, and gravure printing have enabled the low-cost production of flexible sensors directly onto packaging substrates (Chen et al., 2020). Conductive inks containing silver nanoparticles, carbon nanotubes, or graphene are commonly used to print sensor circuits that respond to changes in temperature, humidity, gas concentration, or mechanical stress. These printed sensors are subsequently cured and calibrated to ensure accurate performance under real-world conditions (Kerry et al., 2006).

The processing of biosensor-based smart packaging involves additional complexity due to the presence of biological recognition elements such as enzymes, antibodies, or nucleic acids (Bhatlawande et al., 2024). These components must be immobilized onto sensor surfaces without loss of activity, which requires mild processing conditions and controlled environments. Techniques such as physical adsorption, covalent bonding, and entrapment within polymer matrices are commonly employed for immobilization (Sheibani et al., 2024). During processing, factors such as pH, temperature, moisture, and oxygen exposure are carefully controlled to preserve the biological functionality of biosensors. Although biosensors offer high specificity and sensitivity, their processing challenges, cost, and stability issues currently limit large-scale industrial implementation (Realini & Marcos, 2014).

Another emerging processing strategy in smart food packaging is the use of nanotechnology to enhance sensor sensitivity and functional performance. Nanomaterials such as metal nanoparticles, nanoclays, and nanofibers are incorporated into packaging matrices to improve electrical conductivity, gas sensitivity, and mechanical strength (Drago et al., 2020). Electrospinning is a particularly promising processing technique for producing nanofiber-based sensing layers with high surface area and rapid response times. These nanostructured systems enable more precise detection of spoilage-related compounds, such as ammonia, hydrogen sulphide, and carbon dioxide, which are indicative of food deterioration (Drago et al., 2020). However, the incorporation of nanomaterials necessitates rigorous processing control and safety evaluation to address potential migration and toxicological concerns (Mills, 2005; Realini & Marcos, 2014).

Processing of smart packaging must also consider compatibility with different food categories, as the physicochemical properties of foods influence the performance of smart components (Azeredo & Correa, 2021). High-moisture foods, such as fresh meat and seafood, generate volatile compounds and humidity levels that can affect sensor accuracy and longevity (Ahari & Soufiani, 2021). Conversely, low moisture or dry foods may require more sensitive indicators to detect subtle quality changes. During processing, tailored smart packaging designs are developed by adjusting indicator composition, sensor sensitivity, and package geometry to match specific food systems. Validation studies under simulated storage conditions are essential to ensure reliable performance throughout the product's shelf life (Yam et al., 2005; Kuswandi et al. 2011a, b).

Surface modification and coating technologies play a crucial role in the processing of smart food packaging. Plasma treatment, corona discharge, and chemical grafting are widely used to modify the surface properties of packaging materials, enhancing the adhesion of indicators, sensors, or printed electronics (Robertson, 2019). Coating processes such as dip coating, spray coating, and layer-by-layer assembly allow uniform distribution of functional layers on packaging films. These techniques improve the durability and performance of smart components while maintaining transparency, flexibility, and mechanical integrity of the package (Robertson, 2019).

The integration of smart packaging with modified atmosphere packaging (MAP) systems represents an important processing advancement. Smart indicators that monitor oxygen or carbon dioxide levels are particularly valuable in MAP applications, where gas composition plays a critical role in extending shelf life (Salgado et al., 2021). Processing challenges include ensuring that indicators respond accurately to gas changes without interfering with the controlled atmosphere. Gas-sensitive dyes and electrochemical sensors are typically incorporated into package headspace regions or attached as internal labels. Processing protocols must ensure airtight sealing and precise placement of these components to avoid false readings and ensure consumer trust (Kerry et al., 2006).

An essential processing consideration is the compatibility of smart packaging components with conventional food processing operations, including thermal treatments, high-pressure processing, and modified atmosphere packaging (Salgado et al., 2021). Smart packaging materials must withstand processing conditions without degradation or loss of functionality. For example, indicators and sensors should remain stable during heat sealing, pasteurization, or refrigeration. Processing validation studies are conducted to assess the performance of smart packaging under realistic storage and distribution scenarios, ensuring the reliability and reproducibility of the response signals (Yam et al., 2005; Kerry et al., 2006).

Automation and digitalization are increasingly influencing smart food packaging processing. The integration of Industry 4.0 concepts, such as automated inspection, real-time monitoring, and data analytics, enhances process control and quality assurance (Chen et al., 2020). The integration of data carrier technologies, such as barcodes, QR codes, RFID, and NFC tags, constitutes another main stage in smart packaging processing (Rodrigues et al., 2021). These components are typically applied through printing, embedding, or lamination processes during package fabrication. RFID tags may be incorporated between layers of multilayer packaging films or attached as labels, depending on cost and application requirements. Processing considerations include antenna design, signal readability, and resistance to moisture, temperature, and mechanical stress. The integration of data carriers enables real-time tracking, traceability, and digital communication across the food supply chain (Regattieri et al., 2007; Jones et al., 2005).

With increasing emphasis on sustainability, the processing of smart food packaging also focuses on eco-friendly manufacturing techniques. Solvent-free processing, water-based inks, and energy-efficient curing methods are increasingly adopted to reduce environmental impact (Sheibani et al., 2024). The use of biodegradable substrates combined with compostable indicators and printed electronics aligns smart packaging development with circular economy principles. However, balancing functionality, cost, and environmental performance remains a major processing challenge (Siracusa et al., 2008; Ahari & Soufiani, 2021).

Quality control and regulatory compliance are integral components of smart food packaging processing. All materials and processing aids must comply with food contact regulations and migration limits established by international authorities

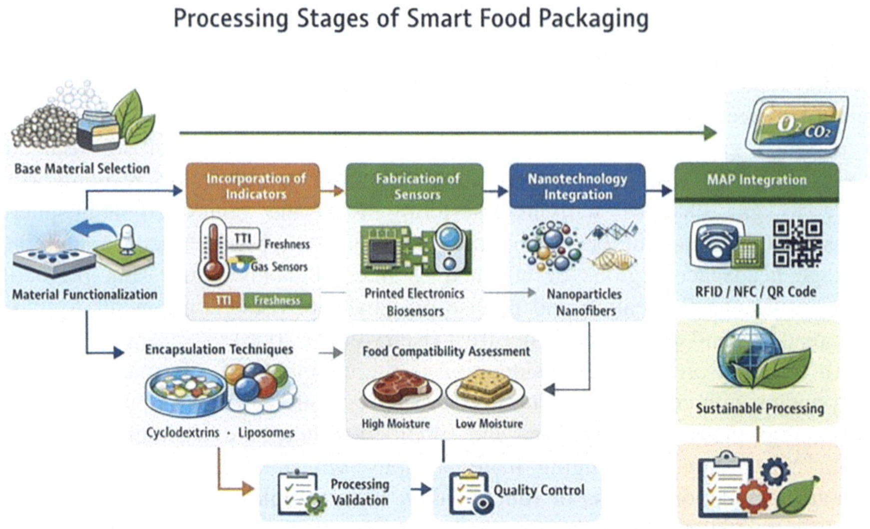

Fig. 9.2 Schematic representation of the main processing stages and technological integration in smart food packaging systems

(Realini & Marcos, 2014). Processing validation includes mechanical testing, barrier property assessment, sensor calibration, and shelf-life studies to ensure safety and effectiveness. Standardization of processing methods and performance evaluation protocols is essential for the commercial adoption of smart packaging technologies (Robertson, 2019).

The processing of smart food packaging is a multidisciplinary and technologically intensive endeavor that extends beyond conventional packaging operations. It involves careful material selection, functionalization, sensor fabrication, surface modification, and seamless integration with food processing, storage, and distribution systems (Sheibani et al., 2024). Figure 9.2 presents a schematic representation of the main processing stages and the technological integration involved in smart food packaging systems, highlighting the sequential and interconnected nature of these operations (Realini & Marcos, 2014). Advances in printing technologies, bio-based and nanostructured materials, and digital integration platforms continue to enhance the feasibility, functionality, and scalability of smart packaging solutions (Sheibani et al., 2024). As consumer demand for food safety, quality transparency, and sustainability continues to increase, optimized and well-coordinated processing strategies will play a pivotal role in the successful implementation and industrial adoption of smart food packaging. Table 9.2 summarizes the main processing stages, associated technologies, and critical considerations in smart food packaging, providing a concise overview of their roles and challenges within the food packaging value chain.

Table 9.2 Processing stages, technologies, and considerations in smart food packaging

Processing stage	Materials/components involved	Processing techniques	Purpose and functional role	Considerations/challenges	References
Base material selection and preparation	Conventional polymers (PE, PP, PET, PVC); bio-based polymers (starch, cellulose, chitosan, PLA, protein films)	Extrusion, solvent casting, compression moulding, coating	Provides mechanical strength, barrier properties, and food contact safety; serves as a substrate for smart components	Compatibility with smart elements, thermal and chemical stability, sustainability, regulatory compliance	Siracusa et al. (2008) and Ahari and Soufiani (2021)
Material functionalization	Polymer matrices, additives, surface modifiers	Plasma treatment, corona discharge, chemical grafting	Enhances surface energy and adhesion for indicators, sensors, and printed electronics	Maintaining transparency, flexibility, and mechanical integrity	Robertson (2019)
Incorporation of indicators	Time temperature indicators, freshness indicators, gas indicators	Blending, surface coating, printing, lamination, and label attachment	Enables visual monitoring of temperature history, spoilage, and package integrity	Indicator stability, signal accuracy, and prevention of food contact	Kuswandi et al. (2011a, b) and Mills (2005)
Integration of natural sensing compounds	Anthocyanins, curcumin, betalains (plant-based pigments)	Solvent casting, electrospinning, nanoencapsulation	Provides eco -friendly, pH, and gas-sensitive responses to food quality changes	Sensitivity to heat, light, and oxygen; color stability	Ahari and Soufiani (2021)
Encapsulation and stabilization of indicators	Cyclodextrins, liposomes, biopolymer matrices	Nanoencapsulation, emulsion-based techniques	Protects active compounds during processing and storage; ensures controlled release	Encapsulation efficiency, migration safety, cost	Bhatlawande et al. (2024)
Fabrication of physical and chemical sensors	Conductive inks (silver nanoparticles, CNTs, graphene)	Inkjet printing, screen printing, gravure printing, and thin film deposition	Enables detection of temperature, humidity, gases, and mechanical stress	Calibration accuracy, curing conditions, and durability	Mills (2005) and Kerry et al. (2006)
Processing of biosensor-based systems	Enzymes, antibodies, nucleic acids	Physical adsorption, covalent bonding, polymer entrapment	High specificity detection of pathogens or metabolites	Sensitivity to processing conditions, cost, and limited industrial scalability	Realini and Marcos (2014)

(continued)

Table 9.2 (continued)

Processing stage	Materials/components involved	Processing techniques	Purpose and functional role	Considerations/challenges	References
Nanotechnology-based enhancement	Metal nanoparticles, nanoclays, nanofibers	Electrospinning, nano-coating, and nanocomposite formation	Improves sensor sensitivity, conductivity, and response speed	Migration risks, toxicological safety, and regulatory approval	Mills (2005) and Realini and Marcos (2014)
Food–package compatibility assessment	High-moisture foods (meat, seafood), low-moisture foods	Tailored indicator design, sensor sensitivity optimization	Ensures accurate performance under specific food environments	Volatile interference, humidity effects, shelf-life variability	Yam et al. (2005) and Kuswandi et al. (2011a, b)
Integration with MAP systems	Oxygen and CO_2 indicators, gas sensors	Internal label placement, headspace integration	Monitors gas composition and package integrity	Airtight sealing, avoidance of false readings	Kerry et al. (2006)
Compatibility with food processing operations	Smart films, indicators, sensors	Heat sealing, pasteurization, and refrigeration testing	Ensures functionality during conventional food processing	Thermal stability, mechanical robustness	Yam et al. (2005) and Kerry et al. (2006)
Data carrier integration	Barcodes, QR codes, RFID, NFC tags	Printing, embedding, and lamination	Enables traceability, real-time monitoring, digital communication	Signal readability, antenna design, environmental resistance	Regattieri et al. (2007) and Jones et al. (2005)
Automation and digitalization	Sensors, RFID, inspection systems	Inline inspection, real-time analytics	Enhances quality control, process reliability, and industry 4.0 integration	Equipment cost, data management	Regattieri et al. (2007)
Sustainable processing approaches	Biodegradable substrates, water-based inks	Solvent-free processing, energy-efficient curing	Reduces environmental impact; supports circular economy	Trade-off between functionality and cost	Siracusa et al. (2008) and Ahari and Soufiani (2021)

Quality control and regulatory validation	Finished smart packaging systems	Migration testing, sensor calibration, shelf-life studies	Ensures food safety, performance reliability, and compliance	Standardization gaps, regulatory harmonization	Robertson (2019) and Realini and Marcos (2014)

9.4 Advantages and Applications of Smart Food Packaging

Smart food packaging has emerged as a transformative innovation in the modern food system, addressing critical challenges related to food safety, quality assurance, shelf-life extension, supply chain efficiency, and sustainability (Chen et al., 2020). Unlike conventional packaging, which primarily functions as a passive barrier to protect food from external environmental factors, smart food packaging integrates advanced materials, indicators, sensors, and information technologies that actively monitor, communicate, and sometimes respond to changes occurring within the packaged food or its surrounding environment (Sheibani et al., 2024). The growing global demand for safe, high-quality, minimally processed foods, coupled with increasing concerns about food waste and environmental impact, has positioned smart food packaging as a vital component of next-generation food preservation and distribution systems (Yam et al., 2005; Ahari & Soufiani, 2021).

One of the most significant advantages of smart food packaging is its ability to enhance food safety (Sheibani et al., 2024). Food-borne illnesses remain a major public health concern worldwide, often resulting from microbial contamination, temperature abuse, or improper storage conditions. Smart packaging systems equipped with freshness indicators, gas sensors, or biosensors can detect early signs of spoilage or contamination by monitoring changes in pH, volatile organic compounds, or pathogenic metabolites (Ahari & Soufiani, 2021). These systems provide real-time or cumulative information about food condition, allowing manufacturers, retailers, and consumers to identify unsafe products before consumption. By offering visible and easily interpretable signals, smart packaging reduces reliance on subjective sensory evaluation and static expiration dates, thereby improving consumer protection and confidence (Kerry et al., 2006; Realini & Marcos, 2014).

Another major advantage of smart food packaging lies in its capacity to extend shelf life while preserving the sensory, nutritional, and microbiological quality of foods (Motelica et al., 2020). Contemporary smart packaging technologies are increasingly designed not only to monitor but also to optimize critical storage and distribution conditions, particularly temperature, relative humidity, and internal package atmosphere, all of which strongly influence food deterioration kinetics (Ahari & Soufiani, 2021). Time-temperature indicators (TTIs), for instance, provide a cumulative record of thermal history and correlate it with quality loss reactions, enabling more precise cold-chain management and dynamic shelf-life prediction. Recent advancements have integrated TTIs with digital sensors, radio frequency identification (RFID), and Internet of Things (IoT) platforms, allowing real-time data acquisition, remote monitoring, and automated decision making across supply chains (Rodrigues et al., 2021). Figure 9.3 presents an overview of the major application areas of smart food packaging and their associated sensing and indicator technologies, illustrating how different sensor systems, such as gas sensors, freshness indicators, and thermal history indicators, are strategically deployed across various food categories and supply chain stages. By enabling early detection of temperature abuse during transportation, storage, or retail display, smart packaging

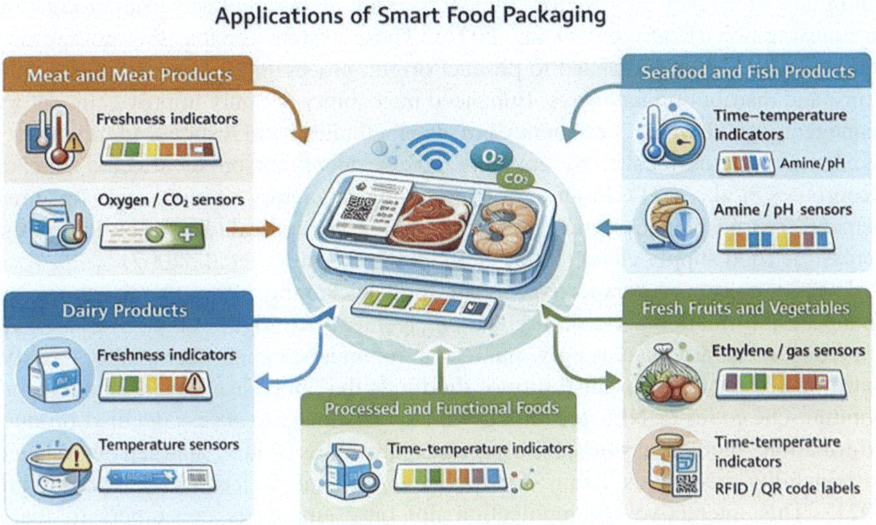

Fig. 9.3 An overview of major application areas of smart food packaging and their associated sensing and indicator technologies

systems help mitigate microbial growth, enzymatic activity, and oxidative reactions that lead to premature spoilage and quality degradation (Bhatlawande et al., 2024). This functionality is particularly critical for highly perishable commodities such as meat, fish, dairy products, and fresh fruits and vegetables, where even short-term deviations from optimal conditions can result in substantial shelf-life reduction and food waste (Taoukis & Labuza, 2003; Robertson, 2019). Moreover, emerging data-driven and predictive models combined with smart packaging technologies are facilitating a shift from static expiry dates to condition-based shelf-life assessment, aligning food safety assurance with sustainability and waste reduction goals.

Smart food packaging also contributes significantly to the reduction of food waste, which is a global sustainability challenge (Azeredo & Correa, 2021). A substantial proportion of food waste occurs due to conservative expiration dates, improper storage, and a lack of real-time information about food quality. Smart packaging systems, particularly freshness indicators and intelligent labels, provide dynamic quality information that reflects the actual condition of the food rather than relying solely on predefined shelf-life estimates (Bhatlawande et al., 2024). This enables consumers and retailers to make informed decisions about food usability, thereby reducing unnecessary disposal of still-safe products. From an environmental and economic perspective, minimizing food waste through smart packaging aligns with sustainable development goals and circular economy principles (Siracusa et al., 2008; Ahari & Soufiani, 2021).

An additional advantage of smart food packaging is its role in improving supply chain transparency and traceability. The integration of data carrier technologies such as barcodes, QR codes, radio frequency identification (RFID), and near-field communication (NFC) allows for efficient tracking of food products from production to consumption (Rodrigues et al., 2021). These systems enable the storage and retrieval of information related to product origin, processing history, storage conditions, and distribution pathways. Enhanced traceability not only improves inventory management and logistics efficiency but also facilitates rapid response to food safety incidents, such as recalls, by enabling precise identification of affected batches (Rodrigues et al., 2021). In an era of increasing regulatory scrutiny and consumer demand for transparency, smart packaging serves as a critical tool for building trust across the food supply chain (Jones et al., 2005; Regattieri et al., 2007).

From a consumer perspective, smart food packaging offers the advantage of enhanced information accessibility and engagement (Rodrigues et al., 2021). Modern consumers are increasingly interested in understanding the quality, safety, nutritional value, and sustainability of the foods they purchase. Smart labels incorporating QR codes or NFC technology allow consumers to access detailed product information, including storage instructions, freshness status, allergen alerts, and sustainability credentials, using smartphones or digital devices (Ahari & Soufiani, 2021). This interactive communication not only empowers consumers to make informed choices but also strengthens brand differentiation and loyalty for food manufacturers (Realini & Marcos, 2014).

In addition to its advantages, smart food packaging has found wide-ranging applications across diverse food categories, reflecting its versatility and adaptability. One of the most prominent application areas is in meat and meat products, where microbial spoilage and oxidation are major quality concerns (Motelica et al., 2020). Freshness indicators that respond to pH changes or volatile amines are commonly used in meat packaging to signal spoilage. Oxygen and carbon dioxide sensors are applied in modified atmosphere packaging (MAP) systems to monitor package integrity and gas composition (Salgado et al., 2021). Furthermore, time temperature indicators are increasingly employed to track cumulative thermal exposure of meat products during cold chain distribution, providing valuable information on potential quality degradation. In some systems, antimicrobial and antioxidant-releasing components are integrated with smart indicators to further inhibit microbial growth and lipid oxidation. These smart packaging solutions help maintain meat quality, reduce spoilage losses, and ensure consumer safety throughout storage and distribution (Kerry et al., 2006; Kuswandi et al. 2011b).

Seafood and fish products represent another important application domain for smart food packaging due to their high perishability and sensitivity to temperature fluctuations (Motelica et al., 2020). Smart packaging systems incorporating TTIs, freshness indicators, and gas sensors are particularly valuable in monitoring the cold chain for seafood products (Salgado et al., 2021). The accumulation of biogenic amines and sulphur-containing compounds during fish spoilage can be detected by smart indicators, providing clear visual signals of quality deterioration. In addition, oxygen sensors integrated into modified atmosphere or vacuum packaging systems assist in detecting package leakage, which is critical for maintaining seafood freshness. Emerging smart labels combined with RFID and IoT platforms further enable real-time traceability and temperature history tracking across the supply chain. These applications not only enhance food safety but also support quality grading and pricing strategies in seafood markets (Yam et al., 2005; Mills, 2005).

In the dairy industry, smart food packaging is applied to monitor freshness, temperature abuse, and package integrity for products such as milk, yogurt, cheese, and cream (Motelica et al., 2020). TTIs and temperature sensors are especially useful for refrigerated dairy products, where maintaining consistent cold storage is essential to prevent microbial growth and quality loss (Salgado et al., 2021). In addition, pH-responsive freshness indicators can detect early signs of fermentation or spoilage in cultured dairy products, while oxygen sensors are used to identify package leakage in cheese and extended-shelf-life milk. Smart labels can also provide consumers with information about storage conditions and remaining shelf life, improving handling practices and reducing spoilage at the household level (Robertson, 2019).

The application of smart packaging in fresh fruits and vegetables focuses on monitoring respiration, ripening, and gas composition within the package (Motelica et al., 2020). Ethylene sensors and gas indicators are used to track ripening processes and detect deviations from optimal storage conditions (Rodrigues et al., 2021). In addition, oxygen and carbon dioxide sensors are incorporated into modified atmosphere packaging systems to regulate respiration dynamics and maintain product quality. Time temperature indicators further assist in identifying temperature abuse during post-harvest handling and cold chain distribution. These smart packaging systems help maintain freshness, delay senescence, and reduce post-harvest losses, which are particularly high for fresh produce. These technologies are increasingly combined with biodegradable materials to support sustainable packaging solutions for fresh foods (Siracusa et al., 2008; Ahari & Soufiani, 2021).

Smart food packaging is also gaining importance in processed and ready-to-eat foods, where convenience, safety, and shelf life are critical factors (Salgado et al., 2021). Intelligent labels and sensors can monitor temperature exposure during transportation and storage, ensuring that products remain within safe limits (Ahari & Soufiani, 2021). Time temperature indicators are particularly valuable for chilled and frozen ready-to-eat meals, while freshness indicators can provide visual cues related to product integrity after opening. In addition, data carrier technologies such as barcodes, QR codes, and RFID facilitate efficient inventory management, batch identification, and real-time traceability across large-scale food service and retail operations. These applications are particularly relevant in the context of globalized food supply chains and growing demand for convenience foods (Regattieri et al., 2007).

Another emerging application area of smart food packaging is in pharmaceutical and nutraceutical food products, including functional foods, fortified foods, and dietary supplements, where product stability, efficacy, and safety are of paramount importance (Motelica et al., 2020). In these applications, smart packaging systems are designed to monitor and control critical environmental factors such as moisture, oxygen concentration, light exposure, and temperature, all of which can significantly influence the degradation of bioactive compounds, vitamins, probiotics, and antioxidants (Motelica et al., 2020). Moisture and oxygen indicators are particularly important for hygroscopic products and oxygen-sensitive formulations, as exposure can lead to reduced potency, oxidation, or microbial contamination. Time temperature indicators are increasingly used to track thermal exposure during storage and distribution, especially for temperature-sensitive nutraceuticals and probiotic-based products. In addition, smart labels incorporating data carriers such as QR codes and RFID enable enhanced traceability, authentication, and anti-counterfeiting measures, which are critical concerns in health-related food and supplement markets. These technologies also facilitate regulatory compliance by supporting accurate shelf-life determination, batch tracking, and quality documentation throughout the supply chain (Mills, 2005). Furthermore, smart packaging can enhance consumer engagement by providing access to detailed product information, dosage instructions, and storage recommendations through digital interfaces. By ensuring product integrity, transparency, and performance, smart food packaging plays a crucial role in strengthening consumer confidence and supporting the growing global demand for functional and health-promoting food products.

Despite its numerous advantages and applications, the adoption of smart food packaging also presents challenges related to cost, technological complexity, and regulatory approval. However, ongoing advances in printable electronics, biodegradable materials, and low-cost sensors are steadily improving the feasibility and scalability of smart packaging systems (Motelica et al., 2020). As sustainability considerations become increasingly important, the integration of smart functionalities with eco-friendly materials is expected to further expand the applications of smart food packaging (Siracusa et al., 2008; Ahari & Soufiani, 2021).

Smart food packaging offers a wide range of advantages, including enhanced food safety, extended shelf life, reduced food waste, improved traceability, and increased consumer engagement (Motelica et al., 2020). Its applications span diverse food categories, ranging from fresh and highly perishable commodities to processed and functional foods. Table 9.3 summarizes the main advantages of smart food packaging along with representative application examples across different food systems, highlighting how specific technologies address safety, quality, and supply chain efficiency challenges (Ahari & Soufiani, 2021). By bridging the gap between conventional packaging systems and digital technologies, smart food packaging plays a crucial role in addressing contemporary challenges within the global food system. Continued research, technological innovation, and regulatory harmonization are expected to further strengthen its role in promoting food quality, safety, transparency, and sustainability in the years to come (Ahari & Soufiani, 2021).

9.5 Conclusion

Smart food packaging represents a significant advancement over conventional packaging systems by integrating sensing, indicating, and communication functionalities into the packaging structure to ensure food quality, safety, and sustainability. As discussed across the preceding sections on the types, processing, advantages, and

Table 9.3 Advantages and representative applications of smart food packaging (References: Chen et al. (2020); Regattieri et al. (2007); Ahari and Soufiani (2021))

Aspect	Description	Applications
Food safety enhancement	Real-time monitoring of spoilage, contamination, and storage conditions using indicators and sensors	Meat, fish, dairy products
Shelf-life extension	Monitoring of temperature, gas composition, and quality changes to prevent premature deterioration	Perishable foods, cold chain products
Food waste reduction	Dynamic freshness information replacing conservative expiration dates	Retail foods, household storage
Supply chain traceability	Integration of RFID, QR codes, and NFC for product tracking and recall management	Global food distribution systems
Consumer engagement	Easy access to quality, safety, and sustainability information through smart labels	Ready-to-eat and branded food products
Quality monitoring in produce	Detection of ripening and respiration through gas and ethylene sensors	Fruits and vegetables
Product integrity assurance	Monitoring environmental factors affecting stability and efficacy	Functional foods, nutraceuticals

applications of smart food packaging, it is evident that this emerging technology has the potential to transform the way food products are preserved, monitored, and delivered throughout the supply chain. The shift from passive packaging toward intelligent systems reflects the growing complexity of modern food distribution networks and the increasing expectations of consumers for transparency, safety, and quality assurance.

The various types of smart food packaging, including indicator-based systems, sensor-based technologies, and data-carrier–enabled packaging, collectively address multiple challenges associated with food spoilage and safety. Indicators such as time temperature indicators, freshness indicators, and gas indicators offer simple yet effective means of communicating quality changes in food products. Sensor-based packaging, including chemical sensors, biosensors, and physical sensors, provides higher accuracy and specificity in detecting environmental and biochemical changes. Data carrier technologies such as RFID, QR codes, and NFC further enhance traceability and information accessibility. Together, these systems enable continuous monitoring and informed decision-making across production, storage, transportation, and consumption stages.

The processing of smart food packaging highlights the multidisciplinary nature of this field, combining material science, food engineering, chemistry, biotechnology, and digital technologies. The successful fabrication of smart packaging systems requires careful selection of base materials, precise incorporation of indicators or sensors, and compatibility with conventional food processing and packaging operations. Advances in printing technologies, nanoencapsulation, biodegradable polymers, and flexible electronics have significantly improved the feasibility and scalability of smart packaging solutions. However, processing challenges related to cost, stability, regulatory compliance, and large-scale industrial integration remain critical considerations for commercial adoption.

From a functional perspective, the advantages of smart food packaging are substantial and multifaceted. Enhanced food safety through early detection of spoilage and contamination is one of the most prominent benefits. By providing real-time or cumulative information on food quality, smart packaging reduces dependence on static expiration dates and subjective sensory evaluation. Shelf-life extension, improved cold-chain management, and reduction of food waste further strengthen the economic and environmental value of these technologies. Additionally, improved traceability and supply chain transparency support regulatory compliance and rapid response to food safety incidents, while interactive smart labels enhance consumer engagement and trust.

The applications of smart food packaging across diverse food sectors demonstrate its versatility and relevance in addressing product-specific challenges. In highly perishable foods such as meat, fish, dairy products, and fresh produce, smart packaging plays a crucial role in maintaining quality and preventing spoilage. In processed and ready-to-eat foods, intelligent systems support safety assurance and logistics efficiency. The growing use of smart packaging in functional foods, nutraceuticals, and high-value products further underscores its importance in ensuring product integrity and consumer confidence. These applications also highlight the role of smart packaging in supporting global food security by reducing losses and improving resource utilization.

Despite its numerous advantages, the widespread implementation of smart food packaging is not without limitations. High production costs, technical complexity, limited consumer awareness, and regulatory uncertainties continue to restrict large-scale commercialization. Additionally, the integration of electronic components raises concerns regarding recyclability and environmental impact. Addressing these challenges requires continued research and innovation focused on cost-effective materials, biodegradable sensing components, standardized performance evaluation methods, and harmonized regulatory frameworks. Consumer education and acceptance will also play a crucial role in the successful adoption of smart packaging technologies.

Looking ahead, the future of smart food packaging lies in the convergence of sustainability, digitalization, and advanced material science. The development of eco-friendly smart packaging systems using biodegradable polymers and natural sensing compounds aligns with global sustainability goals and circular economy principles. Integration with digital platforms, the Internet of Things (IoT), and artificial intelligence is expected to further enhance data-driven decision-making, predictive quality assessment, and supply chain optimization. Such advancements will enable smart packaging to evolve from a monitoring tool into an active participant in intelligent food systems.

In conclusion, smart food packaging represents a paradigm shift in food packaging technology, offering innovative solutions to long-standing challenges related to food safety, quality preservation, waste reduction, and supply chain efficiency. By combining intelligent functionalities with sustainable design, smart packaging has the potential to significantly improve the resilience and transparency of modern food systems. Continued interdisciplinary research, technological advancement, and supportive regulatory policies will be essential to fully realize the benefits of smart food packaging and to ensure its successful integration into the global food industry.

References

Ahari, H., & Soufiani, S. P. (2021). Smart and active food packaging: Insights in novel food packaging. *Frontiers in Microbiology, 12*, 657233.

Azeredo, H. M., & Correa, D. S. (2021). Smart choices: Mechanisms of intelligent food packaging. *Current Research in Food Science, 4*, 932–936.

Azeredo, H. M., Otoni, C. G., Corrêa, D. S., Assis, O. B., de Moura, M. R., & Mattoso, L. H. C. (2019). Nanostructured antimicrobials in food packaging—recent advances. *Biotechnology Journal, 14*(12), 1900068.

Bhatlawande, A. R., Ghatge, P. U., Shinde, G. U., Anushree, R. K., & Patil, S. D. (2024). Unlocking the future of smart food packaging: Biosensors, IoT, and nano materials. *Food Science and Biotechnology, 33*(5), 1075–1091.

Chen, S., Brahma, S., Mackay, J., Cao, C., & Aliakbarian, B. (2020). The role of smart packaging system in food supply chain. *Journal of Food Science, 85*(3), 517–525.

Drago, E., Campardelli, R., Pettinato, M., & Perego, P. (2020). Innovations in smart packaging concepts for food: An extensive review. *Food, 9*(11), 1628.

Heo, W., & Lim, S. (2024). A review on gas indicators and sensors for smart food packaging. *Food, 13*(19), 3047.

Jones, P., Clarke-Hill, C., Hillier, D., & Comfort, D. (2005). The benefits, challenges and impacts of radio frequency identification technology (RFID) for retailers in the UK. *Marketing Intelligence & Planning, 23*(4), 395–402.

Kerry, J. P., O'Grady, M. N., & Hogan, S. A. (2006). Past, current and potential utilisation of active and intelligent packaging systems for meat and muscle-based products. *Meat Science, 74*(1), 113–130.

Kuswandi, B., Jayus, Restyana, A., Abdullah, A., Heng, L. Y., & Ahmad, M. (2011a). A novel colorimetric food package label for fish spoilage based on polyaniline film. *Food Control, 22*(3–4), 596–602.

Kuswandi, B., Wicaksono, Y., Jayus, Abdullah, A., Heng, L. Y., & Ahmad, M. (2011b). Smart packaging: Sensors for monitoring of food quality and safety. *Sensing and Instrumentation for Food Quality and Safety, 5*(3), 137–146.

Mills, A. (2005). Oxygen indicators and intelligent inks for packaging food. *Chemical Society Reviews, 34*(12), 1003–1011.

Mkhari, T., Adeyemi, J. O., & Fawole, O. A. (2025). Recent advances in the fabrication of intelligent packaging for food preservation: A review. *Processes, 13*(2), 539.

Motelica, L., Ficai, D., Oprea, O. C., Ficai, A., & Andronescu, E. (2020). Smart food packaging designed by nanotechnological and drug delivery approaches. *Coatings, 10*(9), 806.

Muncke, J. (2014). Hazards of food contact material: Food packaging contaminants. In Y. Motarjemi (Ed.), *Encyclopedia of food safety* (pp. 430–437). Academic Press. https://doi.org/10.1016/b978-0-12-378612-8.00218-3

Realini, C. E., & Marcos, B. (2014). Active and intelligent packaging systems for a modern society. *Meat Science, 98*(3), 404–419.

Regattieri, A., Gamberi, M., & Manzini, R. (2007). Traceability of food products: General framework and experimental evidence. *Journal of Food Engineering, 81*(2), 347–356.

Robertson, G. L. (2019). *Food packaging: Principles and practice* (3rd ed.). CRC Press.

Rodrigues, C., Souza, V. G. L., Coelhoso, I., & Fernando, A. L. (2021). Bio-based sensors for smart food packaging—Current applications and future trends. *Sensors, 21*(6), 2148.
Salgado, P. R., Di Giorgio, L., Musso, Y. S., & Mauri, A. N. (2021). Recent developments in smart food packaging focused on biobased and biodegradable polymers. *Frontiers in Sustainable Food Systems, 5*, 630393.
Sheibani, S., Jafarzadeh, S., Qazanfarzadeh, Z., Wijekoon, M. J. O., Rozalli, N. H. M., & Nafchi, A. M. (2024). Sustainable strategies for using natural extracts in smart food packaging. *International Journal of Biological Macromolecules, 267*, 131537.
Siracusa, V., Rocculi, P., Romani, S., & Dalla Rosa, M. (2008). Biodegradable polymers for food packaging: A review. *Trends in Food Science & Technology, 19*(12), 634–643.
Taoukis, P. S., & Labuza, T. P. (2003). Time–temperature indicators (TTIs). In R. Ahvenainen (Ed.), *Novel food packaging techniques* (pp. 103–126). Woodhead Publishing.
Yam, K. L., Takhistov, P. T., & Miltz, J. (2005). Intelligent packaging: Concepts and applications. *Journal of Food Science, 70*(1), R1–R10.

Chapter 10
Artificial Intelligence in Food Processing

Abbreviations

AGI	Artificial General Intelligence
AI	Artificial Intelligence
ANN	Artificial Neural Network
API	Application Programming Interface
ATP	Adenosine Triphosphate
BD	Big Data
CNN	Convolutional Neural Network
CO_2	Carbon Dioxide
CV	Computer Vision
DBD	Dielectric Barrier Discharge
DL	Deep Learning
DNC	Differentiable Neural Computer
EIP	Emerging Infections Program
FDA	Food and Drug Administration
GIS	Geographic Information System
GRAS	Generally Recognized as Safe
HACCP	Hazard Analysis and Critical Control Points
IoT	Internet of Things
KPI	Key Performance Indicator
MANN	Memory-Augmented Neural Network
MAP	Modified Atmosphere Packaging
ML	Machine Learning
MSPC	Multivariate Statistical Process Control
NIR	Near Infrared
NLP	Natural Language Processing
NN	Neural Network
PID	Proportional–Integral–Derivative

Y. Kumar, S. Dey, *Emerging Technology in Food Processing*,
https://doi.org/10.1007/978-3-032-24505-2_10

QA	Quality Assurance
QC	Quality Control
RAG	Retrieval-Augmented Generation
RF	Random Forest
RFID	Radio Frequency Identification
RL	Reinforcement Learning
RLHF	Reinforcement Learning from Human Feedback
RMS	Root Mean Square
RNN	Recurrent Neural Network
SEM	Scanning Electron Microscopy
SVM	Support Vector Machine
SVR	Support Vector Regression
USE	Universal Sentence Encoder
WSN	Wireless Sensor Network

10.1 Introduction

The global food-processing industry is experiencing a paradigm shift driven by rapid advancements in digital technologies, particularly artificial intelligence (AI) and machine learning (ML). Traditional food-processing systems were largely dependent on conventional automation, fixed control rules, and human expertise to maintain product quality, safety, and efficiency. While these approaches have supported large-scale standardized production, they struggle to cope with the increasing complexity of modern food systems. Variability in raw materials, dynamic consumer preferences, stringent food safety regulations, sustainability concerns, and the demand for real-time decision-making have highlighted the limitations of conventional automation. As a result, AI and ML have emerged as transformative technologies within the framework of Agri-Food 4.0, enabling intelligent, adaptive, and data-driven food-processing operations (Miranda et al., 2019).

Artificial intelligence refers to the capability of machines and computer systems to perform tasks that typically require human intelligence, such as perception, reasoning, learning, and decision-making. Machine learning, a subfield of AI, focuses on developing algorithms that allow systems to learn patterns and relationships from data and improve performance without being explicitly programmed. In food-processing operations, AI and ML systems analyze large volumes of heterogeneous data generated from sensors, imaging systems, laboratory analyses, production logs, and supply chain records to support predictive modeling, optimization, and autonomous control (Ding et al., 2025).

The importance of AI and ML in food processing lies in their ability to manage the complexity and uncertainty inherent in food materials and processes. Unlike manufactured materials, food raw materials are biologically derived and exhibit natural variability in composition, structure, rheology, color, and microbial load.

These variations significantly influence processing behavior and final product quality. AI-driven models can capture nonlinear and multivariate relationships between raw material properties, processing parameters, and quality attributes, enabling more accurate predictions and robust process control than traditional empirical or rule-based methods (Zhang et al., 2025). Consequently, AI enhances consistency, efficiency, and reliability in food-processing operations.

AI and ML play a critical role in food quality assurance and safety management. Applications such as computer vision-based inspection systems enable real-time, non-destructive evaluation of food products for defects, contamination, and grading. Predictive models assist in hazard detection, microbial risk assessment, and compliance monitoring, reducing reliance on manual inspection and minimizing human error. Furthermore, AI-based shelf-life prediction models integrate environmental, chemical, and microbiological data to estimate product freshness and spoilage more accurately than conventional methods, thereby reducing food waste and improving consumer safety (Nayak et al., 2020).

The scope of AI and ML in food-processing operations extends across the entire food production lifecycle. At the raw material stage, AI supports quality assessment, sorting, and grading of agricultural commodities. During processing, ML algorithms optimize unit operations such as thermal processing, drying, extrusion, fermentation, and mixing by continuously adjusting process parameters based on real-time data. In packaging, AI enables intelligent packaging systems equipped with sensors to monitor product conditions and predict failures or spoilage. Integration with the Internet of Things (IoT) allows continuous data acquisition, while blockchain enhances traceability and transparency across the supply chain (Yan et al., 2025).

Beyond industrial processing, AI-driven systems are increasingly applied in sustainability and waste reduction, supporting energy-efficient operations, predictive maintenance, by-product valorization, and circular economy approaches. Emerging applications such as machine-learning–oriented nanotechnology, personalized nutrition and recommendation systems, and smart kitchen appliances demonstrate how AI is reshaping not only industrial food processing but also consumer-level food management. Collectively, these developments position AI and ML as essential tools for building resilient, sustainable, and intelligent food-processing systems capable of meeting future global food demands (Liakos et al., 2025).

This chapter of AI and ML represent a fundamental shift from static, deterministic automation to learning-based, adaptive, and intelligent food-processing operations. Their integration with sensors, robotics, IoT, and data analytics forms the backbone of next-generation food manufacturing systems, ensuring improved quality, safety, efficiency, and sustainability.

10.2 Principle and Mechanism of Artificial Intelligence (AI)

Artificial intelligence (AI) is a multidisciplinary field of computer science that aims to create systems capable of performing tasks that typically require human intelligence, such as learning, reasoning, perception, decision-making, and problem-solving. The concept of AI is often debated because it attempts to replicate or simulate aspects of human brain architecture and cognition, which are still not fully understood (Dodiya & Shah, 2021). As a result, AI has historically been perceived as complex and even speculative, resembling science fiction to non-experts (Fig. 10.1).

In practice, AI systems are implemented through computer programs supported by appropriate hardware. Programming languages such as LISP (List and Symbol Processing) and PROLOG (Logic Programming) were among the earliest tools developed to support symbolic reasoning and logical inference in AI systems (Abiodun et al., 2019). Modern AI, however, has expanded far beyond symbolic approaches and now integrates statistical learning, neural computation, and data-driven models.

The fundamental principle of AI lies in enabling machines to mimic human-like cognitive capabilities. These capabilities include reasoning, learning from experience, recognizing patterns, adapting to new situations, and making decisions under uncertainty. AI mechanisms become particularly valuable as the volume, complexity, and repetition of data increase, making manual human analysis impractical.

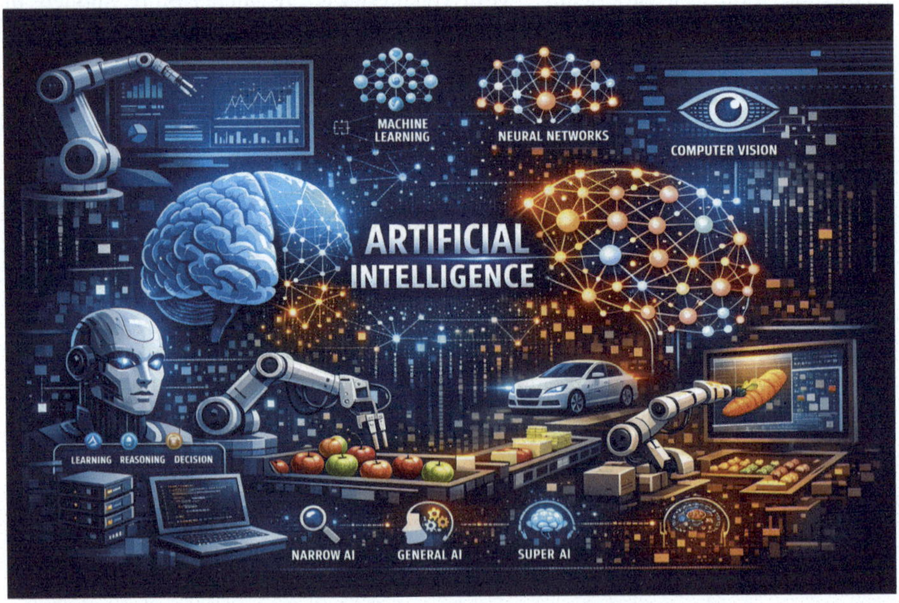

Fig. 10.1 Fundamentals of artificial intelligence

Thus, AI systems are designed to process large datasets efficiently while continuously improving their performance through learning mechanisms.

AI studies are broadly divided into two major categories: classical (symbolic) artificial intelligence and artificial neural networks (ANNs). Classical AI focuses on rule-based reasoning and logic, whereas ANNs are inspired by biological neural systems and emphasize learning from data (Abiodun et al., 2019). Together, these paradigms form the foundation upon which modern AI systems operate (Table 10.1).

10.2.1 Component of AI Systems

An AI system consists of several interconnected components that work together to simulate intelligent behavior. These components enable the system to perceive its environment, process information, learn from data, and take appropriate actions.

1. *Data and Knowledge Base*

   Data is the foundation of any AI system. It may include structured data (numerical values, sensor readings) or unstructured data (images, text, audio). In classical AI systems, knowledge bases store facts, rules, and relationships that allow logical reasoning and inference.
2. *Learning Mechanism*

   Learning is a core component of AI and refers to the system's ability to improve its performance over time based on experience. Artificial neural networks learn by adjusting internal parameters (weights) in response to input data and feedback. This learning mechanism enables pattern recognition and prediction.
3. *Reasoning and Inference Engine*

   Reasoning allows AI systems to draw conclusions from available information. In symbolic AI, inference engines apply logical rules to stored knowledge, while in neural systems, reasoning emerges from learned representations. Practical reasoning becomes increasingly important as data complexity increases.
4. *Decision-Making Module*

   AI systems incorporate decision-making components that select optimal actions based on inputs, objectives, and constraints. This is especially critical in autonomous systems where human supervision is limited.
5. *Hardware and Software Infrastructure*

   AI systems rely on appropriate computing hardware (processors, sensors, memory) and software frameworks to execute algorithms efficiently. Advances in computing power have significantly accelerated AI development.

Table 10.1 Role of AI, machine learning, deep learning, and computer vision concepts in food processing

Section	Description	Applications in food processing	References
Principle and mechanism of AI	AI enables machines to mimic human cognitive abilities such as learning, reasoning, perception, and decision-making. AI systems integrate symbolic reasoning and data-driven learning approaches to handle complex and large-scale data efficiently.	Intelligent decision-making, adaptive control systems, automation beyond fixed rules	Dodiya and Shah (2021), Abiodun et al. (2019)
Components of AI systems	Core components include data/knowledge base, learning mechanisms, reasoning and inference engine, decision-making module, and supporting hardware/software infrastructure.	Autonomous process control, real-time monitoring, intelligent manufacturing	Abiodun et al. (2019)
Types of artificial intelligence	Classified into narrow AI (task-specific), general AI (human-level intelligence), and super AI (beyond human intelligence). Current applications are primarily narrow AI.	Industrial automation, inspection systems, predictive analytics	Hintze (2016), Dodiya and Shah (2021), and Joshi (2025)
AI vs. conventional automation	AI systems learn, adapt, and make context-aware decisions, while conventional automation follows fixed, rule-based instructions without learning capability.	Adaptive processing, intelligent robotics, smart manufacturing	Frankó et al. (2020) and Joshi (2025)
Machine learning in food Pprocessing	ML enables systems to learn from processing data and model nonlinear, multidimensional relationships between inputs and outputs without a full mechanistic understanding.	Process optimization, quality prediction, automated control	Khan et al. (2022) and Menichetti et al. (2023)
Supervised learning	Uses labeled datasets to train models for regression and classification tasks. Performance depends on data quality and representative samples.	Quality grading, defect detection, shelf-life prediction	Khan et al. (2022)

(continued)

Table 10.1 (continued)

Section	Description	Applications in food processing	References
Unsupervised learning	Identifies hidden patterns and structures in unlabeled data. Useful for exploratory analysis and anomaly detection.	Raw material clustering, process deviation detection	Sharma et al. (2026)
Reinforcement learning	Learns optimal actions through interaction with the environment using rewards and penalties.	Dynamic process control, adaptive parameter optimization	Wu et al. (2024)
Artificial neural networks (ANNs)	Inspired by biological neurons, capable of modeling complex nonlinear relationships using layered structures and backpropagation learning.	Quality prediction, process modeling, shelf-life estimation	Nayak et al. (2020), Khan et al. (2022), Gbashi and Njobeh (2024)
Convolutional neural networks (CNNs)	Specialized deep learning models for image data using convolutional layers to extract spatial features hierarchically.	Surface defect detection, contamination detection, packaging inspection	Ding et al. (2023), Ding et al. (2025), Gbashi and Njobeh (2024)
Recurrent neural networks (RNNs)	Designed for sequential and time-series data with memory mechanisms. Advanced forms include LSTM and GRU.	Spoilage prediction, microbial risk forecasting, cold-chain monitoring	Ding et al. (2023, 2025)
Computer vision principles	Combines imaging systems with AI algorithms to extract visual features such as color, shape, texture, and size for decision-making.	Automated inspection, grading, quality control	Guo et al. (2020) and Liakos et al. (2025)
Image acquisition and processing	Uses CCD/CMOS cameras, RGB, and hyperspectral imaging; preprocessing improves image quality and feature extraction.	Non-destructive quality assessment, contamination detection	Guo et al. (2020) and Liakos et al. (2025)
Quality inspection and grading	Vision systems classify food products based on visual quality attributes using deep learning models.	Fruit grading, packaging inspection, labeling compliance	Liakos et al. (2025) and Rojas Santelices et al. (2025)
Defect detection and sorting systems	CNN-based systems detect defects and integrate with automated actuators for high-speed sorting.	Removal of defective products, waste reduction	Liakos et al. (2025), Guo et al. (2020), and Hemamalini et al. (2022)

10.2.2 Types of Artificial Intelligence

Artificial intelligence can be classified based on its capability and level of intelligence. The most widely accepted classification includes narrow AI, general AI, and super AI.

1. *Narrow AI (Weak AI)*

   Narrow AI refers to AI systems designed to perform specific tasks within a limited domain. These systems operate under predefined constraints and cannot generalize their intelligence beyond their programmed function. Examples include internet search engines, facial recognition systems, voice assistants, autonomous driving modules, and industrial automation tools. Most AI systems currently in use belong to this category. Narrow AI excels at executing repetitive and data-intensive tasks with high accuracy and efficiency. However, it lacks consciousness, self-awareness, and general reasoning ability (Dodiya & Shah, 2021). Arend Hintze classified many Narrow AI systems into Type 1 (Reactive Machines) and Type 2 (Limited Memory Systems). Reactive machines, such as IBM's Deep Blue chess system, can analyze current situations but cannot use past experiences. Limited memory systems, such as autonomous vehicles, use historical data temporarily to make decisions (Hintze, 2016).
2. *General AI (Strong AI)*

   Artificial general intelligence (AGI) refers to AI systems capable of understanding, learning, and applying knowledge across a wide range of tasks at a level comparable to human intelligence (Joshi, 2025). Unlike narrow AI, AGI would demonstrate flexible, general-purpose intelligence, allowing it to transfer knowledge between domains and adapt to unfamiliar situations. AGI is often considered the "holy grail" of AI research, yet it remains theoretical. Significant debate exists regarding its precise definition and the path to achieving it (Joshi, 2025). Current AI systems lack true reasoning, consciousness, and common-sense understanding, which are essential for AGI. According to Hintze's classification, Type 3 (Theory of Mind) AI would represent a step toward AGI. Such systems would understand human intentions, emotions, and beliefs.
3. *Super AI*

   Super AI refers to a hypothetical form of intelligence that surpasses human cognitive abilities in all domains, including creativity, decision-making, emotional intelligence, and problem-solving. These systems would possess self-awareness and consciousness, corresponding to Type 4 (Self-Aware AI) in Hintze's classification (Hintze, 2016). Super AI remains purely speculative and raises ethical, philosophical, and safety concerns. Issues such as loss of human control, goal misalignment, and unintended consequences dominate discussions around this concept. Currently, no real-world implementation of Super AI exists.

10.2.3 AI Versus Conventional Automation

AI is frequently confused with conventional automation and robotics, but the two concepts differ significantly in principle and functionality. Automation refers to the use of machines or systems to perform repetitive, predefined tasks using fixed rules and programmed instructions. Traditional automation systems lack adaptability and cannot learn from new data. Robotics, for example, focuses on building machines that can perform physical tasks independently using sensors and programmed logic. These machines may appear intelligent, but operate strictly within their predefined rules (Frankó et al., 2020). In contrast, AI emphasizes intelligence rather than repetition. AI systems can learn from experience, adapt to changing environments, and make decisions under uncertainty. While automation follows "if-then" rules, AI systems analyze patterns and improve performance over time. Artificial intelligence (AI) differs fundamentally from conventional automation in several significant aspects. AI systems possess the ability to learn from data and adapt their behavior over time, whereas conventional automation systems operate strictly according to preprogrammed rules and do not improve through experience. In terms of decision-making, AI can analyze contextual information and make informed, situation-specific decisions, while automation systems simply execute fixed instructions without understanding the surrounding context. AI systems also demonstrate greater flexibility, as they can generalize their capabilities within a domain and handle variations in input, whereas automated systems are typically designed for specific, repetitive tasks. Furthermore, AI exhibits cognitive capabilities such as reasoning, perception, and pattern recognition that resemble human intelligence, which are absent in traditional automation. Consequently, while conventional automation primarily enhances efficiency and consistency in repetitive operations, AI enables intelligent behavior and autonomy. Modern industrial systems increasingly combine AI with automation to improve overall performance, safety, and adaptability in complex and dynamic environments (Joshi, 2025).

10.3 Machine Learning in Food Processing

The rapid advancement of artificial intelligence has significantly influenced modern food manufacturing systems, particularly through machine learning (ML). Within the Agri-Food 4.0 paradigm, machine learning plays a central role in enabling intelligent, adaptive, and automated food-processing systems that can handle the inherent variability of raw food materials and evolving consumer demands (Khan et al., 2022). Unlike conventional automation, which relies on predefined rules and fixed control strategies, machine learning allows systems to learn from data, identify patterns, and make data-driven decisions in complex and dynamic processing environments.

Food processing operations involve multiple interacting variables, such as raw material composition, processing parameters, and environmental conditions, all of which influence final product quality. These relationships are often nonlinear, multidimensional, and difficult to model using traditional mechanistic approaches. Machine learning offers a powerful alternative by extracting meaningful insights directly from processing data and linking inputs to outputs without requiring complete mechanistic understanding (Menichetti et al., 2023). Consequently, ML has emerged as a critical enabling technology for automated food processing, process optimization, and quality control.

10.3.1 Definition and Principles of Machine Learning

Machine learning is a subfield of artificial intelligence that focuses on developing computational models capable of learning from data and improving their performance over time without explicit programming. In food processing, machine learning systems analyze data collected from sensors, images, spectra, and historical production records to support decision-making, prediction, and control (Jiménez-Carvelo et al., 2022).

The fundamental principle of machine learning is pattern recognition. ML algorithms identify relationships between input variables (e.g., temperature, moisture content, ingredient composition) and output variables (e.g., texture, color, shelf life, or microbial stability). Once trained, these models can predict outcomes for new data and assist in real-time process control (Zhang et al., 2025).

A typical machine learning workflow in food processing includes four important steps:

(i) Data acquisition: Collection of relevant data through inline sensors, laboratory analyses, images, or spectra.
(ii) Feature selection or engineering: Extraction of meaningful variables from raw data, such as texture descriptors or spectral features.
(iii) Model training: Application of learning algorithms to establish relationships between inputs and outputs.
(iv) Model validation and testing: Evaluation of model performance using independent data to ensure robustness and reliability (Zhang et al., 2025).

Machine learning builds upon earlier statistical approaches such as multivariate statistical process control (MSPC), which has been used for monitoring food-processing operations such as pasteurization (Gao et al., 2025). However, unlike MSPC, machine learning is better suited to handling high-dimensional data and variability arising from raw materials and consumer customization. This makes ML particularly valuable for food processes where mechanistic models are incomplete or impractical.

10.3.2 Types of Machine Learning

Machine learning algorithms can be broadly classified into supervised learning, unsupervised learning, and reinforcement learning, based on how they learn from data. Each type has distinct applications in food processing.

1. *Supervised Learning*

   Supervised learning is the most widely used machine learning approach in food technology. In this method, models are trained using labeled datasets, where the desired output is known for each input. The algorithm learns to map inputs to outputs by minimizing prediction errors (Khan et al., 2022). In food processing, supervised learning is commonly applied to tasks such as quality prediction, defect detection, shelf-life estimation, and process optimization. For example, regression models can predict textural properties of extruded products based on processing parameters, while classification models can identify defective food products from images or sensor data (Khan et al., 2022). Supervised learning is particularly effective when historical data are available, and product quality attributes can be quantified. However, its performance depends heavily on data quality and the availability of representative labeled samples (Khan et al., 2022).
2. *Unsupervised Learning*

   Unsupervised learning deals with unlabeled data, where the algorithm seeks to identify inherent structures or patterns without predefined output variables. This approach is useful for exploratory data analysis and anomaly detection in food processing (Sharma et al., 2026). In industrial food manufacturing, unsupervised learning methods such as clustering are used to group similar raw materials, processing conditions, or product quality profiles. These techniques help identify sources of variability and detect abnormal processing behavior (Sharma et al., 2026). Unsupervised learning is particularly valuable in early stages of process development, where labeled data may not yet exist. However, interpretation of results often requires expert knowledge (Sharma et al., 2026).
3. *Reinforcement Learning*

   Reinforcement learning (RL) involves learning through interaction with an environment. The algorithm learns by trial and error, receiving feedback in the form of rewards or penalties based on its actions. Over time, the system learns optimal control strategies (Wu et al., 2024). In food processing, reinforcement learning is well suited for dynamic process control, where real-time decisions are required under changing conditions. For example, RL can be used to optimize temperature, pressure, or flow rate adjustments in response to sensor feedback during processing operations (Wu et al., 2024). Although reinforcement learning holds significant potential for adaptive food-processing systems, its industrial application is still emerging due to challenges related to data requirements, system safety, and computational complexity (Wu et al., 2024).

10.3.3 Common Machine Learning Algorithms Used in Food Technology

A wide range of machine learning algorithms have been applied in food processing and quality assessment. The choice of algorithm depends on the nature of the data, processing objectives, and system complexity.

1. *Artificial Neural Networks (ANNs)*

   Artificial neural networks are inspired by biological neural systems and are widely used in food technology for modeling complex, nonlinear relationships. ANNs have been applied to predict food quality attributes, optimize processing parameters, and model operations such as extrusion and drying (Nayak et al., 2020; Khan et al., 2022). Simple ANNs, such as single-layer perceptrons, are a form of machine learning algorithm.

   ANNs are particularly effective when mechanistic models are difficult to establish, as they can capture empirical relationships directly from data. However, they are often considered "black-box" models with limited interpretability (Khan et al., 2022).
2. *Support Vector Machines (SVMs)*

   Support vector machines are supervised learning algorithms used for classification and regression tasks. In food processing, SVMs have been applied to classify food quality grades, detect adulteration, and analyze spectral data (Zhu & Spachos, 2021). SVMs perform well with high-dimensional data and limited sample sizes.
3. *Decision Trees and Random Forests*

   Decision trees use hierarchical rules to make predictions and are easy to interpret. Random forests, which combine multiple decision trees, improve prediction accuracy and robustness. These algorithms have been used for quality classification, fault detection, and process monitoring in food manufacturing (Vidyarthi et al., 2020).
4. *Deep Learning Models*

   Deep learning, a subset of machine learning, uses multilayer neural networks to process complex data such as images and spectra. Deep learning models are particularly effective for computer vision applications, including detection of fruit damage, color changes, and surface defects during processing and storage (Nayak et al., 2020).
5. *Hybrid and Gray-Box Models*

   Hybrid or gray-box models combine mechanistic knowledge with machine learning. This approach leverages first-principle understanding to constrain data-driven models, reducing data requirements and improving generalization (Zhang et al., 2019). Gray-box models are especially useful in complex food processes such as extrusion, where partial mechanistic understanding exists but full modeling is impractical.

10.4 Deep Learning and Neural Networks

Deep learning is a subfield of artificial intelligence (AI) that focuses on learning hierarchical representations of data through multilayered neural network architectures. Unlike traditional machine learning methods that rely heavily on manual feature engineering, deep learning models automatically extract relevant features from raw data, enabling superior performance in complex tasks such as image recognition, time-series prediction, and pattern discovery. In recent years, deep learning has gained remarkable attention in food processing and food safety due to its ability to handle large-scale, nonlinear, and multimodal data generated throughout the food supply chain (Ding et al., 2023).

Artificial neural networks (ANNs), convolutional neural networks (CNNs), and recurrent neural networks (RNNs) form the foundation of modern deep learning systems. Each architecture is designed to address specific data characteristics, such as static tabular data, spatial image data, or sequential time-dependent data. Their integration into food processing has enabled rapid quality inspection, contamination detection, process optimization, and predictive risk assessment, surpassing the limitations of traditional analytical methods (Ding et al., 2025).

10.4.1 Basics of Artificial Neural Networks (ANNs)

Artificial neural networks (ANNs) are computational models inspired by the biological neural structure of the human brain. An ANN consists of interconnected processing units called neurons, organized into three primary layers: the input layer, one or more hidden layers, and the output layer. Each neuron computes a weighted sum of inputs, applies a nonlinear activation function, and transmits the result to the next layer (Gbashi & Njobeh, 2024).

The learning process of ANNs typically relies on backpropagation, an optimization algorithm that adjusts connection weights by minimizing the prediction error using gradient descent. This mechanism enables the network to iteratively learn complex nonlinear mappings between inputs and outputs (Gbashi & Njobeh, 2024).

In food processing, early ANN models were applied to tasks such as quality grading, sensory evaluation, and process parameter prediction. However, conventional fully connected ANNs suffer from several limitations. They require extensive manual feature engineering, struggle with high-dimensional data, and exhibit reduced performance when handling spatial or temporal dependencies. These drawbacks have led to the evolution of specialized deep learning architectures such as CNNs and RNNs, which overcome many of these challenges (Ding et al., 2023).

Despite their limitations, ANNs remain important as foundational models and are still used in combination with deep architectures for regression and classification problems in food engineering, such as predicting moisture content, shelf life, and physicochemical properties of food products.

10.4.2 Convolutional Neural Networks (CNNs)

Convolutional neural networks (CNNs) are deep learning models specifically designed to process grid-structured data, particularly images. CNNs introduce convolutional layers that apply learnable filters to extract local spatial features such as edges, textures, and shapes. These local features are then hierarchically combined through deeper layers to form high-level representations (Ding et al., 2025).

A typical CNN architecture includes convolutional layers, activation functions (e.g., ReLU), pooling layers for dimensionality reduction, fully connected layers, and an output layer for classification or regression tasks. This structure significantly reduces the number of trainable parameters compared to fully connected networks, improving computational efficiency and generalization performance (Ding et al., 2023).

CNNs have demonstrated exceptional performance in food-processing applications, particularly in image-based food safety inspection. They are widely used for detecting surface defects, contamination, adulteration, and packaging faults. By preserving spatial relationships within images, CNNs outperform traditional machine vision and statistical models in recognizing subtle visual patterns that indicate spoilage or foreign matter (Gbashi & Njobeh, 2024).

Advanced CNN variants such as AlexNet, VGGNet, ResNet, and DenseNet have further enhanced detection accuracy. These architectures enable deeper networks, improved gradient flow, and multi-scale feature extraction, making them suitable for complex food images captured under varying lighting and environmental conditions. The integration of CNNs with hyperspectral imaging (HSI) systems has further expanded their capability by enabling simultaneous analysis of spatial and spectral information, allowing non-destructive detection of chemical residues and internal defects in food products (Ding et al., 2023).

In industrial food-processing lines, CNN-based systems enable real-time inspection, automated sorting, and quality grading, reducing reliance on manual labor and minimizing human error.

10.4.3 Recurrent Neural Networks (RNNs)

Recurrent neural networks (RNNs) are specialized neural networks designed to process sequential and time-dependent data. Unlike feedforward networks, RNNs incorporate feedback connections that allow information to persist across time steps. This memory mechanism enables RNNs to capture temporal dependencies and dynamic patterns within sequences (Ding et al., 2023).

In food processing, time-series data such as temperature, humidity, pressure, microbial growth trends, and logistics conditions play a crucial role in safety and quality control. RNNs are particularly effective in modeling these temporal patterns and predicting future states based on historical observations (Ding et al., 2023).

However, standard RNNs often suffer from vanishing or exploding gradient problems when handling long sequences. To address these limitations, advanced variants such as long short-term memory (LSTM) networks and gated recurrent units (GRU) were developed. LSTM networks introduce memory cells and gating mechanisms (input, forget, and output gates) that regulate information flow, enabling stable learning of long-term dependencies (Ding et al., 2025).

In food-processing systems, RNNs and LSTMs are extensively used for risk prediction and early warning systems. By analyzing historical and real-time data from sensors and monitoring devices, these models can forecast spoilage risks, microbial contamination probability, and cold-chain failures. Compared to traditional statistical methods such as ARIMA or regression models, RNN-based approaches exhibit superior performance in capturing nonlinear relationships and complex interactions among multiple variables (Gbashi & Njobeh, 2024).

10.5 Computer Vision in Food Processing

Computer vision has emerged as a transformative technology in the food-processing industry, enabling rapid, non-destructive, and objective assessment of food quality and safety. By integrating digital imaging systems with artificial intelligence (AI) and machine learning (ML) algorithms, computer vision systems can automatically acquire, process, and interpret visual information from food products. These systems are increasingly replacing traditional manual inspection methods, which are time-consuming, subjective, and prone to human error. In modern food-processing plants, computer vision plays a vital role in quality inspection, grading, defect detection, and automated sorting, contributing significantly to food safety, regulatory compliance, and production efficiency (Liakos et al., 2025).

10.5.1 Principles of Computer Vision

Computer vision is an interdisciplinary field that enables machines to "see" and interpret visual information in a manner similar to human vision. In food processing, the primary objective of computer vision is to extract meaningful information from images of food products and use it for decision-making tasks such as classification, grading, and defect detection. A typical computer vision system consists of image acquisition, preprocessing, feature extraction, and decision-making or classification stages (Guo et al., 2020).

The principle of computer vision relies on the assumption that visual attributes such as color, shape, texture, and size correlate with food quality parameters. For example, surface color and texture can indicate ripeness, freshness, or spoilage, while shape and size are used for grading fruits and vegetables. Advanced computer vision systems utilize machine learning and deep learning algorithms, particularly convolutional neural networks (CNNs), to automatically learn these features from data rather than relying on manually designed rules (Liakos et al., 2025).

In food quality control, computer vision systems operate under controlled illumination and imaging conditions to ensure consistency and repeatability. Proper lighting is essential to minimize shadows and reflections, which may interfere with accurate feature extraction. The captured images are then converted into digital signals that can be processed by algorithms to identify patterns associated with acceptable or defective products (Rojas Santelices et al., 2025).

10.5.2 Image Acquisition and Processing Techniques

Image acquisition is the first and most critical step in any computer vision system. In food-processing applications, images are commonly captured using charge-coupled device (CCD) or complementary metal–oxide–semiconductor (CMOS) cameras mounted above conveyor belts. Depending on the application, different imaging modalities may be used, including RGB imaging, grayscale imaging, and hyperspectral imaging (HSI). RGB imaging is widely used due to its low cost and simplicity, whereas hyperspectral imaging provides both spatial and spectral information, enabling deeper analysis of internal composition and contamination (Liakos et al., 2025).

Once images are acquired, they undergo preprocessing to improve quality and remove noise. Preprocessing techniques include image resizing, contrast enhancement, noise filtering, and background subtraction. These steps are essential to ensure that the relevant features of the food product are clearly visible and distinguishable from the background (Guo et al., 2020).

Image processing techniques are then applied to extract meaningful features. Traditional image processing methods focus on color histograms, texture descriptors, and shape analysis. However, modern systems increasingly rely on deep learning-based feature extraction, where CNNs automatically learn hierarchical features ranging from simple edges and textures to complex shapes and defect patterns. Such approaches have demonstrated superior performance in handling variations in lighting, orientation, and product appearance (Liakos et al., 2025).

10.5.3 Quality Inspection and Grading of Food Products

Quality inspection and grading are essential operations in the food industry, as they directly influence consumer acceptance, pricing, and regulatory compliance. Computer vision systems provide a fast, objective, and non-destructive means of evaluating food quality attributes. In fruits and vegetables, grading is commonly based on size, color, shape, and surface appearance. For example, apples and citrus fruits are graded according to color uniformity and the absence of surface blemishes, while grains are assessed based on shape and size distribution (Rojas Santelices et al., 2025).

The use of CNN-based vision systems has significantly improved the accuracy of quality grading. According to Liakos et al. (2025), supervised deep learning models dominate food quality applications, achieving high classification and regression accuracy across various products such as fruits, eggs, confectionery, and packaged foods. These systems can classify products into quality grades in real time, reducing labor costs and ensuring consistent standards (Guo et al., 2020).

In the beverage and processed food industries, computer vision is also used to inspect packaging integrity, label correctness, and fill levels. Automated vision systems can detect mislabeling, improper sealing, or under-filled containers at high speeds, ensuring compliance with safety and labeling regulations. By integrating computer vision with Industry 4.0 concepts, food processors can achieve continuous monitoring and data-driven quality control (Liakos et al., 2025).

10.5.4 Defect Detection and Sorting Systems

Defect detection is one of the most impactful applications of computer vision in food processing. Defects may include bruises, cracks, discoloration, contamination, or foreign objects, all of which pose risks to food safety and consumer satisfaction. Traditional manual inspection methods are often inconsistent and inefficient, especially in high-throughput processing lines. Computer vision-based defect detection systems address these limitations by providing rapid and accurate inspection (Rojas Santelices et al., 2025).

Recent advances in deep learning have enabled highly reliable defect detection systems. CNN-driven models have been successfully applied to detect surface damage in eggs, defects in gummy candies, bruises in fruits, and packaging flaws in processed foods. For instance, Liakos et al. (2025) report that CNN-based inspection systems achieved accuracy levels exceeding 98% in egg surface defect detection, demonstrating their suitability for real-time industrial applications (Guo et al., 2020).

Sorting systems integrate defect detection algorithms with automated actuators, such as air jets or robotic arms, to separate defective products from acceptable ones. These systems operate at high speeds and can handle large volumes of food products with minimal human intervention. In addition, anomaly detection techniques, such as autoencoders and student–teacher networks, are increasingly used when labeled defect data are scarce. These methods learn the normal appearance of products and flag deviations as potential defects (Liakos et al., 2025).

The adoption of computer vision-based sorting systems improves product consistency, reduces waste, and enhances overall processing efficiency. However, challenges remain, including variability in product appearance, lighting conditions, and the detection of subtle or internal defects. Ongoing research focuses on integrating hyperspectral imaging, sensor fusion, and explainable AI to overcome these limitations and improve system transparency and reliability (Hemamalini et al., 2022).

10.6 AI Applications in Food-Processing Operations

Artificial intelligence (AI) has become a transformative technology in modern food-processing operations, enabling intelligent decision-making, automation, and predictive capabilities across the entire food value chain. By integrating machine learning (ML), deep learning (DL), computer vision, big data analytics, and Internet of Things (IoT) technologies, AI systems address critical challenges related to food quality, safety, efficiency, sustainability, and consumer satisfaction. Recent research demonstrates that AI-driven systems outperform traditional rule-based methods by learning complex, nonlinear relationships from large and heterogeneous datasets generated during food production and distribution (Ding et al., 2023) (Table 10.2).

10.6.1 Raw Material Quality Assessment

Raw material quality assessment is a critical step in food processing because the physical, chemical, and biological variability of raw ingredients directly affects process performance and final product quality. Traditional quality inspection relies heavily on manual sorting, laboratory testing, and sensory evaluation, which are time-consuming, subjective, and unsuitable for large-scale industrial operations. AI-based systems overcome these limitations by enabling rapid, non-destructive, and objective quality evaluation using machine learning and computer vision techniques (Ding et al., 2023).

In modern food-processing facilities, AI-driven raw material assessment integrates imaging systems (RGB, hyperspectral, multispectral), spectroscopy (NIR, FTIR), and sensor data to evaluate attributes such as size, shape, color, surface defects, moisture content, firmness, fat content, and chemical composition. Convolutional neural networks (CNNs) are widely used to analyze visual and spectral data, enabling accurate classification and grading of agricultural commodities such as fruits, vegetables, grains, meat, and seafood. These models can identify subtle defects, bruises, contamination, and maturity stages that may not be visible to human inspectors (Ding et al., 2023).

Machine learning models such as support vector machines (SVMs), artificial neural networks (ANNs), and random forests are also used to predict internal quality parameters, including sugar content, protein concentration, moisture level, and microbial load. By correlating sensor signals with laboratory reference data, AI systems provide real-time quality predictions that support intelligent

Table 10.2 AI applications in food-processing operations

AI application area	Functions and contributions in food processing	References
Raw material quality assessment	Non-destructive evaluation of physical, chemical, and biological attributes; grading and classification using computer vision, spectroscopy, CNNs, ANNs, SVMs, and random forests; real-time prediction of internal quality parameters; improved traceability and waste reduction	Ding et al. (2023) and Liakos et al. (2025)
Process optimization and control	Data-driven modeling of nonlinear processing behavior; real-time parameter optimization; reinforcement learning–based adaptive control; soft sensing for difficult-to-measure quality attributes; energy and resource efficiency improvement	Ding et al. (2023) and Liakos et al. (2025)
Predictive maintenance of equipment	Sensor-based monitoring of vibration, temperature, and power consumption; fault diagnosis and remaining useful life prediction using ML, RNNs, and autoencoders; reduced downtime and contamination risk	Ding et al. (2023) and Liakos et al. (2025)
Shelf-life prediction and spoilage detection	Dynamic shelf-life estimation using ML and DL; integration of environmental and microbial data; intelligent packaging with real-time freshness monitoring; improved inventory management and waste reduction	Navina et al. (2026)
Food safety and hazard detection	Early risk prediction using large heterogeneous datasets; AI-supported inspection prioritization; integration of external surveillance data (FoodNet, recalls); detection of biological, chemical, and processing-induced hazards	Tajkarimi (2020)
Food quality assurance and control	Continuous, non-destructive quality monitoring; CNN-based visual inspection; spectroscopic analysis of internal attributes; anomaly detection and digital quality records for compliance and traceability	Agrawal et al. (2025)
Intelligent packaging systems	AI-driven inspection of packaging integrity; prediction of gas composition and shelf life; ANN-based packaging optimization; sensor-integrated and sustainable packaging solutions	Liakos et al. (2025)
Smart sensors and IoT integration	Real-time monitoring of processing parameters; AI-based anomaly detection and autonomous control; predictive maintenance; enhanced traceability across processing and distribution	Ding et al. (2023)
Food supply chain and traceability	AI-powered grading and defect detection; IoT-enabled real-time monitoring; blockchain-integrated data integrity; fraud prevention and rapid recall management	Liakos et al. (2025)
Machine-learning–oriented nanotechnology	ML analysis of nanosensor signals; detection of contaminants and spoilage; SERS–CNN	Yan et al. (2025)

(continued)

Table 10.2 (continued)

AI application area	Functions and contributions in food processing	References
	systems for trace-level hazard detection; intelligent packaging and process monitoring	
Personalized nutrition and recommendation systems	AI-driven dietary recommendations; integration of health, genetic, and behavioral data; computer vision–based food recognition; support for customized food product development	Tsolakidis et al. (2024)
Food waste reduction and sustainability	AI-based process optimization; predictive quality management; demand forecasting; by-product valorization; reduced energy, water, and material waste; circular economy support	Malefors et al. (2022)

decision-making in raw material acceptance and preprocessing. This ensures consistent product quality while minimizing waste and processing losses (Liakos et al., 2025).

AI-based raw material assessment improves traceability and transparency by generating digital quality records that can be integrated into supply chain management systems. These records support regulatory compliance, supplier evaluation, and risk management. Overall, AI enhances efficiency, accuracy, and reliability in raw material quality assessment, forming the foundation for intelligent food-processing operations (Liakos et al., 2025).

10.6.2 Process Optimization and Control

Process optimization and control represent one of the most impactful applications of AI in food processing. Food manufacturing processes such as thermal treatment, fermentation, drying, extrusion, mixing, and baking involve complex, nonlinear interactions between processing parameters and product quality attributes. Conventional control systems, including PID controllers and rule-based automation, struggle to adapt to variability in raw materials and changing operational conditions. AI-based control systems address these challenges by learning process behavior directly from data (Ding et al., 2023).

Machine learning models such as ANNs, decision trees, and ensemble learning methods are trained using historical process data to predict quality outcomes such as texture, moisture content, color, flavor, and nutritional value. These predictive models are integrated into advanced process control frameworks that enable real-time adjustment of operating parameters. Reinforcement learning algorithms further enhance process optimization by learning optimal control strategies through continuous interaction with the process environment (Ding et al., 2023).

AI-driven process control improves production efficiency, reduces energy consumption, and enhances product consistency. For example, in thermal processing

Fig. 10.2 AI in food process optimization and control

operations, AI models can dynamically optimize temperature-time profiles to ensure microbial safety while minimizing nutrient degradation. In fermentation processes, AI systems monitor microbial activity and environmental conditions to achieve consistent product quality despite biological variability (Liakos et al., 2025).

Moreover, AI enables soft sensing, where difficult-to-measure quality parameters are estimated indirectly from easily accessible sensor data. This reduces reliance on expensive analytical equipment and enables continuous monitoring. Research indicates that AI-based process optimization significantly outperforms traditional control methods in terms of robustness, adaptability, and economic performance (Liakos et al., 2025) (Fig. 10.2).

10.6.3 Predictive Maintenance of Food-Processing Equipment

Predictive maintenance is a data-driven approach that uses AI to anticipate equipment failures before they occur. Food-processing equipment operates under harsh conditions involving moisture, temperature fluctuations, mechanical stress, and hygienic constraints. Equipment failures can lead to production downtime, product losses, and food safety risks. Traditional maintenance strategies, such as reactive or scheduled maintenance, are inefficient and costly. AI-based predictive maintenance provides a proactive solution (Liakos et al., 2025).

Sensors installed on food-processing equipment continuously collect data related to vibration, temperature, pressure, noise, and power consumption. Machine learning algorithms analyze these data streams to detect anomalies and degradation patterns indicative of impending failures. Models such as SVMs, recurrent neural networks (RNNs), and autoencoders are commonly used for fault diagnosis and remaining useful life prediction (Liakos et al., 2025).

AI-driven maintenance systems reduce unplanned downtime, extend equipment lifespan, and improve operational safety. In hygienic food-processing environments, predictive maintenance also minimizes contamination risks by ensuring timely servicing of critical components. Studies show that AI-based maintenance strategies lead to significant cost savings and improved reliability compared to conventional approaches (Ding et al., 2023).

10.6.4 Shelf-Life Prediction and Spoilage Detection

Shelf-life prediction and spoilage detection are critical aspects of food-processing operations, as they directly influence food quality, safety, consumer satisfaction, and economic sustainability. Artificial intelligence (AI) and machine learning (ML) technologies have significantly transformed conventional approaches to shelf-life estimation by enabling data-driven, dynamic, and highly accurate predictions. Traditional shelf-life determination methods rely on fixed expiration dates derived from empirical testing under controlled conditions, which often fail to capture real-time variations in storage, transportation, and handling environments. In contrast, AI-based models continuously analyze large volumes of data to provide adaptive shelf-life predictions (Navina et al., 2026).

AI and ML algorithms process complex patterns associated with food degradation by integrating multiple variables such as temperature, humidity, oxygen concentration, light exposure, and microbial growth. Advanced models, including artificial neural networks and deep learning architectures, can learn non-linear relationships between these factors and quality deterioration indicators, resulting in more precise spoilage detection and shelf-life estimation. Intelligent packaging systems equipped with sensors further enhance this capability by monitoring food conditions in real time and transmitting data to predictive models that assess freshness and safety throughout the supply chain (Navina et al., 2026).

The application of AI in shelf-life prediction supports sustainable food processing by reducing overproduction, minimizing postharvest losses, and decreasing food waste. Accurate spoilage detection enables timely interventions, improved inventory management, and optimized distribution strategies. Moreover, AI-driven shelf-life models can be integrated with analytical techniques such as spectroscopy and chromatography to strengthen quality assessment frameworks. Overall, the adoption of AI and ML in shelf-life prediction and spoilage detection represents a major advancement in food-processing operations, promoting efficiency, safety, and sustainability in modern food systems (Navina et al., 2026).

10.6.5 Food Safety and Hazard Detection

Food safety and hazard detection are among the most critical responsibilities of the food-processing industry, as failures can lead to serious public health consequences, regulatory actions, and economic losses. Artificial intelligence (AI) has emerged as a powerful tool to enhance food safety management systems by enabling data-driven risk prediction, early hazard detection, and proactive decision-making. Unlike traditional food safety approaches that rely heavily on periodic inspections and reactive responses, AI-based systems analyze large volumes of heterogeneous data to identify patterns and predict risks before incidents occur (Tajkarimi, 2020).

A notable real-world application of AI in food safety is demonstrated by the Chicago Department of Public Health. The department uses AI algorithms to analyze extensive datasets related to food safety violations, including time since last inspection, duration of operation, nearby sanitation complaints, recent temperature data, crime statistics, and licensing information. By processing these diverse inputs through Python-based AI software, the system classifies restaurants and public establishments into high-risk and low-risk categories. The results support targeted inspections and efficient allocation of regulatory resources, significantly improving food safety outcomes. The system's design and results are publicly available through the City of Chicago's Food Inspection Reporting System and GitHub, highlighting transparency and reproducibility (Tajkarimi, 2020).

Effective AI-based food safety systems depend on the availability and integration of both external and internal data sources. External data include product recalls, foodborne outbreak records, electronic health records, and surveillance data collected by governmental and nongovernmental organizations. A prominent example is the Foodborne Diseases Active Surveillance Network (FoodNet), managed by the Centers for Disease Control and Prevention (CDC) in collaboration with state health departments and laboratories. FoodNet compiles, monitors, and interprets data on foodborne illnesses such as listeriosis, salmonellosis, campylobacteriosis, and infections caused by Shiga toxin–producing *Escherichia coli*. These datasets provide valuable insights into emerging risks and trends in food safety (Tajkarimi, 2020).

Internal food safety data generated within food-processing facilities are equally important. These include sanitation verification records, pest control data, supplier audits, consumer complaints, HACCP and preventive control monitoring, laboratory test results, and environmental pathogen testing. AI models can integrate these datasets to identify correlations between processing conditions and safety outcomes. Furthermore, AI systems are increasingly used to assess chemical and biological hazards, including antimicrobial resistance, zoonotic diseases, chemical contaminants such as pesticides and dioxins, and processing-induced hazards like acrylamide formation (Tajkarimi, 2020).

For successful implementation, three main elements are essential: identification of relevant food safety and quality key performance indicators (KPIs), application of AI-based experimental design and predictive modeling, and conversion of analytical results into actionable decisions. Automated sensors, laboratory analyses, and

Fig. 10.3 AI in food safety and hazard detection

predictive tools such as the Pathogen Modeling Program and ComBase feed data into AI models that support real-time hazard detection and risk mitigation. Ultimately, AI-driven food safety and hazard detection systems provide a holistic, proactive, and efficient approach to protecting public health and ensuring food quality across the processing chain (Tajkarimi, 2020) (Fig. 10.3).

10.6.6 Food Quality Assurance and Control

Food quality assurance and control are fundamental components of food-processing operations, ensuring that products consistently meet defined standards for safety, sensory attributes, nutritional value, and regulatory compliance. With increasing production volumes, product diversity, and consumer expectations, traditional quality control methods based on manual inspection and end-point testing are often insufficient. Artificial intelligence (AI) has therefore emerged as a powerful enabler of intelligent, continuous, and data-driven quality assurance systems in modern food-processing environments (Agrawal et al., 2025).

AI-based food quality assurance systems integrate machine learning algorithms, computer vision, spectroscopy, and sensor technologies to monitor quality attributes throughout the processing lifecycle. Unlike conventional quality control, which is typically reactive and sample-based, AI enables real-time and nondestructive monitoring of products during processing. Machine learning models are trained using historical production data, laboratory analyses, and sensory evaluation results to establish relationships between processing conditions and quality outcomes. These models can then predict deviations from quality standards and trigger corrective actions before defects occur (Agrawal et al., 2025).

Computer vision is one of the most widely adopted AI tools in food quality control. Vision systems combined with convolutional neural networks (CNNs) are used to evaluate appearance-related attributes such as color, size, shape, surface texture, and uniformity. These systems are extensively applied in the inspection of fruits, vegetables, bakery products, meat, and packaged foods, where visual quality strongly influences consumer acceptance. AI-based vision systems outperform human inspectors in speed, consistency, and accuracy, especially in high-throughput production lines (Agrawal et al., 2025).

Spectroscopic techniques such as near-infrared (NIR), hyperspectral imaging, and Raman spectroscopy further enhance AI-driven quality assurance by providing information on internal and chemical properties of food products. Machine learning algorithms analyze spectral data to predict moisture content, fat and protein levels, sugar concentration, and freshness indicators. This allows processors to ensure compliance with nutritional labeling requirements and product specifications without destructive testing (Agrawal et al., 2025).

AI also plays a critical role in statistical quality control and anomaly detection. Advanced algorithms continuously analyze process data to identify abnormal patterns that may indicate quality deterioration, equipment malfunction, or raw material variability. Early detection of such deviations reduces rework, waste, and product recalls. Moreover, AI-supported quality assurance systems generate digital quality records that improve traceability, audit readiness, and regulatory compliance (Agrawal et al., 2025).

Overall, AI-based food quality assurance and control systems represent a shift from reactive inspection to proactive quality management. By enabling continuous monitoring, predictive analytics, and intelligent decision-making, AI improves product consistency, enhances consumer trust, and supports the efficient and sustainable operation of modern food-processing industries (Agrawal et al., 2025).

10.6.7 Intelligent Packaging Systems

Intelligent packaging systems represent a rapidly advancing application of artificial intelligence (AI) in food-processing operations, combining machine learning (ML), deep learning (DL), and sensor technologies to enhance food quality control, safety monitoring, and sustainability. Traditional packaging systems primarily serve as passive barriers for protection and containment, whereas intelligent packaging integrates sensing, data analysis, and predictive capabilities to monitor product condition in real time and support informed decision-making throughout the food supply chain (Liakos et al., 2025).

Machine learning plays a crucial role in predictive quality control (QC) and packaging performance evaluation. Advanced ML models are increasingly being applied to predict packaging failures, optimize package design, and automate inspection processes such as defect detection, seal integrity verification, and label compliance assessment. Deep learning–based computer vision systems and sensor fusion approaches enable high-speed, non-destructive inspection on production lines, ensuring consistent packaging quality under industrial conditions (Liakos et al., 2025).

Research demonstrates the effectiveness of AI-driven modeling in postharvest packaging applications. For example, studies investigating dielectric barrier discharge (DBD) cold plasma combined with modified atmosphere packaging (MAP) have shown that machine learning models such as artificial neural networks (ANNs) can accurately predict in-package gas composition, including carbon dioxide and ethylene levels, which are key indicators of fruit respiration and shelf life. By using physicochemical and mechanical quality parameters as inputs, ANN models have outperformed traditional regression techniques, providing robust predictions that support optimized packaging treatments and extended shelf life. These findings highlight the potential of AI integration in packaging digitalization strategies (Liakos et al., 2025).

AI also contributes to sustainable packaging design through the integration of wireless sensor networks (WSNs) and ANN-based optimization frameworks. Sensor-acquired data related to energy consumption, environmental impact, and packaging performance can be analyzed by AI models to minimize pollution, reduce energy usage, and optimize material selection across the packaging life cycle. Such AI-driven approaches have demonstrated superior performance compared to conventional machine learning and IoT-based methods, emphasizing their value in developing green and low-carbon food packaging solutions (Liakos et al., 2025).

In addition, deep learning–based image recognition systems enable real-time packaged food identification and safety verification. Mobile platforms using advanced object detection architectures allow consumers and regulators to instantly access product information, recall status, and safety data through simple image capture, improving transparency and consumer trust (Liakos et al., 2025).

Overall, intelligent packaging systems powered by AI represent a transformative shift in food-processing operations. By integrating sensors, predictive modeling, and deep learning–based inspection, these systems enhance quality assurance, improve shelf-life management, support sustainability goals, and enable smarter, more responsive food packaging solutions in modern agri-food industries.

10.6.8 Smart Sensors and IoT Integration

Smart sensors and Internet of Things (IoT) integration form the technological backbone of intelligent food-processing systems under the Agri-Food 4.0 framework. The increasing demand for food quality, safety, traceability, and sustainability has driven the adoption of AI-enabled sensor networks capable of real-time data acquisition, analysis, and decision-making. By combining smart sensors with artificial intelligence (AI) and machine learning (ML) algorithms, food-processing operations are transformed from conventional automated systems into adaptive, self-optimizing production environments (Ding et al., 2023).

Smart sensors are deployed throughout food-processing facilities to continuously monitor critical parameters such as temperature, humidity, pressure, pH, gas composition, vibration, and microbial indicators. These sensors generate large volumes of high-frequency data that are transmitted through IoT platforms to centralized or cloud-based systems. AI algorithms analyze these data streams to detect anomalies, predict process deviations, and optimize operational performance. Unlike traditional monitoring systems, AI-driven sensor networks can learn from historical and real-time data, enabling proactive control and rapid response to changing processing conditions (Ding et al., 2023).

In food processing operations, IoT-integrated smart sensors play a vital role in quality control, process optimization, and equipment health monitoring. For example, sensor data related to thermal processing conditions can be analyzed by machine learning models to ensure microbial safety while preserving nutritional and sensory quality. In fermentation and drying processes, AI systems interpret sensor signals to regulate environmental conditions and maintain product consistency despite raw material variability. Similarly, vibration and acoustic sensors combined with AI models support predictive maintenance by identifying early signs of equipment wear or failure (Ding et al., 2023).

Smart sensors and IoT integration also enhance traceability and transparency across the food supply chain. Sensor-equipped storage units and transportation systems monitor environmental conditions during distribution, while AI models predict shelf life and spoilage risks. This information supports informed decision-making, reduces food waste, and improves compliance with food safety regulations. Furthermore, IoT-enabled AI platforms facilitate seamless data sharing among stakeholders, strengthening coordination and accountability (Ding et al., 2023).

Overall, the integration of smart sensors and IoT with AI represents a critical advancement in food-processing operations. By enabling continuous monitoring, intelligent analytics, and autonomous decision-making, these technologies improve efficiency, safety, and sustainability, paving the way for resilient and digitally connected food systems (Fig. 10.4).

10.6.9 Food Supply Chain and Traceability

Food supply chains have become increasingly complex due to globalization, extended distribution networks, and rising consumer demands for transparency, safety, and authenticity. Traditional traceability systems, which often rely on manual documentation and isolated databases, are limited in their ability to provide real-time visibility and rapid response to food safety incidents. Artificial intelligence (AI), in combination with machine learning (ML), Internet of Things (IoT), blockchain, and big data analytics, is transforming food supply chain management by enabling

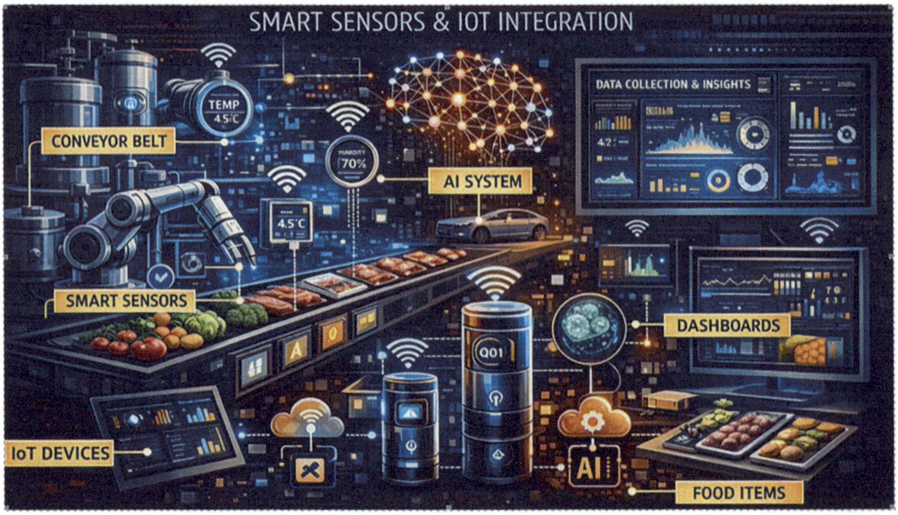

Fig. 10.4 AI in food smart sensors and IoT integration

data-driven traceability, provenance verification, and transparency across all stages from farm to consumer (Liakos et al., 2025).

AI-based traceability systems, often described as part of the "Traceability 4.0" paradigm, integrate multiple digital technologies to enhance food authenticity, safety, and quality. Machine learning models and neural networks are used to analyze large volumes of heterogeneous data collected along the supply chain, including production records, environmental sensor data, logistics information, and quality inspection results. In fruit and vegetable supply chains, AI-powered computer vision systems have been applied for automated grading of products such as blueberries and apples, while ML-based models support quality assessment of kiwifruit, carrots, and other horticultural commodities. Neural network models have also demonstrated high accuracy in defect detection for dried fruits and vegetables, improving consistency and reducing reliance on manual inspection (Liakos et al., 2025).

IoT plays a critical role in enabling real-time traceability by deploying sensor networks, RFID tags, and QR codes to continuously monitor environmental parameters such as temperature, humidity, carbon dioxide concentration, and product location throughout storage and transportation. AI algorithms analyze these data streams to detect deviations from optimal conditions, predict spoilage risks, and support timely interventions. When integrated with blockchain technology, IoT-generated data are recorded in immutable ledgers, ensuring data integrity, transparency, and trust among stakeholders. Such blockchain–AI systems enhance fraud prevention, support regulatory compliance, and enable rapid recall initiation in the event of safety issues (Liakos et al., 2025).

Big data analytics further strengthens AI-driven traceability by enabling predictive modeling for supply chain optimization, waste reduction, and resource efficiency. Studies report significant improvements in yield monitoring, classification accuracy for origin authentication, and enhanced fraud detection when AI is combined with advanced analytical techniques. In addition, natural language processing (NLP)–based AI models have been applied to analyze large bodies of literature and reports related to food waste interventions across global supply chains. These supervised learning models efficiently classify stakeholders and intervention strategies, supporting evidence-based decision-making and policy development (Liakos et al., 2025).

Overall, AI applications in food supply chain traceability emphasize supervised learning and classification tasks, reflecting the structured and categorical nature of traceability decisions such as authenticity verification and compliance assessment. Despite challenges related to implementation cost, interoperability, and technical expertise, AI-driven traceability systems represent a significant advancement toward transparent, resilient, and sustainable food supply chains, aligning with modern regulatory and consumer expectations (Liakos et al., 2025).

10.6.10 Machine-Learning–Oriented Nanotechnology

The rapid evolution of food-processing technologies has created an urgent need for advanced, sensitive, and real-time analytical tools to ensure food quality, safety, and authenticity. Machine-learning (ML)–oriented nanotechnology has emerged as a transformative approach that integrates nanomaterial-based sensing with data-driven intelligence. This convergence enables food-processing systems to handle complex food matrices, high-dimensional datasets, and dynamic production environments more efficiently than conventional analytical methods (Yan et al., 2025).

Nanotechnology contributes to food processing primarily through nanosensors and nanomaterials capable of generating highly sensitive optical, electrochemical, or spectroscopic signals. Optical nanomaterials such as gold and silver nanoparticles, quantum dots, carbon dots, metal nanoclusters, and metal–organic frameworks—are particularly valuable due to their strong signal amplification, rapid response, and high specificity toward target analytes. However, these nanosensors often produce complex and voluminous datasets that are difficult to interpret using traditional statistical techniques. Machine learning overcomes this limitation by enabling automated pattern recognition, feature extraction, and predictive modeling from large datasets (Yan et al., 2025).

In food-processing operations, ML-driven nanotechnology plays a vital role in food quality assessment. Nanomaterial-based sensors interact with food components or contaminants and generate multidimensional optical signals related to freshness, adulteration, nutrient composition, or spoilage. Machine learning algorithms such as artificial neural networks (ANNs), support vector machines (SVMs), random forests (RFs), and deep learning models analyze these signals to classify food quality grades, predict contamination levels, or estimate shelf life with high accuracy. This approach significantly reduces reliance on time-consuming laboratory analyses like GC–MS or HPLC, enabling rapid and on-site decision-making (Yan et al., 2025).

Another important application lies in food safety monitoring during processing. ML-oriented nanotechnology enables the detection of pesticide residues, antibiotics, heavy metals, and pathogenic microorganisms at trace levels. For example, surface-enhanced Raman scattering (SERS) substrates combined with convolutional neural networks (CNNs) allow accurate identification and quantification of contaminants even in complex food matrices. These intelligent systems improve sensitivity, minimize false positives, and support continuous monitoring along processing lines (Yan et al., 2025).

Machine learning also facilitates the rational design and optimization of nanomaterials used in food processing. Traditional nanomaterial development relies heavily on trial-and-error experimentation, which is resource-intensive and inefficient. ML models can predict nanomaterial properties, optimize synthesis parameters, and assess performance based on structural and physicochemical features. This accelerates the development of high-performance nanosensors with improved stability, selectivity, and safety profiles, while reducing experimental costs and environmental impact (Yan et al., 2025).

In addition, ML-oriented nanotechnology supports intelligent packaging and process monitoring. Nanomaterial-based indicators embedded in packaging can respond to changes in temperature, gas composition, or microbial activity. Machine learning algorithms interpret these signals to provide real-time information on food freshness and storage conditions. When integrated with smart devices and IoT platforms, such systems enable end-to-end monitoring of food processing, storage, and distribution (Yan et al., 2025).

Despite its advantages, several challenges remain. High-quality and standardized datasets are essential for reliable machine learning models, yet data collection in food systems is often fragmented and heterogeneous. The long-term safety and environmental impact of nanomaterials also require thorough evaluation before large-scale adoption. Moreover, the interpretability of complex ML models must be improved to enhance trust and regulatory acceptance in food-processing applications (Yan et al., 2025).

The machine-learning–oriented nanotechnology represents a powerful paradigm for modern food-processing operations. By combining the sensitivity of nanomaterials with the analytical intelligence of machine learning, this approach enables rapid, accurate, and real-time quality control, safety assurance, and process optimization. As data infrastructure, sensor design, and regulatory frameworks continue to evolve, ML-driven nanotechnology is expected to play a central role in building intelligent, sustainable, and resilient food-processing systems (Yan et al., 2025).

10.6.11 Personalized Nutrition and Recommendation Systems

Personalized nutrition and recommendation systems represent one of the most transformative applications of artificial intelligence (AI) and machine learning (ML) in modern food processing and food systems. Traditional nutrition guidelines often follow a generalized "one-size-fits-all" approach, which fails to account for individual variability in genetics, metabolism, lifestyle, health status, and food preferences. AI-driven personalized nutrition systems overcome these limitations by leveraging large-scale data, advanced algorithms, and intelligent decision-support frameworks to deliver customized dietary recommendations tailored to individual needs (Tsolakidis et al., 2024).

At the core of personalized nutrition systems are recommendation engines powered by ML models such as artificial neural networks, decision trees, clustering algorithms, and deep learning architectures. These systems integrate heterogeneous data sources, including anthropometric data (age, sex, height, weight), physiological indicators (BMI, blood glucose, cholesterol), dietary habits, food preferences, allergies, and physical activity levels. More advanced platforms also incorporate genetic information, gut microbiota profiles, and real-time sensor data from wearable devices. By analyzing these multidimensional datasets, AI models can identify complex, nonlinear relationships between diet and health outcomes, enabling more precise and adaptive nutrition advice than conventional methods (Tsolakidis et al., 2024).

In food-processing operations, personalized nutrition systems support the development of targeted food products and customized meal plans. AI-driven analytics allow food manufacturers to design formulations that cater to specific population groups, such as individuals with diabetes, cardiovascular diseases, lactose intolerance, or obesity. Recommendation systems can suggest optimized combinations of ingredients, portion sizes, and nutrient compositions that align with individual health goals while maintaining sensory quality and consumer acceptability. This capability is particularly valuable for functional foods, fortified products, and specialized diets, where precise nutrient balancing is essential (Tsolakidis et al., 2024).

A significant advancement in this domain is the integration of computer vision and natural language processing (NLP) into nutrition recommendation platforms. Image-based food recognition systems use convolutional neural networks to identify food items from photographs, estimate portion sizes, and calculate nutrient intake automatically. NLP techniques analyze recipe descriptions, ingredient lists, and user feedback to refine recommendations further. These technologies reduce reliance on manual data entry, improve accuracy, and enhance user engagement, making personalized nutrition solutions more practical for large-scale adoption in food systems (Tsolakidis et al., 2024).

Personalized nutrition systems are increasingly deployed through mobile and web-based applications, serving as an interface between consumers, food processors, and healthcare professionals. Such platforms provide real-time feedback, adaptive meal planning, and behavioral nudges that encourage healthier eating habits. In food-processing operations, aggregated and anonymized data from these platforms can be used to forecast demand for personalized products, optimize production schedules, and reduce waste by aligning output with consumer-specific needs. This data-driven approach enhances both operational efficiency and sustainability within the food industry (Tsolakidis et al., 2024).

Despite their benefits, personalized nutrition and recommendation systems face challenges related to data quality, privacy, interoperability, and model interpretability. High-quality labeled datasets are required to train robust AI models, while ethical considerations surrounding sensitive health and genetic data must be carefully managed. Nevertheless, ongoing advances in explainable AI, secure data-sharing frameworks, and sensor technologies are steadily addressing these limitations (Tsolakidis et al., 2024).

The AI-enabled personalized nutrition and recommendation systems are redefining food-processing operations by bridging the gap between individual dietary needs and large-scale food production. By combining data-driven intelligence with advanced computational models, these systems support healthier diets, innovative food product development, and more responsive, consumer-centric food-processing ecosystems (Tsolakidis et al., 2024).

10.6.12 Food Waste Reduction and Sustainability

Food waste reduction and sustainability have become central challenges in modern food processing due to increasing global food demand, resource constraints, and environmental concerns. A significant proportion of food loss occurs during processing, storage, and distribution because of inefficiencies, poor quality control, inaccurate demand forecasting, and suboptimal decision-making. Artificial intelligence (AI) and machine learning (ML) offer powerful tools to address these issues by enabling predictive, adaptive, and data-driven approaches that enhance sustainability across food-processing operations (Malefors et al., 2022).

One of the major contributions of AI to food waste reduction lies in process optimization. Machine learning models analyze large volumes of production data, including raw material characteristics, processing parameters, and yield outcomes, to identify inefficiencies and sources of loss. By learning complex, nonlinear relationships between inputs and outputs, AI systems can recommend optimal processing conditions that maximize yield while minimizing waste. For example, AI-driven optimization of thermal processing, drying, extrusion, and fermentation has been shown to reduce overprocessing, improve product consistency, and decrease rejection rates. Such intelligent control systems contribute directly to sustainable manufacturing by lowering energy consumption and raw material wastage (Malefors et al., 2022).

AI also plays a crucial role in predictive quality management, which is essential for preventing unnecessary disposal of food products. Computer vision and deep learning models are widely used to assess visual quality attributes such as color, size, shape, and surface defects in real time. Unlike traditional rule-based inspection systems, AI-based vision systems can adapt to natural variability in food materials and distinguish between acceptable variations and true defects. This capability reduces false rejections and ensures that only genuinely unsafe or unacceptable products are removed from the processing line, thereby reducing food waste without compromising quality or safety (Malefors et al., 2022).

Another important application of AI in sustainability is shelf-life prediction and spoilage forecasting. Machine learning models integrate data from sensors measuring temperature, humidity, gas composition, and microbial indicators to predict product degradation more accurately than conventional empirical methods. By providing reliable estimates of remaining shelf life, AI systems enable better inventory management, dynamic pricing, and improved distribution planning. This reduces premature disposal of food products that are still safe for consumption and supports more efficient utilization of resources throughout the supply chain (Malefors et al., 2022).

AI-driven demand forecasting and supply chain optimization further enhance sustainability in food processing. Advanced ML algorithms analyze historical sales data, seasonal trends, consumer behavior, and external factors such as weather or market fluctuations to generate accurate demand predictions. These insights help food processors align production volumes with actual market needs, reducing overproduction and associated waste. When combined with IoT and real-time data streams, AI systems enable responsive and flexible production planning, which is essential for minimizing losses in perishable food systems (Malefors et al., 2022).

In addition, AI supports the valorization of food by-products and waste streams. Machine learning models are increasingly used to identify opportunities for converting processing waste into value-added products, such as functional ingredients, animal feed, bioactive compounds, or bioenergy. By analyzing compositional data and processing feasibility, AI can assist in selecting appropriate technologies and pathways for waste utilization, contributing to circular economy principles in the food industry (Malefors et al., 2022).

From an environmental perspective, AI contributes to sustainability by reducing energy and water usage in food processing operations. Intelligent monitoring and control systems optimize equipment performance, detect anomalies, and support predictive maintenance, thereby preventing resource-intensive breakdowns and inefficiencies. AI-based life cycle assessment tools also help evaluate the environmental impact of different processing strategies, guiding decision-making toward more sustainable options (Malefors et al., 2022).

Overall, AI applications for food waste reduction and sustainability represent a transformative shift from reactive to proactive food-processing management. By integrating machine learning, sensor technologies, and data analytics, food processors can reduce losses, improve resource efficiency, and lower environmental impacts while maintaining product quality and safety. As digitalization advances, AI-driven sustainability strategies are expected to play a critical role in achieving resilient, efficient, and environmentally responsible food-processing systems.

10.7 Conclusion

The application of artificial intelligence (AI) and machine learning (ML) in food processing has marked a significant transition from conventional, rule-based automation to intelligent, data-driven, and adaptive systems. As outlined in this chapter, the integration of AI begins with a foundational understanding of its principles and mechanisms, including the core components of AI systems and the distinction between narrow, general, and super AI. While current industrial applications largely rely on narrow AI, its ability to learn, reason, and adapt has already demonstrated clear advantages over traditional automation, particularly in handling variability, complexity, and uncertainty inherent in food materials and processes.

Machine learning plays a central role in modern food processing by enabling systems to extract meaningful insights from large datasets generated across production lines. Supervised, unsupervised, and reinforcement learning approaches support diverse tasks ranging from quality classification and anomaly detection to real-time process optimization. Common machine learning algorithms such as artificial neural networks, support vector machines, decision trees, and ensemble methods have proven effective for modeling nonlinear relationships between processing conditions and product quality attributes. These capabilities are further enhanced through deep learning and neural network architectures, including convolutional neural networks for image-based inspection and recurrent neural networks for temporal process modeling.

Computer vision has emerged as a powerful tool in food processing, enabling automated quality inspection, grading, defect detection, and sorting of food products. By combining advanced imaging techniques with AI-driven analysis, food processors can achieve non-destructive, high-speed, and consistent evaluation of products. These technologies reduce dependence on manual inspection while improving accuracy and throughput.

The broad spectrum of AI applications in food-processing operations highlights its transformative impact across the value chain. AI-driven systems support raw material quality assessment, process optimization and control, predictive maintenance of equipment, shelf-life prediction, and spoilage detection. Moreover, AI enhances food safety and hazard detection, strengthens food quality assurance systems, enables intelligent packaging, and integrates smart sensors and IoT for real-time monitoring. At the supply chain level, AI improves traceability and transparency, while emerging fields such as machine learning–oriented nanotechnology and personalized nutrition demonstrate the expanding scope of intelligent food systems. Additionally, AI-driven strategies for food waste reduction and sustainability contribute to more efficient use of resources and reduced environmental impact.

Looking toward the future, trends such as digital twins, autonomous food-processing units, AI–IoT–blockchain integration, and explainable AI are expected to further advance the food industry. However, challenges remain, including data availability and quality, model interpretability, high implementation costs, cybersecurity risks, and the need for skilled interdisciplinary professionals. Addressing these challenges through standardization, regulatory alignment, and collaborative research will be essential for widespread adoption. Overall, AI and ML are poised to become foundational technologies in food processing, shaping resilient, safe, and sustainable food systems for the future.

References

Abiodun, O. I., Jantan, A., Omolara, A. E., Dada, K. V., Umar, A. M., Linus, O. U., et al. (2019). Comprehensive review of artificial neural network applications to pattern recognition. *IEEE Access, 7*, 158820–158846.

Agrawal, K., Goktas, P., Holtkemper, M., Beecks, C., & Kumar, N. (2025). AI-driven transformation in food manufacturing: A pathway to sustainable efficiency and quality assurance. *Frontiers in Nutrition, 12*, 1553942.

Ding, H., Tian, J., Yu, W., Wilson, D. I., Young, B. R., Cui, X., et al. (2023). The application of artificial intelligence and big data in the food industry. *Foods, 12*(24), 4511.

Ding, H., Hou, H., Wang, L., Cui, X., Yu, W., & Wilson, D. I. (2025). Application of convolutional neural networks and recurrent neural networks in food safety. *Foods, 14*(2), 247.

Dodiya, M., & Shah, M. (2021). A systematic study on shaping the future of solar prosumage using deep learning. *International Journal of Energy and Water Resources, 5*(4), 477–487.

Frankó, A., Vida, G., & Varga, P. (2020). Reliable identification schemes for asset and production tracking in industry 4.0. *Sensors, 20*(13), 3709.

Gao, J., Wang, S., Erdogdu, F., Marra, F., Sarghini, F., & Chen, L. (2025). Machine learning-driven innovations in food processing: A systematic review of applications, challenges, and future developments. *Trends in Food Science & Technology, 164*, 105204.

Gbashi, S., & Njobeh, P. B. (2024). Enhancing food integrity through artificial intelligence and machine learning: A comprehensive review. *Applied Sciences, 14*(8), 3421.

Guo, Z., Zhang, M., Lee, D. J., & Simons, T. (2020). Smart camera for quality inspection and grading of food products. *Electronics, 9*(3), 505.

Hemamalini, V., Rajarajeswari, S., Nachiyappan, S., Sambath, M., Devi, T., Singh, B. K., & Raghuvanshi, A. (2022). Food quality inspection and grading using efficient image segmentation and machine learning-based system. *Journal of Food Quality, 2022*(1), 5262294.

Hintze, A. (2016). Understanding the four types of artificial intelligence. *Consulté le, 10*(11), 2021.

Jiménez-Carvelo, A. M., Cruz, C. M., Cuadros-Rodríguez, L., & Koidis, A. (2022). Machine learning techniques in food processing. In *Current developments in biotechnology and bioengineering* (pp. 333–351). Elsevier.

Joshi, S. (2025). Comprehensive review of Artificial General Intelligence AGI and Agentic GenAI: Applications in business and finance. Available at SSRN 5250611.

Khan, M. I. H., Sablani, S. S., Nayak, R., & Gu, Y. (2022). Machine learning-based modeling in food processing applications: State of the art. *Comprehensive Reviews in Food Science and Food Safety, 21*(2), 1409–1438.

Liakos, K. G., Athanasiadis, V., Bozinou, E., & Lalas, S. I. (2025). Machine learning for quality control in the food industry: A review. *Foods, 14*(19), 3424.

Malefors, C., Secondi, L., Marchetti, S., & Eriksson, M. (2022). Food waste reduction and economic savings in times of crisis: The potential of machine learning methods to plan guest attendance in Swedish public catering during the Covid-19 pandemic. *Socio-Economic Planning Sciences, 82*, 101041.

Menichetti, G., Ravandi, B., Mozaffarian, D., & Barabási, A. L. (2023). Machine learning prediction of the degree of food processing. *Nature Communications, 14*(1), 2312.

Miranda, J., Ponce, P., Molina, A., & Wright, P. (2019). Sensing, smart and sustainable technologies for Agri-Food 4.0. *Computers in Industry, 108*, 21–36.

Navina, B., Huthaash, K. K., & Samuel, V. R. (2026). Shelf-life prediction through AI and ML to artificial intelligence in food science. In *Artificial intelligence in food science* (pp. 377–388). Academic Press.

Nayak, J., Vakula, K., Dinesh, P., Naik, B., & Pelusi, D. (2020). Intelligent food processing: Journey from artificial neural network to deep learning. *Computer Science Review, 38*, 100297.

Rojas Santelices, I., Cano, S., Moreira, F., & Peña Fritz, Á. (2025). Artificial vision systems for fruit inspection and classification: Systematic literature review. *Sensors, 25*(5), 1524.

Sharma, P., Bisht, A., Agarwal, G., & Rana, M. (2026). Unsupervised learning techniques in food science: Clustering and dimensionality reduction. In *Artificial intelligence in food science* (pp. 81–95). Academic Press.

Tajkarimi, M. (2020). Food safety and quality data management using artificial intelligence. *Food Protection Trends, 40*(6), 464–467.

Tsolakidis, D., Gymnopoulos, L. P., & Dimitropoulos, K. (2024). Artificial intelligence and machine learning technologies for personalized nutrition: A review. *Informatics, 11*(3), 62. MDPI.

Vidyarthi, S. K., Tiwari, R., Singh, S. K., & Xiao, H. W. (2020). Prediction of size and mass of pistachio kernels using random Forest machine learning. *Journal of Food Process Engineering, 43*(9), e13473.

Wu, M., Liu, W., & Zheng, S. (2024). Intelligent food safety: A prediction model based on attention mechanism and reinforcement learning. *Applied Artificial Intelligence, 38*(1), 2379731.

Yan, Y., Li, J., Yu, Y., Ren, Z., Huang, Y., & Li, Z. (2025). Machine learning-driven optical nanomaterials typical applications in food quality assessment: Present obstacles and future horizons. *Food Science and Human Wellness, 15*. https://doi.org/10.26599/FSHW.2025.9250661.

Zhang, X., Zhou, T., Zhang, L., Fung, K. Y., & Ng, K. M. (2019). Food product design: A hybrid machine learning and mechanistic modeling approach. *Industrial & Engineering Chemistry Research, 58*(36), 16743–16752.

Zhang, L., Boom, R. M., & Ma, Y. (2025). Machine learning in automated food processing: A mini review. *Annual Review of Food Science and Technology, 16*(1), 25–37.

Zhu, L., & Spachos, P. (2021). Support vector machine and YOLO for a mobile food grading system. *Internet of Things, 13*, 100359.

Chapter 11
Food Waste Reduction Technology

Abbreviations

ABTS	2,2′-Azinobis-(3-ethylbenzothiazoline-6-sulfonic acid)
AI	Artificial Intelligence
BET	Brunauer–Emmett–Teller (surface area analysis)
BOD	Biological Oxygen Demand
CO	Carbon Monoxide
CO_2	Carbon Dioxide
COD	Chemical Oxygen Demand
DDGS	Dried Distiller's Grains with Solubles
DPPH	2,2-Diphenyl-1-picrylhydrazyl
DSC	Differential Scanning Calorimetry
EAE	Enzyme-Assisted Extraction
ES	Eggshell
ESM	Eggshell Membrane
FAO	Food and Agriculture Organization
FSC	Food Supply Chain
FSCW	Food Supply Chain Waste
FTIR	Fourier Transform Infrared Spectroscopy
FW	Food Waste
GCMS	Gas Chromatography Mass Spectrometry
GHG	Greenhouse Gas
HHP	High Hydrostatic Pressure
HPLC	High-Performance Liquid Chromatography
HTC	Hydrothermal Carbonization
ILE	Ionic Liquid Extraction
IoT	Internet of Things
KV/cm	Kilovolt per centimeter
LC-MS	Liquid Chromatography–Mass Spectrometry

Y. Kumar, S. Dey, *Emerging Technology in Food Processing*,
https://doi.org/10.1007/978-3-032-24505-2_11

MAE	Microwave-Assisted Extraction
MPa	Megapascal
NMR	Nuclear Magnetic Resonance
PEF	Pulsed Electric Field
PLE	Pressurized Liquid Extraction
PUFA	Polyunsaturated Fatty Acids
SDG	Sustainable Development Goals
SEM	Scanning Electron Microscopy
SFE	Supercritical Fluid Extraction
STE	Stirred-Tank Extraction
STE/UP	Stirred-Tank Extraction with Ultrasonication Pretreatment
TCR	Total Carotenoid Recovery
TFC	Total Flavonoid Content
TPC	Total Phenolic Content
TTI	Time Temperature Indicator
UAE	Ultrasonic-Assisted Extraction
UAHE	Ultrasound-Assisted Heating Extraction
UNEP	United Nations Environment Programme
UV-Vis	Ultraviolet–Visible Spectroscopy
VFAs	Volatile Fatty Acids.

11.1 Introduction

Due to the industrial revolution, there was a significant increase in waste output around the sixteenth century as a result of the large-scale migration of people from rural to urban locations. The increase in population that followed this urban mitigation also brought about a notable growth in the quantity and kind of garbage generated within city limits (Amasuomo & Baird, 2016). The word "waste" refers to the discarded results of human activity that still retain components from the original, useful product, but were deemed undesirable. Waste may be categorised using a number of criteria, and it can take on many different forms. Commonly used characteristics in waste classification include physical states, physical attributes, biodegradability, possibility for reuse, source of generation, and degree of environmental damage. Waste may be generically classified into three categories based on their physical characteristics: gaseous, solid, and liquid. While various countries employ different classifications, the following are the most often used ones: commercial waste, food waste, agricultural waste, demolition and construction waste, mining waste, industrial waste, and household/domestic waste (Amasuomo & Baird, 2016).

Food waste (FW) includes leftovers, waste, and by-products produced in homes, restaurants, hotels, catering companies, and other food-related businesses. FW is the residual material that must be disposed of because its economic value is less than the cost of collection and recovery (Kavitha et al., 2020). It encompasses everything, whether edible or not, that may be recycled or disposed of after being abandoned from the Food Supply Chain (FSC). This includes managing food waste through a range of methods, such as drainage disposal, landfilling, burning to produce bioenergy, open dumping, and disposal into aquatic bodies. Food waste can occur at every stage of the food supply chain as a result of inefficiencies or less-than-ideal procedures by different parties involved, as noted by Foresight (2010).

Food losses can happen in a number of ways during the production, processing, packaging, handling, distribution, and marketing phases of the food supply chain. These losses comprise leftovers from meals cooked in commercial places like hotels, restaurants, and cafes that lack efficient social utilization practices, as well as inedible items like leaves, stalks, and skins. Food processing and preparation frequently provide by-products, which are valuable and marketable commodities or services generated from stages other than the original production (Amasuomo & Baird, 2016). Animal organs and body parts are among the food derivatives that add to the food waste stream, even though edible remnants are sometimes diverted for use as animal feed. Food waste arises across the entire food supply chain, from production to consumption, and includes materials from both plant and animal sources. It is broadly categorized by the FAO into groups such as fruits and vegetables, cereals, dairy, meat, poultry, and seafood. Fruits and vegetables account for the highest losses due to perishability, often discarded in both pre- and post-consumer stages. Their waste, rich in fibers, antioxidants, and phytochemicals, can be valorized into dietary supplements, natural preservatives, and bioactive compounds. Cereal by-products like bran and germ are reservoirs of dietary fiber, proteins, and phenolics, repurposed in food, pharmaceutical, and nutraceutical applications. Dairy waste, particularly whey and effluents, is utilized to produce high-protein powders, biofuels, and bioplastics. Poultry waste, including feathers, offal, and eggshells, is processed into animal feed, fertilizers, collagen, and calcium supplements. Similarly, seafood waste comprising bones, skins, and shells is converted into collagen, fish oil, chitosan, and bioactive peptides. These valorization pathways not only reduce environmental burdens but also contribute to sustainable development by recovering valuable nutrients and promoting circular economy practices. Integrating innovative technologies for extraction and processing enhances the utility of these wastes, transforming them into functional ingredients and eco-friendly products across food, agriculture, pharmaceutical, and energy sectors.

The valorization of food waste through advanced extraction technologies presents a sustainable and efficient pathway for recovering high-value bioactive compounds with potential applications in the food, pharmaceutical, and cosmetic industries. Modern extraction techniques such as microwave-assisted extraction (MAE), supercritical fluid extraction (SFE), ultrasonic-assisted extraction (UAE), and pressurized liquid extraction (PLE) have emerged as green and efficient alternatives to conventional methods, offering higher yields, reduced solvent usage, and shorter processing times. Techniques such as enzyme-assisted extraction (EAE), ionic liquid extraction (ILE), and high hydrostatic pressure (HHP) enhance cell wall disruption, facilitating the release of bioactives such as polyphenols, flavonoids, peptides, and dietary fibers. Meanwhile, novel physical techniques such as pulsed electric fields (PEF) and fermentation processes further improve extraction efficiency and bioactivity by promoting biochemical transformations and microbial metabolite generation. Additionally, thermal and thermochemical processes like anaerobic digestion, pyrolysis, gasification, and hydrothermal carbonization (HTC) enable the conversion of food waste into energy carriers and nutrient-rich by-products, adding value to otherwise discarded materials. Together, these technologies support the circular bioeconomy model by transforming food waste into functional ingredients and bioenergy, emphasizing resource recovery, sustainability, and innovation in waste management. Comprehensive characterization of the extracted bioactives ensures their quality, functionality, and potential for industrial application. This chapter discusses the generation, classification, and valorization of food waste from various sources, including fruits and vegetables, cereals, dairy, poultry, and seafood. It highlights the composition and potential of each waste category for the recovery of valuable bioactive compounds, such as dietary fibers, proteins, antioxidants, enzymes, and minerals. The increasing challenge of food waste has led to the development of emerging technologies aimed at improving efficiency and sustainability across the food supply chain. Technologies such as smart packaging and sensors, artificial intelligence and digital monitoring, Internet of Things (IoT) applications, and blockchain-based traceability systems play a crucial role in preventing spoilage, optimizing inventory management, and enhancing transparency. These innovations enable real-time monitoring of food quality, storage conditions, and movement from production to consumption. Collectively, they support informed decision-making, reduce unnecessary losses, and promote sustainable food waste reduction practices (Liegeard & Manning, 2020; da Costa et al., 2022; Ahmadzadeh et al., 2023). This chapter emphasizes sustainable approaches and emerging technologies for transforming food waste into value-added products, supporting environmental sustainability, resource efficiency, and circular economy practices.

11.2 Origin of Food Waste

Food loss during production, undeniable food waste during consumption, and superfluous food waste also occurring during consumption, such as banana peels or fruit cores, are the three categories into which the European Commission (2014) divides food waste (FW) according to its point of occurrence within the food supply chain (FSC). According to Gustavsson et al. (2011), these categories are further divided into several FSC stages, such as cultivation, postharvest management, storage, distribution, and ultimate consumption.

According to Parfitt et al. (2010), inefficiencies in seller and customer behavior during the FSC's marketing and consuming stages are the main cause of food loss. FW is defined by Lipinski et al. (2013) as edible goods that are valuable enough to be consumed by humans but are thrown away either before or after they begin to deteriorate. Food supply chain waste (FSCW) includes human food items with a high organic content that are thrown away mostly during the production and marketing phases, including losses from spoiling or insect damage. Though there are differences across the FSC, FW generation is common, especially at the merchandising and consumer stages. Unbelievably, according to FAO data, 51% of food produced worldwide is wasted before and after consumption, posing a serious environmental and social problem (Parfitt et al., 2010; Gustavsson et al., 2011).

Nearly one-third of the world's food output is wasted, according to the FAO. As a result, this issue has gained more attention and is now a particular sub-goal of Agenda 2030's SDG 12. According to recent UNEP statistics, families produced an astounding 61% of the anticipated 931 million tonnes of firewood that were produced in 2019 (Giordano & Franco, 2021). Regardless of where it comes from, FW is frequently disposed of in landfills or treated chemically or biologically to lessen its negative effects on the environment. Notably, the main contributors to global FW are Brazil, the United States, China, and India (Sinha & Tripathi, 2021). While traditional waste management practices like landfilling and composting are used in developing countries, they are not sustainable due to issues including gas emissions, unpleasant odors, and environmental pollution (Carmona-Cabello et al., 2020).

FW is pervasive across the FSC, starting with agriculture production and being made worse by things like poor quality crops or insignificantly profitable co-products. Businesses are using novel FW reutilization strategies and sustainable practices more often in order to lower costs and improve operational efficiency (Kavitha et al., 2020). Utilizing important biomolecules that are abundantly found in FW, such as proteins, polysaccharides, lipids, and phenolic compounds, these processes turn FW into high-value goods, including flavoring agents, oils, and nutritional meals (Sinha & Tripathi, 2021).

Considering the seriousness of the problem, developing methods for value estimate might spur improvements in FW management and promote more sustainable food supply chains. These developments tackle important issues facing the food industry, such as minimizing greenhouse gas emissions, reducing the amount of trash dumped in landfills, preventing inorganic leachates from contaminating water supplies, and optimizing conventional waste treatment techniques like composting (Sinha & Tripathi, 2021). These efforts are demonstrated by creative closed-loop supply chain models, which illustrate the possibility of thorough FW recycling across the FSC (Kavitha et al., 2020). Food loss may be caused by a variety of factors, including changes in the climate and environment, as well as economic factors like laws and quality requirements (Kader, 2009). The complex character of FW highlights the necessity of comprehensive approaches that cover its causes, care, and management throughout the full FSC.

11.3 Types of Food Waste and Utilization

Food waste includes resources abandoned throughout different phases of production, processing, storage, transportation, and consumption from both plant and animal sources (Zhu et al., 2023). Food waste is classified by the Food and Agriculture Organisation (FAO) according to food groups. The results show that cereals, pulses, and oilseeds make up about 30% of food waste, followed by meat, poultry, and fish (35%), dairy and dairy products (20%), and fruits and vegetables (45–50%), which have the highest rates of food waste (Zhu et al., 2023). Disturbingly, estimates from FAO data indicate that food waste in families worldwide amounts to 570 million tonnes yearly, or 74 kilograms per person annually (FAO, 2023). Economically speaking, food waste in industrialized and rich nations amounts to an astounding $680 billion, exceeding the $310 billion attributable to underdeveloped countries (FAO, 2023). A systematic method that involves grouping food waste and then identifying critical components using different techniques is necessary to realize the potential value addition of food waste (Table 11.1) (Fig. 11.1).

Fruit and vegetable waste constitutes a significant portion of global food loss and is a pressing concern due to its nutritional potential and environmental impact. This waste is broadly categorized into pre-consumer and post-consumer waste, with further classification based on its nature as avoidable, partially avoidable, and unavoidable waste (Wang et al., 2022). Pre-consumer waste includes waste generated during agricultural production, harvesting, post-harvest handling, processing, transportation, and retail. This often arises from overproduction, cosmetic imperfections, improper storage, and spoilage. Aesthetic standards set by retailers frequently lead to rejection of perfectly edible produce. In developing countries, significant losses occur due to inadequate infrastructure and storage systems. Post-consumer waste, on the other hand, is generated at the household or foodservice level due to

Table 11.1 Essential components derived from food waste sources

Food waste sources	Cereal, pulses, and nuts	Milk products	Meat, poultry, eggs, and seafood
Essential components	Vitamin E, Phytates, insoluble dietary fiber, phenols, γ-Oryzanol, tocopherols, carbohydrates, lipid-soluble vitamins, folic acid, Phytosterols, amino acids, oligosaccharides, peptides, activated carbon, proteins, lipids, fatty acids, minerals	Prebiotics, biodiesel, ethanol, whey protein, lactose, galactooligosaccharides, baker's yeast, minerals, ethanol, single cell protein, animal feed, whey protein, lactose	Feather meal, lactic acid, animal feed, blood meal, meat and bone meal, probiotics, chitin, calcium carbonate, protein, astaxanthin, chitinase

Papageorgiou and Skendi (2018); Wanyo et al. (2014); Balandrán-Quintana et al. (2015); Kiran et al. (2014); Ben-Othman et al. (2020); Hua et al. (2010); Lappa et al. (2019); Yaakob et al. (2019); Marques et al. (2016); Ashayerizadeh et al. (2017)

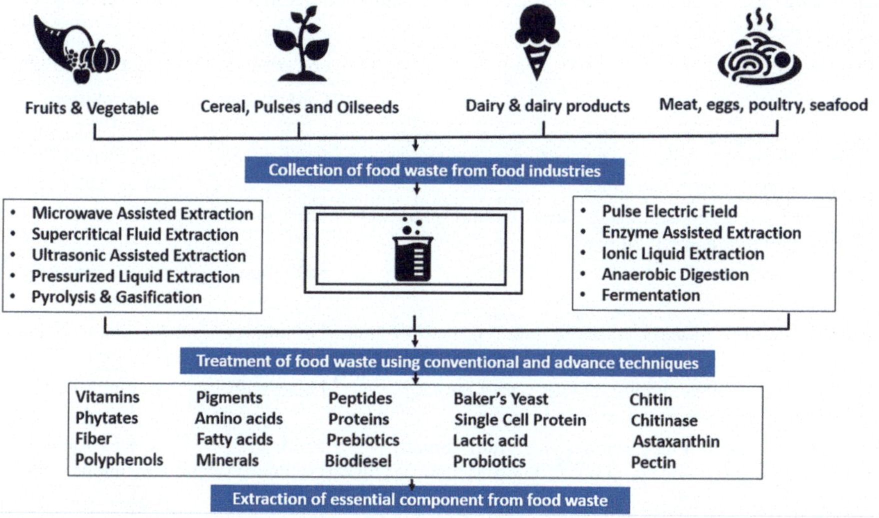

Fig. 11.1 Extraction of essential components from food waste using various techniques

over-purchasing, poor planning, improper storage, or plate waste (Wang et al., 2022).

Fruit and vegetable waste can also be categorized based on its physical form: Peels and skins (e.g., banana peels, citrus rinds, potato skins); Pomace and pulp (from juice and jam industries); Stems, stalks, and cores; Leaves and trimmings; Spoiled or overripe produce. Each of these components holds significant value. For example, peels and skins are rich in dietary fiber, antioxidants, and bioactive compounds such as polyphenols, flavonoids, and essential oils. Pomace, a by-product of juice processing, contains fibers, vitamins, and minerals. Leaves and trimmings, though often discarded, may contain micronutrients and phytochemicals beneficial for human health (Shukla et al., 2024). The valorization of fruit and vegetable waste into value-added products is an emerging solution that supports sustainable food systems and circular economies. One of the most promising areas of utilization is the extraction of natural food additives and supplements. Peels and skins can be used to derive colorants, flavor enhancers, and natural preservatives due to their antimicrobial and antioxidant properties. Dietary fibers extracted from fruit and vegetable waste can be incorporated into bakery products, beverages, and snacks to enhance nutritional quality (Shukla et al., 2024). Additionally, waste components are used in the production of phytochemicals, prebiotics, and functional food ingredients through microbial fermentation. These compounds are increasingly sought after for their roles in improving gut health, boosting immunity, and managing chronic diseases such as diabetes and cardiovascular conditions. For instance, carrot or beetroot pomace is used in bakery products, while apple and citrus peels are incorporated into cereal bars or supplements for their pectin content (Wang et al., 2022). Fruit and vegetable waste is also a viable source of natural enzymes, essential

Table 11.2 Fruits and vegetables waste contains essential components and therapeutic activity

Food waste	Essential component	Beneficial effects
Shells of coffee	Caffeic acid	Antibacterial, antioxidant
Peels of apple, artichoke, date, cauliflower, banana, mango, orange, carrot and onion	Polyphenols	Antimicrobial, antioxidant
Pomace of grape, pomegranate, carrot; tomato seeds and skin; persimmon and lemon peel	Carotenoids	Antioxidant
Broccoli seeds	Sulforaphane	Anticarcinogenic; antibacterial
Peels of Satsuma, orange, lemon, banana; grape seeds and skin	Flavonoids	Antibacterial, antioxidant
Peel of grape, orange, lemon, potato; carrot pomace, tamarind seeds, watermelon rinds	Dietary fiber	Antiflatulence
Peel of citrus, pumpkin; pistachio green hull, tomato wastes, sugar beet	Pectins	Help in bowling movement
Tomato seeds; pistachio pomace	Unsaturated fatty acids	Antioxidant
Peels of pineapple, papaya; kinnow mandarin waste	Amino acids and proteins	Increase bone density
Sapota seeds	Saponins	Antibacterial
Banana stem, apple peels	Glycosides	Anticancer, antioxidant

Zhu et al. (2023); Andrés et al. (2017); Bellur Nagarajaiah and Prakash (2015); Hwang et al. (2011); Xing et al. (2018); Elbadrawy and Sello (2016); Abdolshahi et al. (2015); El-Safy et al. (2012); Kothari and Seshadri (2010); Milner and Romagnolo (2010)

oils, and organic acids for the pharmaceutical and cosmetic industries. Furthermore, the conversion of waste into biodegradable packaging, compost, animal feed, or biofuels supports environmental sustainability by reducing landfill load and greenhouse gas emissions. Finally, the development of nutrient-enriched products targeted at vulnerable populations such as children, the elderly, or individuals in food-insecure regions offers a dual benefit of reducing food waste and improving public health. Recovery systems must, however, be carefully monitored to ensure food safety and quality of re-circulated products (Augustin et al., 2020). Table 11.2 offers in-depth information on fruit and vegetable waste and utilization.

Waste derived from grains includes by-products produced at different technological phases. While bran is produced after the extraction of certain chemicals or during grain polishing during the rice processing process, cleaning techniques produce waste materials like straw and cob (Viganó et al., 2015). Cereal milling and processing industries generate a variety of by-products, many of which are rich in bioactive compounds and functional ingredients. Key milling by-products include bran, a primary residue from dry milling of cereals like wheat, rice, and oats. Bran consists of the aleurone layer and the seed coat and is particularly rich in dietary fibers and phenolic compounds. Germ, another by-product, is separated during milling and serves as a valuable source of oil and micronutrients. Screenings, comprising grains and seeds that do not meet quality standards, are often repurposed for animal feed or further processed for bioactives (Cassano & Galanakis, 2018). From fermentation processes, spent grains emerge as the predominant by-product of barley-based beer production, while Dried Distiller's Grains with Solubles (DDGS) result from maize-based ethanol manufacturing and are increasingly investigated for their functional polysaccharides. Additional valuable derivatives include rice bran oil, extracted after polishing brown rice, and maize germ cake, a protein- and fiber-rich residue following oil extraction (Skendi et al., 2020).

These by-products are reservoirs of bioactive compounds whose recovery is increasingly aligned with green chemistry and sustainability goals. Dietary fibers, both soluble (e.g., β-glucans, arabinoxylans) and insoluble (e.g., cellulose, hemicellulose, lignin), are primarily sourced from wheat bran, maize fiber, and oat hulls. They function as prebiotics and gut health enhancers and are commonly extracted via micronization, microfluidization, enzymatic hydrolysis, or air classification. Phenolic compounds and antioxidants, including ferulic acid, flavonoids, and polyphenols, are abundant in rice bran, maize bran, and spent grains. Their extraction is facilitated by enzymatic treatments (e.g., carbohydrases), solvent systems, or supercritical CO_2, enhancing their bioavailability and efficacy as natural antioxidants or anti-inflammatory agents. Proteins and amino acids, particularly from wheat gluten, rice, and maize fractions, find application in bakery enrichment and bioactive peptide generation (Skendi et al., 2020). Extraction typically involves dry or wet milling, alkaline solubilization, and membrane-based separation such as ultrafiltration. Lipid fractions, notably from rice bran oil and maize germ oil, are rich in unsaturated fatty acids (oleic, linoleic), tocopherols, γ-oryzanol, and phytosterols, and are obtained using cold pressing, solvent extraction, or supercritical CO_2 techniques (Diwan et al., 2018; Sherazi & Mahesar, 2016) (Table 11.3). Furthermore, cereal by-products are a source of functional polymers and gums such as heteroxylans from maize bran and arabinoxylans from DDGS, which are utilized in hydrogel formation, bioactive delivery systems, and food stabilization (Skendi et al., 2020). These are typically extracted via aqueous or alkali processes followed by precipitation or encapsulation methods. Altogether, cereal-based by-products represent a promising, sustainable source of high-value compounds with wide-ranging applications in food, pharmaceutical, and nutraceutical industries (Liu et al., 2023).

The dairy industry is a significant contributor to global food supply but also a major producer of waste, with an estimated 128 million tonnes of milk lost annually due to inadequate handling, microbial spoilage, and inefficient processing. The sector generates a diverse array of wastes broadly categorized into solid, liquid, and gaseous forms, each demanding specific treatment and valorization strategies to mitigate environmental impact and enhance sustainability (Chaudhary et al., 2023). Liquid wastes, primarily in the form of wastewater, constitute the most voluminous category. These effluents are rich in biodegradable organic matter, including lactose, proteins, fats, and salts. If untreated, they pose a serious environmental threat due to their high chemical oxygen demand (COD) and biological oxygen demand (BOD). However, advancements in anaerobic digestion allow for the conversion of this

Table 11.3 Characterization of bioactive using the extraction methods of cereal-based food waste

Cereal type	Type of food waste (by-product)	Extractable bioactives	Extraction methods/ Techniques
Wheat	Bran, germ, grain screenings, gluten	Dietary fibers (arabinoxylans, β-glucans) Proteins Lipids Polyphenols (ferulic acid)	Milling Enzymatic hydrolysis Microfluidization Air classification
Rice	Bran, germ, bran oil, broken Rice	Dietary fibers Antioxidants (-γ-oryzanol, vitamin E) Phenolic acids Essential fatty acids (oleic, linoleic)	Enzymatic treatment (to enhance extractability of phenolics) Supercritical CO_2 extraction (for oil) Micro/nano-encapsulation
Maize (corn)	Bran (Pericarp), germ cake, distiller's grains (DDGS)	Arabinoxylans Ferulic acid Dietary fibers Lipids (from germ) Proteins (zein)	Dry and wet milling Solvent extraction Fermentation Hydrogel formation Micronization
Barley	Spent grains (brewer's grains), bran	Enzymes (lipases, trypsin) carriers Proteins Fibers	Enzyme immobilization Milling and malting Solid-state fermentation
Rye	Bran	Dietary fibers Phenolic compounds	Milling Size reduction (micron technology)
Oats	Bran, hulls, flaking residues	β-glucans Dietary fibers Lipids	Milling Fractionation Particle size reduction
Sorghum/millet/ triticale	Bran, germ, residues	Polyphenols Proteins Dietary fibers	Enzymatic treatment Solvent extraction Micronization
Pseudocereals (quinoa, Amaranth, buckwheat, etc.)	Husk, bran, Saponin-rich outer layers	Saponins Polyphenols Proteins Fibers	Solvent extraction Fermentation Enzymatic hydrolysis

Viganó et al. (2015); Cassano and Galanakis (2018); Skendi et al. (2020); Diwan et al. (2018); Sherazi and Mahesar (2016); Liu et al. (2023)

wastewater into biogas, a renewable energy source that reduces both the organic load and operational costs of waste treatment. Membrane-based separation technologies, such as ultrafiltration and reverse osmosis, have also emerged as effective tools for recovering valuable components from dairy effluents, further aligning with the principles of resource recovery and circular economy (Al-Tayawi et al., 2023).

Solid dairy wastes include sludge, cow dung, leftover curd, and cream residues. These materials are nutrient-rich and can be composted to produce organic fertilizers, thereby supporting agricultural sustainability. A notable by-product in this category is whey, generated during cheese and paneer production. Whey contains high levels of lactose, proteins, and vitamins, making it suitable for diverse applications. It can be transformed into high-protein powders, nutritional beverages, or fermented products, and also serves as an excellent substrate for microbial fermentation to yield biofuels (bioethanol, biomethane), bioplastics (e.g., lactic acid polymers), and single-cell proteins (Kushwaha et al., 2011). Additionally, whey is utilized as animal feed and a microbial growth medium, offering several revenue-generating opportunities. Nonedible dairy wastes such as expired or contaminated milk, scum, whey permeate, and curd clumps are also significant in volume and pollution potential. These wastes can undergo aerobic or anaerobic microbial treatment, leading to the production of environmentally beneficial outputs such as biogas and organic acids. Dairy wastewater, often high in nitrogen and phosphorus compounds, can also be biologically treated to extract nutrients, contributing to both pollution reduction and nutrient recycling (Al-Tayawi et al., 2023). Gaseous emissions, although less discussed, are an important aspect of dairy waste. Emissions from boilers and other processing equipment primarily include CO_2 and other combustion gases, which can be reduced through improved fuel combustion technologies and the installation of air filtration systems (Al-Tayawi et al., 2023). Lastly, packaging waste such as plastic and cardboard poses disposal challenges but offers opportunities for recycling when adequately sorted (Parashar et al., 2016). Implementing integrated waste management systems across the dairy production chain not only reduces environmental burden but also supports economic growth through resource recovery. This holistic approach promotes clean, green, and sustainable industrial practices, making the dairy industry more resilient and aligned with global sustainability goals.

The poultry industry generates substantial amounts of waste that pose significant environmental challenges if not effectively managed. Poultry waste can be broadly classified into four categories: solid waste, liquid waste, gaseous emissions, and blood-derived by-products. Each of these waste streams contains valuable nutrients or bioactive compounds that can be repurposed, aligning with sustainable waste management and circular economy principles. Solid waste includes feathers, manure, litter, offal, and hatchery discards. Poultry manure and litter are rich in nitrogen and phosphorus and have long been used as organic fertilizers or compost materials to enhance soil fertility. Feathers, primarily composed of keratin, can be hydrolyzed into feather meal for animal feed or further processed into bioplastics and amino acid-rich hydrolysates (Cheng et al., 2024) (Table 11.4). Offal and hatchery waste are often rendered into protein meals that serve as components in aquaculture and pet food, demonstrating their nutritional and commercial value. Liquid waste primarily originates from slaughterhouses and poultry processing facilities (Cheng et al., 2024). This wastewater contains high levels of organic matter, including

Table 11.4 Characterization of bioactive using the extraction methods of meat, poultry, egg, and seafood-based food waste

Type of food waste	Extractable bioactives/valuable components	Extraction techniques
Feathers	Keratin, amino acids	Hydrothermal processing, enzymatic hydrolysis, chemical hydrolysis
Manure and litter	Nitrogen, phosphorus	Composting, anaerobic digestion, vermicomposting
Offal and hatchery discards	Proteins, fats, micronutrients	Rendering, enzymatic hydrolysis
Slaughterhouse liquid waste	Proteins, fats, blood residues	Anaerobic digestion, membrane filtration, centrifugation
Blood (Hemoglobin & Plasma)	Bioactive peptides, flavor enhancers, natural colorants	Enzymatic hydrolysis, membrane separation, freeze drying
Gaseous emissions (e.g., ammonia)	Bioenergy (e.g., methane, hydrogen)	Gas capture and storage, biofiltration, chemical scrubbing
Meat co-products (liver, heart)	Proteins, amino acids, vitamins, minerals	Cooking/preprocessing, canning, emulsification
Bones, connective tissue	Collagen, gelatin, calcium	Acid/alkali treatment, enzymatic hydrolysis, heat extraction
Animal fats	Fatty acids, energy-rich lipids	Rendering, filtration, transesterification (for biodiesel)
Eggshell (ES)	Calcium carbonate	Washing, drying, grinding, calcination (for enhanced solubility)
Eggshell membrane (ESM)	Collagen-like proteins, glycosaminoglycans, hyaluronic acid	Separation from shell, enzymatic hydrolysis, solvent extraction, ultrafiltration
Fish heads, skins, bones, offcuts	Proteins, collagen, gelatin, calcium, hydroxyapatite	Enzymatic hydrolysis, acid-base extraction, thermal treatment
Viscera	Fish oil, bioactive peptides, enzymes (e.g., trypsin, pepsin)	Solvent extraction, enzymatic digestion, centrifugation
Shells (shrimp, crab, lobster)	Chitin, chitosan	Deproteinization, demineralization, chemical hydrolysis
Squid skin	Chitosan, collagen, melanin	Enzymatic hydrolysis, chemical extraction
Processing effluent	Proteins, nitrogen, phosphorus, polyunsaturated fatty acids (PUFAs)	Membrane filtration, flocculation, centrifugation, microbial recovery
Blood residues	Hemoglobin, peptides, minerals	Freeze drying, enzymatic hydrolysis, isoelectric precipitation
Skins (fish, squid), scales	Collagen, gelatin, bioactive peptides	Enzymatic hydrolysis, acid solubilization, ultrafiltration
Bones	Hydroxyapatite, calcium, phosphates	Acid/alkali treatment, calcination, thermal decomposition
Fish oil from viscera or heads	Omega-3 fatty acids (EPA, DHA), carotenoids	Wet pressing, supercritical CO_2 extraction, molecular distillation
Shell waste	Chitin, chitosan	Alkali deacetylation, acid treatment

(continued)

Table 11.4 (continued)

Type of food waste	Extractable bioactives/valuable components	Extraction techniques
Any carbon-rich seafood by-products	Carbon-based nanomaterials, biosorbents	Pyrolysis, carbonization, microwave-assisted synthesis

Aditya et al. (2021); Shirsath and Henchion (2021); Adhikari et al. (2018); Marques et al. (2016); Cheng et al. (2024); Prameela et al. (2017); Valcarcel et al. (2017); Akhila et al. (2024)

proteins, fats, and blood residues, making it a significant pollutant if left untreated. However, through anaerobic digestion, this organic load can be converted into biogas, a renewable energy source that reduces reliance on fossil fuels. The treated effluent from these processes can be reused for irrigation or incorporated into nutrient recovery systems, thus reducing the industry's overall water footprint. Gaseous emissions such as ammonia, methane, and hydrogen sulfide are released during the decomposition of manure and other organic waste. These gases contribute to air pollution and odor issues, but can be captured for use in bioenergy production. Additionally, emissions can be minimized through biofiltration and improved manure management practices, further supporting environmental sustainability (Adhikari et al., 2018; Marques et al., 2016). Blood-derived by-products also present opportunities for value addition. Hemoglobin and blood plasma are rich in bioactive peptides and proteins. Enzymatic hydrolysis of these components can yield functional ingredients for food products, such as flavor enhancers and natural colorants. Moreover, these compounds have pharmaceutical potential, serving as antioxidants and inhibitors for various biological targets. These applications underscore the potential of blood waste in creating high-value products that promote health and nutrition (Shirsath & Henchion, 2021). Another major area of poultry waste is meat co-products and offal, often referred to as the "fifth quarter." These by-products include edible parts like liver, heart, kidney, and tongue, which are rich in amino acids, vitamins, and minerals. These items are consumed in traditional diets or incorporated into processed meat products to enhance their nutritional profile. Overcoming negative consumer perceptions through preprocessing and product development can increase their acceptance in mainstream markets. In the pharmaceutical and biomedical fields, co-products like bones, blood, and connective tissue are valuable sources of bioactive compounds. For instance, collagen and gelatin derived from animal tissues are essential in wound dressings, surgical materials, and medical capsules. Additionally, animal fats are processed into industrial products such as biodiesel, soaps, and biolubricants, providing renewable alternatives to petroleum-based chemicals. In animal feed and pet food sectors, low-risk by-products are rendered into meat and bone meals and fat-rich additives, supporting nutrient recycling. Furthermore, rendered products like bone meal and blood meal serve as efficient fertilizers, especially in organic farming, thus closing the nutrient loop and reducing dependence on synthetic inputs. A critical yet often overlooked form of poultry waste is egg waste, primarily composed of eggshell (ES) and eggshell membrane (ESM). These components are typically discarded in large quantities by households and egg processing industries. Eggshells constitute about 10% of an egg's total weight and are composed predominantly of calcium carbonate (~94%). This makes them an excellent and affordable source of bioavailable calcium. Recent studies have demonstrated that eggshell-derived calcium can be used to fortify food products such as yogurt, biscuits, sausages, cakes, and coffee, thus offering a natural and sustainable alternative to synthetic calcium supplements. In agriculture, eggshells can be applied as soil conditioners to enhance soil pH and nutrient availability. Moreover, eggshell powder can be incorporated into animal feed as a calcium supplement and has potential industrial applications, including use as a catalyst in biodiesel production. The eggshell membrane, although less studied, contains collagen-like proteins and other bioactive compounds that offer significant pharmaceutical and cosmetic potential (Aditya et al., 2021). It has been explored for use in wound healing materials and anti-inflammatory supplements. Valorizing eggshell and eggshell membrane reduces the environmental burden associated with landfill disposal and adds value to what is conventionally considered waste (Aditya et al., 2021). This not only promotes a zero-waste approach but also supports the development of functional, cost-effective ingredients for various industrial applications. Valorizing these by-products contributes to reducing environmental pollution, conserves natural resources, and supports circular bioeconomy models. Developing effective processing technologies and increasing consumer and industrial awareness will be key to unlocking the full potential of poultry and egg waste, transforming them from burdensome refuse into valuable assets for food, agriculture, pharmaceuticals, and energy sectors.

Seafood processing generates vast quantities of waste that are often discarded despite being rich in valuable nutrients. These wastes can be broadly categorized into solid waste (such as fish heads, skins, bones, viscera, and shells), liquid waste (effluents containing organic matter, nitrogen, phosphorus, and residual antibiotics), and bioactive-rich by-products (such as collagen, peptides, and lipids). Solid waste forms the bulk of seafood discards and includes biologically potent materials like fish skin, bones, and offcuts, which are excellent sources of proteins, collagen, and calcium (Prameela et al., 2017; Valcarcel et al., 2017). Shellfish and squid by-products, including shells and skin, contain high levels of chitin and chitosan, useful in biotechnology and pharmaceuticals. Liquid waste from processing units, although often overlooked, is rich in proteins, polyunsaturated fatty acids (PUFAs), and carotenoids, which can be recovered through secondary processing. The loss of these nutrients not only contributes to environmental degradation but also represents a significant economic loss. To address this, enzymatic, chemical, and microbial processing techniques are being employed to convert these waste streams into high-value compounds (Valcarcel et al., 2017). These include bioactive peptides with antimicrobial and antihypertensive properties, fish oil for nano-encapsulation in drug delivery, and fishmeal and fish feed additives for the aquaculture industry (Table 11.4). Additionally, the extraction of hydroxyapatite from fish bones supports biomedical applications such as bone repair, while collagen and gelatin from skins are used in cosmetics and pharmaceuticals. Chitosan, derived from shell waste, serves as a biodegradable support for enzyme immobilization in industrial processes.

Emerging uses also include producing carbon-based nanomaterials from seafood waste for advanced technological applications. Overall, valorizing seafood waste not only mitigates pollution but also creates sustainable economic opportunities across food, healthcare, agriculture, and industrial sectors, aligning with global circular economy goals (Akhila et al., 2024).

11.4 Processing of Food Waste Reduction Technology

The food waste contains a number of essential components that can be characterized from the various processing techniques (Table 11.5).

11.4.1 Microwave-Assisted Extraction (MAE)

MAE directly affects polar materials by applying an electric field, magnetic field, and microwave heating. By use of an ion conduction mechanism and dipole rotation, it turns into heat. Pressure and temperature increases cause the solute molecules to split from the sample matrix, allowing the solvent to diffuse and liberate itself. MAE directly affects polar materials by applying an electric field, magnetic field, and microwave heating. By use of an ion conduction mechanism and dipole rotation, it turns into heat. Solvent diffuses and is released when solute molecules are separated from the sample matrix by a rise in temperature and pressure. MAE is used to extract the caffeine and polyphenols from green tea leaves, pectin from beet and orange peel, resveratrol from grape and peanuts waste, and antioxidant from mango peel (Dorta et al., 2013; Fishman et al., 2008; Guo et al., 2012; Pan et al., 2003; Liu et al., 2023). Studies have shown that MAE applied to food wastes like citrus peels and apple pomace can yield pectin of high quality, while preserving its structural integrity and gelling properties. Additionally, MAE reduces degradation of thermolabile compounds and can be fine-tuned for specific temperature and time profiles, enabling selective extraction. Its scalability and compatibility with green solvents make it a promising sustainable alternative in the valorization of agro-industrial waste streams for high-value pectin recovery (Marić et al., 2018).

11.4.2 Supercritical Fluid Extraction (SFE)

SFE makes use of its increased diffusivity and lower viscosity, which allow it to diffuse more readily in solid materials and increase extraction yield. Carbon dioxide, the primary solvent in SFE, has a critical temperature and pressure of 31.1 °C and 7.4 MPa, respectively. SFE technique is used to extract the polyphenol and antioxidants from canola seed, lycopene, β-carotene from tomato peels/seeds, apricot

Table 11.5 Characterization of food waste using extraction techniques

Extraction technique	Bioactive Compounds	Food waste source	Advantages
Microwave-assisted extraction (MAE)	Caffeine, polyphenols, pectin, resveratrol, antioxidants	Green tea leaves, beet peel, orange peel, grape waste, peanut skin, mango peel, citrus peels, apple pomace	Rapid heating, reduced solvent use, preserves thermolabile compounds, selective extraction
Supercritical fluid extraction (SFE)	Lycopene, β-carotene, polyphenols, antioxidants, resveratrol, chlorogenic acid	Tomato peels/seeds, apricot pomace, carrot pulp, grape seeds, broccoli stems, custard apple seeds	Eco-friendly, high-purity extracts, low solvent residue, CO_2-based system with ethanol CO-solvents
Ultrasound-assisted extraction (UAE)	Flavanones, hesperidin, catechins, pectin, carotenoids, anthocyanins, flavonoids	Grape seed/skin, cranberry pomace, carrot waste, citrus peel, spent tea leaves, apple peel, watermelon rind	Enhanced mass transfer, low temperature, minimal solvent use, preserves compound functionality
Pressurized liquid extraction (PLE)	Procyanidins, quercetin, polyphenols, flavonoids, hydroxytyrosol, tyrosol, oleuropein	Red grape pomace, grape skin, tomato waste, olive pomace, seaweed, saffron waste	Fast, automated, minimal solvent, tunable parameters, compatible with green solvents
High hydrostatic pressure (HHP)	Pectin, polyphenols, sulforaphane, proteins, peptides	Pomelo peels, broccoli, tomato peel waste	Non-thermal, improves solubility, high bioactivity retention, and effective microbial inactivation
Pulsed electric field (PEF)	Phenolics, pectin, betalains, carotenoids, starch, α-tocopherol, proteins, polysaccharides	Grape seed, red beet root, apple pomace, potato and sweet potato waste, tomato, carrot, cocoa shells, seafood by-products	Increases permeability, enhances yield, preserves nutritional value, is eco-friendly and scalable process
Enzyme-assisted extraction (EAE)	Oils, pectin, aroma compounds, bioactive peptides	Mango kernel, rice bran, apple pomace, soy meal, banana peel	High specificity, mild conditions, preserves bioactivity, eco-friendly
Ionic liquid extraction (ILE)	Polyphenols, carotenoids, flavonoids, organic acids	Fruit peels, rice bran, berry pomace, cereal bran	Highly selective, low volatility, reusable solvents, operates under mild conditions
Anaerobic digestion	Biogas, volatile fatty acids (VFAs), organic acids	Onion, pineapple, potato peels, mixed food waste	Renewable energy production, waste stabilization, reduction of pathogens
Fermentation	Ethanol, lactic acid, GOS, peptides, dietary fibers	Citrus peel, dairy/meat by-products, okara, banana/potato peels	Produces valuable biochemicals, improves digestibility, valorizes carbohydrates

(continued)

Table 11.5 (continued)

Extraction technique	Bioactive Compounds	Food waste source	Advantages
Pyrolysis	Bio-oil, syngas, biochar	Nut shells (almond, pistachio), dried food waste	Converts waste to fuel, fast process, handles heterogeneous waste
Gasification	Syngas (H_2, CO), biochar	Lignocellulosic food residues	Clean fuel production, high thermal efficiency
Hydrothermal carbonization (HTC)	Hydrochar, water-soluble antioxidants, humic substances	High-moisture waste (fruit peels, citrus waste, olive pomace)	Handles wet waste, produces solid fuel and antioxidants, low emissions

Dorta et al. (2013); Fishman et al. (2008); Guo et al. (2012); Pan et al. (2003); Liu et al. (2023); Marić et al. (2018); Hassas-Roudsari et al. (2009); Sabio et al. (2003); Wijngaard et al. (2012); de Andrade Lima et al. (2019); Afraz et al. (2023); Zhu et al. (2023); Barp et al. (2023); Monrad et al. (2010); Katsinas et al. (2021); Pappas et al. (2021); Perez-Vazquez et al. (2023); Grassino et al. (2020); Guo et al. (2017); Marciniak et al. (2018); Boateng (2022); Tomas et al. (2024); Chatzimitakos et al. (2023); Boulila et al. (2015); Manasa et al. (2013); Zuorro et al. (2011); Luo et al. (2025); Peron et al. (2024); Xing et al. (2023); Ren et al. (2018); Hunter et al. (2021); El Sheikha & Ray (2022); Pontonio et al. (2024); Faria et al. (2023); Niksiar and Nasernejad (2017); Wang et al. (2018); Czerski (2022); Petrović et al. (2024)

pomace, carrot pulp (Hassas-Roudsari et al., 2009; Sabio et al., 2003; Wijngaard et al., 2012). Tomato waste from the vegetable processing industry has been used to extract the lycopene and pectin (Liu et al., 2023). Green, yellow, and red pepper wastes, as well as the flesh and peels of sweet potatoes, tomatoes, apricots, pumpkins, and peaches, were all exposed to SFE under optimal circumstances (59 °C, 350 bar, 15 g/min CO_2, 15.5% (v/v) ethanol as co-solvent, 30 min of extraction time). Total carotenoid concentration, total carotenoid recovery (TCR), and antioxidant activity were measured in the resulting extracts (de Andrade Lima et al., 2019). Numerous studies demonstrate the valorization of agri-food waste using SFE-CO_2. For instance, tomato pomace, grape seeds, broccoli stems, and custard apple seeds have yielded high concentrations of antioxidants, including lycopene, β-carotene, resveratrol, and chlorogenic acid. Co-solvents like ethanol enhance the extraction of polar compounds, extending SFE's applicability. Compared to conventional methods (e.g., hexane or dichloromethane-based extraction), SFE offers superior safety, lower solvent residues, and reduced environmental impact. The technology's success is evident in the commercial decaffeination of coffee and hop oil extraction, inspiring widespread research into its application for waste valorization. As the demand for natural additives and functional ingredients increases, SFE represents a green, scalable solution for converting food waste into high-value bioactives. This aligns with circular economy principles, offering environmental and economic benefits while promoting sustainable food system innovation (Afraz et al., 2023).

11.4.3 Ultrasonic-Assisted Extraction (UAE)

UAE uses ultrasound's cavitation impact to speed up the release of active compounds and encourage extraction. The UAE high-frequency sound wave is sufficiently powerful to cause cavitation that is easily absorbed by cell walls. Mechanical vibrations, which may be delivered into a solid, liquid, or gas at frequencies higher than 20 kHz, are the basis of sound waves. Since humans can only detect sounds between 16 Hz and 16 kHz, ultrasound is defined as sound that is above their range of hearing. It is primarily defined by wavelength and frequency (kHz to MHz range). This method is utilized for a number of food-processing operations, such as enzyme and microbe inactivation, extraction, drying, emulsification, and homogenization. According to several studies, the extraction process is facilitated by the disruption of cells caused by the collapse of cavitation bubbles near cell walls, which is induced by ultrasound. In addition, the ultrasound wave facilitated hydration and swelling, which led to the occurrence of bigger holes in the plant's cell wall and structural alterations in the plant tissue. In this way, mass transfer is enhanced, and the dispersion of the substance inside the cell is encouraged (Marić et al., 2018). UAE may promote cell growth and boost extraction effectiveness. UAE technique is used for the extraction of flavanones, hesperidin, catechins, carotenoids, pectin, and lycopene in fruits and vegetables (Zhu et al., 2023). Grape seed/skin, blueberry waste, cranberry pomace, carrot waste, citrus peel, spent tea leaves, and apple peel are utilized to produce the cyanidin, catechin, and flavan-3-ol (Liu et al., 2023). When applied to food wastes such as citrus peel, watermelon rind, or sugar beet pulp, UAE enhances pectin yield and maintains its functional quality. The mechanical effects of cavitation improve solvent penetration and diffusion, allowing better recovery without strong acids or high temperatures. UAE, combined with mild heating (UAHE), further optimizes extraction efficiency while preserving pectin's physicochemical properties, such as esterification degree. This method is energy-efficient, eco-friendly, and adaptable, making it highly beneficial for sustainable biorefining processes involving diverse plant-based wastes and by-products (Marić et al., 2018).

11.4.4 Pressurized Liquid Extraction (PLE)

To keep the solvent liquid throughout the extraction process, PLE uses solvent extraction at high temperatures and pressures that are always below the corresponding critical points. Under these specific pressure and temperature circumstances, the physicochemical characteristics of the extraction solvent alter, facilitating deeper and easier penetration of the solid or semi-solid matrix to be extracted. Specifically, the mass transfer rate rises in tandem with an increase in analyte solubility and a reduction in solvent viscosity and surface tension. These circumstances lead to a quick extraction method that uses little solvent and produces large yields. Furthermore, automated instrumentation enhances repeatability and facilitates the development of labor-saving techniques (Barp et al., 2023). Through the application of pressure, PLE raises the temperature over the typical solvent boiling point. The benefit of PLE is that it uses less solvent and takes less time. This technique is used to extract procyanidin and quercetin from red grape pomace, grape skin, and tomato waste (Monrad et al., 2010; Liu et al., 2023) and polyphenols (hydroxytyrosol, tyrosol, oleuropein), and flavonoids from olive pomace (Katsinas et al., 2021). In a comparative study focused on saffron processing waste, pressurized-liquid extraction (PLE), conventional stirred-tank extraction (STE), and stirred-tank extraction with ultrasonication pretreatment (STE/UP) were assessed using aqueous solutions of citric and lactic acids as environmentally friendly solvents. Although PLE enabled rapid extraction, it yielded extracts with lower concentrations of polyphenols and pigments and exhibited comparatively weaker antioxidant activity. Conversely, STE and STE/UP produced richer extracts with higher ferric-reducing power and antiradical capacity. Among the methods, STE/UP with 1% lactic acid emerged as the most effective, as determined by hierarchical cluster analysis, yielding extracts with superior polyphenol content and enhanced antioxidant potential. Despite its lower initial performance, PLE's advantage lies in its significantly reduced extraction time, highlighting its potential for further optimization. Modifications to pressure, temperature, solvent composition, or extraction cycles may enhance its efficacy, enabling it to rival or surpass conventional methods. Importantly, PLE aligns with principles of green chemistry by minimizing energy input, solvent use, and environmental impact (Pappas et al., 2021). Studies on seaweed, for instance, reveal its rich composition of antioxidants, essential minerals, and pigments, making it a highly valuable matrix for PLE application. Although seaweeds are more commonly consumed in Asia, their increasing popularity in Western markets presents an opportunity for wider utilization of their processing waste using advanced extraction methods such as PLE. Recent investigations highlight the advantages of PLE in extracting compounds from seaweeds and other food waste matrices. Optimization of parameters such as solvent type, temperature, pressure, extraction time, and number of cycles is crucial for maximizing yields. Moreover, combining PLE with other techniques, such as supercritical fluid extraction (SFE) using CO_2, has demonstrated improved recovery efficiencies and economic feasibility, particularly for scale-up scenarios. This hybrid approach can produce concentrated extracts with high levels of bioactive compounds, suitable for use in nutritional supplements, functional foods, and food packaging materials. PLE has shown particular promise in applications involving food-grade green solvents, such as aqueous solutions of citric or lactic acid, aligning with sustainable extraction protocols. These solvents are not only safer for food applications but also enhance the recovery of hydrophilic bioactives like polyphenols. Although PLE often yields lower concentrations of target compounds compared to some conventional or ultrasound-assisted extractions under non-optimized conditions, its drastically shorter extraction time and lower environmental impact are compelling advantages. The food industry is exploring various applications for bioactives extracted from food waste via PLE. These include the development of edible coatings, bioplastics, and bio-nanocomposites to enhance packaging materials, as well as the enrichment of food and animal feed products. Despite limited experimental data on edible seaweeds specifically, analogous studies on similar matrices suggest that PLE holds substantial potential for broader food applications (Perez-Vazquez et al., 2023).

11.4.5 High Hydrostatic Pressure (HHP)

To improve food safety and shelf life, high hydrostatic pressure (HHP) treatment is a useful method for eliminating microbes and inactivating enzymes under high pressure. Based on the principles of mass transfer, HHP is often applied in the region of 100 to 1200 MPa as a non-thermal approach. According to phase behavior theory, the applied pressure increases plant cell permeability, which results in cell component diffusivity; in other words, the compound becomes more soluble as pressure increases (Grassino et al., 2020). This technique is used to extract the sulforaphane from raw broccoli, pectin from pomelo peels (Guo et al., 2017), and increase 15% recovery of pectin and polyphenols from tomato peel waste as compared to conventional methods (Grassino et al., 2020). Characterization of HHP-treated food waste includes protein structural analysis, enzyme-substrate interactions, peptide yield and profiling, bioactivity assessments (e.g., antioxidant, antihypertensive), and nutritional and safety evaluations. These analyses help optimize enzymatic hydrolysis efficiency, enhance peptide functionality, and support the sustainable conversion of food by-products into valuable bioactive peptide ingredients (Marciniak et al., 2018).

11.4.6 Pulse Electric Field (PEF)

By applying an external electric field (10–80 KV/cm), it causes the cell membranes to experience a critical potential, which impacts pore formation, ruptures, and increases the permeability of the membrane. This technique is used to extract the phenols, betalains, and pectin from grape seed, red beet root, and apple pomace (Boateng, 2022). Potato and sweet potato waste from vegetable processing industries is used to extract the starch, steroidal alkaloids, α-tocopherol, cellulolytic enzymes, and biopolymer films (Liu et al., 2023). It enhances carotenoid recovery from food waste by increasing cell membrane permeability, enabling efficient extraction with minimal thermal damage. It preserves nutritional quality, reduces solvent use, and shortens processing time. PEF is a green, scalable technique ideal for valorizing carotene-rich waste into functional food and nutraceutical ingredients (Tomas et al., 2024). It enhances bioactive compound extraction from food waste such as tomato, potato, carrot, corn, onion, asparagus roots, chicory, coffee, cocoa, and seafood by-products by increasing cell membrane permeability. It improves yields of carotenoids, polyphenols, proteins, and polysaccharides, while preserving antioxidant activity and structural integrity, offering a sustainable, non-thermal valorization method (Chatzimitakos et al., 2023).

11.4.7 *Enzyme-Assisted Extraction (EAE)*

Enzyme-assisted extraction (EAE) is a pretreatment that is thought to be environmentally beneficial for extracting oils and bioactive compounds. This technique is used to extract oils from mango kernel, soybean, and rice bran. Along with this, lycopene extracted from tomato pulp, gingerol from ginger, and essential oils from bay leaves (Boulila et al., 2015; Manasa et al., 2013; Zuorro et al., 2011). Apple by-products have been utilized to extract the pectin, lactic acid, citric acid, aroma compounds, butanol, fructooligosaccharides, and pectinases (Liu et al., 2023). Food waste, particularly from animal processing by-products, contains valuable enzymes such as proteases and lipases that can be extracted for industrial use. Enzyme extraction techniques involve isolating these bioactive proteins from tissues like liver, stomach, and blood, using methods such as affinity chromatography and gel filtration. These techniques allow for efficient recovery while maintaining enzyme activity and purity. Characterizing food waste through enzyme extraction reveals its potential as a resource rather than waste, enabling valorization into functional ingredients that improve food quality, enhance sustainability, and reduce environmental impact in the food industry (Luo et al., 2025). It has been successfully applied to apple pomace, banana peels, and other fibrous wastes. EAE improves extract purity and reduces co-extraction of unwanted components such as lignin and phenolics. It also allows customization of pectin properties by selecting appropriate enzyme types and concentrations. While slower than MAE or UAE, EAE is highly selective, environmentally benign, and supports green chemistry principles. Its main limitations are cost and enzyme recovery challenges, but when integrated into biorefinery systems, it offers an effective and sustainable approach for pectin recovery from agricultural and food processing residues (Marić et al., 2018).

11.4.8 *Ionic Liquid Extraction (ILE)*

ILE is an extraction technique where a solvent is an organic liquid. These liquids have a simple solubility and high boiling point system because they are strong ion conductors. The ability to dissolve hydrophilic and hydrophobic compounds at room temperature during the extraction process is the biggest benefit of using ionic liquids as solvents. This technique has been used to extract the polyphenols, tannins, and catechins from fruit and vegetable wastes (Zhu et al., 2023). In cereal, pulses, and oilseeds waste (barley by-product, rice bran and husk, wheat, legumes) has been used to extract the vitamin E, phytates, insoluble dietary fiber, phenols, γ-oryzanol, tocopherols, carbohydrates, lipid soluble vitamins, folic acid, phytosterols, amino acids, oligosaccharides, and peptides (Liu et al., 2023). Ionic extraction utilizes ionic liquids to efficiently extract bioactive compounds from food waste like berry pomace. These solvents enable the selective recovery of phenolics, flavonoids, and pigments under mild conditions, preserving bioactivity. The method offers high yield, low toxicity, and sustainability, making it ideal for valorizing agro-industrial residues (Peron et al., 2024). Food waste, rich in valuable bioactives such as polyphenols, flavonoids, carotenoids, alkaloids, and organic acids, presents a promising feedstock for value-added recovery. When ILs are employed as extraction solvents, these compounds can be efficiently separated due to ILs' unique tunable properties, such as adjustable polarity, hydrophobicity, and hydrogen-bonding capabilities. The characterization of these bioactives typically involves quantitative and qualitative analysis using advanced analytical techniques such as high-performance liquid chromatography (HPLC), gas chromatography-mass spectrometry (GC–MS), Fourier transform infrared spectroscopy (FTIR), nuclear magnetic resonance (NMR), and UV–vis spectroscopy. These methods allow for a detailed profiling of the chemical nature, structural features, and concentration of extracted components. Polyphenols, one of the most abundant antioxidant classes in food waste, are effectively recovered using ILs due to their ability to form hydrogen bonds with IL cations or anions. Studies show that imidazolium- and phosphonium-based ILs can selectively enhance the recovery of catechins, quercetin, gallic acid, and rutin from fruit peels and vegetable residues. Characterization of these extracts reveals high antioxidant capacity, as measured by DPPH or ABTS assays, and strong radical-scavenging activity. FTIR spectra of IL-extracted phenolic fractions typically exhibit prominent absorption bands corresponding to hydroxyl (-OH) stretching, carbonyl (C=O) vibrations, and aromatic ring structures, confirming the phenolic nature of the extract. Moreover, total phenolic content (TPC) and total flavonoid content (TFC) analyses further validate the enrichment of bioactives post-IL extraction. Similarly, carotenoids such as lycopene and β-carotene, which are hydrophobic and sensitive to degradation, benefit from the mild and non-volatile nature of ILs. ILs with hydrophobic anions, such as $[PF_6]^-$ or $[NTf_2]^-$, show superior solubilization capacity for carotenoids from tomato peels or carrot waste. The recovered fractions display strong UV–vis absorption in the 400–500 nm range, and NMR analyses confirm the structural integrity of carotenoids after extraction, suggesting minimal degradation during processing. In terms of organic acids and small bioactive molecules like volatile fatty acids (VFAs), ILs such as [P666,14][Cl] have shown effectiveness in extracting short-chain acids, particularly at higher pH levels where conventional extraction solvents fail. These VFAs are characterized using HPLC equipped with refractive index detectors or GC–MS following derivatization, which confirms the presence and purity of acids like acetic, propionic, and butyric acids. IL-based extraction preserves the acid profile and improves selectivity, especially for longer-chain VFAs. Furthermore, bioactive alkaloids and glycosides, though present in lower concentrations, are also efficiently extracted by functionalized ILs with targeted affinity, and their characterization using mass spectrometry-based techniques reveals detailed fragmentation patterns and molecular weights that support compound identification (Xing et al., 2023).

11.4.9 Anaerobic Digestion

The waste of onion, potato, carrot, pineapple, and other fruits and vegetables consists of 75% sugar and hemicellulose, 9% cellulose, and 5% lignin, which is biodegradable and can be used for the production of biomethane gas using methane-producing bacteria (Ren et al., 2018). Hunter et al. (2021) reported various methods used in anaerobic digestion to extract different value-added products from various food waste containing carbohydrates, protein, and lipids. In the methanogenesis process, biogas is produced, acetogenesis produces biohydrogen, acidogenesis produces biofuels, volatile fatty acids, carboxylic acids, hydrocarboxylic acids, dicarboxylic acids, furans, hydrolysis produces sugars, amino acids, phytochemicals, phenols, flavonoids, carotenoids, vitamins, and terpenes (Hunter et al., 2021). Food waste (FW), rich in organic matter, is a promising substrate for aerobic digestion due to its high biodegradability and moisture content. In aerobic digestion, microorganisms decompose FW in the presence of oxygen, producing stabilized compost, heat, and carbon dioxide. This process is rapid, reducing the volume and mass of waste significantly while eliminating pathogens and odors. FW's low lignin and high sugar, starch, and protein content facilitate efficient microbial breakdown. Proper aeration, moisture balance, and carbon-to-nitrogen (C/N) ratio optimization enhance the process. Aerobic digestion offers an environmentally friendly alternative for FW management, yielding nutrient-rich compost for agricultural applications (Pilarska et al., 2023).

11.4.10 Fermentation

By using a fermentation method, bioethanol can be produced from different methods, such as alkali, acid, or enzymatic pretreatment from the fruit and vegetable wastes (banana peel, potato peel, citrus wastes, and domestic wastes) (El Sheikha & Ray, 2022). *Saccharomyces cerevisiae* was the culture generally utilized in fermentation, which can only consume hexose sugars. It is possible to produce ethanol using pentose sugars in conjunction with other fermentative organisms (Zhu et al., 2023). Dairy industry waste is utilized to extract the prebiotics, lactose, whey protein, biodiesel, ethanol, minerals, baker's yeast, and galactooligosaccharides (Liu et al., 2023). Meat and seafood processing industries wastes is used to develop fertilizer, chitin, animal feed, and protein, astaxanthin, chitinase, lactic acid is extracted (Liu et al., 2023). The food industry produces substantial waste, especially from cereal milling, yet by-products like bran and germ retain high nutritional value. Sourdough fermentation, either spontaneous or with selected strains, is an effective, sustainable method to upcycle these materials. It improves the bioavailability of fibers, proteins, and phenolic compounds by breaking down cell wall matrices and reducing antinutritional factors like phytic acid. Additionally, it enhances sensory properties and shelf life, enabling their reintroduction into food products. Beyond nutritional improvements, sourdough fermentation contributes to food safety by reducing mycotoxins and stabilizing lipids, making it a valuable tool in circular food production systems (Pontonio et al., 2024). Fermented food waste is increasingly recognized as a promising source of bioactive compounds with significant health and industrial potential. Various fermented food residues, such as those from vegetables, dairy products, bread, brewing, and wineries, contain a wide array of primary and secondary metabolites, including polyphenols, peptides, polysaccharides, flavonoids, and oligosaccharides. These compounds exhibit notable biological activities, such as antioxidant, antimicrobial, anti-inflammatory, anticancer, antihypertensive, anti-obesity, and immunomodulatory effects, making them suitable for applications in nutraceuticals, functional foods, food packaging, and therapeutic development. Vegetable-based residues such as cabbage and radish powders from kimchi, and soybean dregs (okara), are rich in prebiotic dietary fibers and proteins. These contribute to gut health, enhance satiety, and provide potential anti-obesity benefits. For example, insoluble residues from okara were found to contain phenolic compounds and proteins that demonstrated in vivo anti-obesity effects. Sauerkraut juice, another by-product of fermented cabbage, was processed into a powder rich in phenols, vitamins, and minerals, offering potential as a natural salt substitute without compromising sensory qualities. Dairy fermentation by-products, such as acid whey from Greek yogurt and buttermilk, are also rich in bioactive components. Deproteinized whey contains calcium phosphate, which is useful in milk fortification and baking applications. Buttermilk, on the other hand, contains milk fat globule membrane (MFGM) components that have shown anti-hypertensive effects in human trials, pointing to its functionality in cardiovascular health management. In the brewing and winery industries, significant quantities of residues like grape marc and brewer's spent grain are generated. These by-products are valuable for their high content of flavonoids, phenolics, proteins, and fibers. Grape seed proanthocyanidins, for example, have demonstrated a satiating effect and the ability to stimulate GLP-1, a hormone linked to insulin secretion. These compounds also show strong antioxidant and antimicrobial properties, which are critical in food preservation and human health. Fermentation residues also offer industrially relevant compounds. In bread waste, fermentable sugars and proteins can be used to produce alcoholic beverages, although challenges remain in enzyme extraction and flavor quality. Bread waste has also shown promise as a raw material in extrusion cooking, offering enhanced dietary fiber content compared to traditional flour-based extrudates. Soy sauce residue, when chemically treated, provides high levels of insoluble fiber and proteins useful in seasonings and food additives. When biologically treated, it can serve as a methane gas substrate, highlighting its dual potential in both food and energy sectors. Peptides from yeast biomass and tofu whey wastewater (e.g., lunasin) have been recovered using membrane filtration and chromatography, exhibiting anti-inflammatory and anticancer potential through modulation of nitric oxide and cytokine production (Faria et al., 2023).

11.4.11 Pyrolysis and Gasification

One thermal method that produces bio-oil (major) and gas (syngas) from fruit and vegetable wastes in an oxygen-free atmosphere is called pyrolysis. It operates at temperatures between 400 and 800 °C. At a relatively high temperature range of 800 to 900 °C, gasification partly oxidizes organic wastes to create a mixture of flammable gases. Various reactor designs result in various products because of the reaction temperatures involved in catalytic pyrolysis and gasification. Fruit and vegetable wastes are broken down by pyrolysis, producing tar, bio-oil, char, and syngas. Gasification has the benefit of reducing the amount of dioxins produced by conventional incineration, as it breaks down hydrogenated syngas. However, carbon-based fruit and vegetable wastes may be processed using both gasification and pyrolysis. Pyrolysis and gasification are used to produce low-calorific value gas from the fruit and vegetables (almond shells, pistachio shells, oil palm shells) (Niksiar & Nasernejad, 2017; Wang et al., 2018). Food waste typically contains a high proportion of organic matter, with moisture content often ranging between 60% and 80%, which directly influences its thermal behavior and energy potential. After drying, its elemental composition generally consists of carbon (40–50%), hydrogen (5–7%), oxygen (35–45%), and minor amounts of nitrogen and sulfur. The ash content varies, depending on the presence of components like eggshells, bones, or fibrous plant material. On a dry basis, food waste has a heating value of approximately 15–20 MJ/kg, which is comparable to other organic wastes and low-grade biomass, making it a viable feedstock for thermal conversion. During thermochemical treatment, food waste is transformed into three primary product streams: solid (biochar or ash), liquid (bio-oil or tar), and gaseous (syngas) fractions. The solid residue, commonly referred to as biochar when obtained via pyrolysis, is characterized by an increased fixed-carbon content and improved fuel properties, including higher heating value and lower H/C and O/C ratios. These changes indicate a more stable and aromatic structure, which can enhance its applicability in soil amendment, carbon sequestration, and pollutant adsorption. Biochar also demonstrates improved porosity and surface area, attributes that depend strongly on the thermal treatment temperature and residence time. Additionally, functional groups such as carboxyl, hydroxyl, and carbonyl remain on the surface of biochar, contributing to its reactivity and potential environmental applications. The gaseous products resulting from food waste gasification mainly consist of carbon monoxide (CO), hydrogen (H_2), methane (CH_4), and carbon dioxide (CO_2). The relative proportions of these gases depend on the process parameters and the type of gasifying agent used, such as air, steam, or oxygen. Typically, the produced syngas has a calorific value in the range of 4–8 MJ/Nm^3. The composition and energy content of the gas are influenced by the feedstock's chemical makeup and the operating conditions, including temperature, pressure, and heating rate. Moreover, impurities such as tar, particulates, and trace metals may be present in the gas, requiring efficient cleaning systems to ensure its suitability for combustion or chemical synthesis. The liquid fraction, or bio-oil, obtained from pyrolysis contains a complex mixture of oxygenated organic compounds, including acids, alcohols, ketones, aldehydes, and phenols. This bio-oil typically has low pH, high viscosity, and poor stability due to its high oxygen content and tendency for polymerization. While it holds potential as a renewable fuel or chemical feedstock, significant upgrading is usually required. Inorganic analysis of food waste reveals elements such as potassium, sodium, calcium, magnesium, and phosphorus, along with occasional trace metals. These components can impact the efficiency and safety of thermal systems by contributing to fouling, slagging, or corrosion. Overall, the characterization of food waste and its thermal conversion products provides essential insights into optimizing process design, improving energy recovery, and supporting the development of sustainable waste management practices (Czerski, 2022).

11.4.12 Hydrothermal Carbonization (HTC)

HTC is a thermal process that is capable of handling wastes with high moisture contents (80–90%), particularly those derived from fruits and vegetables. HTC is a type of wet processing that operates between 180 and 350 °Celsius under pressure for a period of 0.2 to 120 h. Many academics are currently paying close attention to HTC. HTC has the ability to treat huge volumes of trash in a short length of time (a few hours) without releasing an odor. It can also transform precipitated fruit and vegetable wastes into bio-energy at relatively low temperatures. In order to significantly reduce the reaction time, Niksiar and Nasernejad (2017) investigated the adsorption capability of copper ions in composting water following HTC with rice husk, citrus waste, and olive pomace at various temperatures. The valorization of food waste through hydrothermal carbonization (HTC) offers a sustainable pathway for the recovery and characterization of bioactive compounds. Food waste, including fruit peels, vegetable residues, and agro-industrial by-products, is a rich source of bioactive compounds such as polyphenols, flavonoids, organic acids, and antioxidants. These compounds, although often underutilized, can be effectively extracted, stabilized, or transformed during HTC into value-added materials with diverse functionalities. The biochemical composition of food waste significantly influences the nature of the bioactive compounds present in the resulting hydrochar. Common bioactives include gallic acid, catechins, caffeic acid, ferulic acid, and flavonoid glycosides, many of which originate from fruit-based residues such as citrus peels, apple pomace, or grape skins. During HTC, partial hydrolysis and depolymerization reactions occur, promoting the release of bound phenolics and degradation of macromolecules into smaller, bioactive fragments that can either be retained in the solid phase or dissolve into the process water. Characterization of the hydrochar reveals a significant concentration of surface-bound bioactive compounds. Fourier transform infrared spectroscopy (FTIR) is commonly employed to detect functional groups such as hydroxyl, carbonyl, and carboxyl moieties, which are characteristic of phenolic and organic acid structures. Thermogravimetric analysis (TGA) and differential scanning calorimetry (DSC) can also be used to analyze the thermal stability and composition of hydrochars, revealing the preservation or transformation of organic components. Liquid chromatography–mass spectrometry (LC-MS) and high-performance liquid chromatography (HPLC) are standard techniques for quantifying specific bioactives in the liquid and solid HTC fractions. Studies on food waste-derived hydrochars report high retention of antioxidant phenolics on the carbon surface, often linked to their interaction with lignocellulosic fragments or the formation of humic-like substances. These compounds enhance the antioxidant capacity of hydrochars, as confirmed through DPPH (2,2-diphenyl-1-picrylhydrazyl) and ABTS (2,2′-azinobis-3-ethylbenzothiazoline-6-sulfonic acid) assays. Additionally, the HTC process water contains a wide spectrum of water-soluble bioactive molecules. Analytical profiling reveals the presence of sugars, organic acids, furans, aldehydes, and low-molecular-weight phenolics, which arise from the degradation of polysaccharides and lignin during hydrothermal treatment. These compounds can be recovered through downstream processes such as solvent extraction, membrane filtration, or adsorption onto resins, providing additional routes for value recovery. The antioxidant capacity and reducing power of HTC-derived hydrochars and process water are of great interest. Antioxidant assays demonstrate that bioactive-rich hydrochars show substantial radical-scavenging activity, often correlating with the type of food waste and HTC temperature used. Lower HTC temperatures tend to preserve more native phenolic structures, while higher temperatures promote the formation of novel, aromatic polymeric structures with antioxidant potential. The presence of nitrogen-containing compounds, often derived from protein-rich waste such as food processing sludge or legumes, contributes to the formation of nitrogen-doped hydrochars. These materials exhibit enhanced electrochemical activity and antioxidant behavior due to the stabilization of nitrogen functional groups within the carbon matrix. Surface characterization via scanning electron microscopy (SEM) and Brunauer–Emmett–Teller (BET) analysis reveals morphological and textural changes that impact the accessibility of bioactives on the hydrochar. Mesoporous structures and increased surface area facilitate interactions with environmental pollutants or biological systems, enabling potential applications in nutraceuticals, pharmaceuticals, or as bioactive fillers in packaging materials. Overall, HTC provides a platform not only for carbonizing food waste but also for concentrating and stabilizing valuable bioactive compounds. These compounds, retained in hydrochar or liberated into process water, can be characterized and potentially recovered, transforming food waste into functional materials with antioxidant, antimicrobial, or chemical reactivity properties relevant to environmental, agricultural, or food sectors (Petrović et al., 2024).

11.5 Emerging Technologies in Food Waste Reduction

The increasing magnitude of food waste across the global food supply chain has accelerated the adoption of emerging technologies aimed at prevention, monitoring, and efficient utilization of food resources. Traditional waste management approaches are no longer sufficient to address losses occurring during production, processing, distribution, and consumption. Recent technological advancements, such as smart packaging, artificial intelligence (AI), Internet of Things (IoT), and blockchain, have demonstrated significant potential in reducing food waste by improving traceability, decision-making, and shelf-life management (Ahmadzadeh et al., 2023).

11.5.1 Smart Packaging and Sensors

Smart packaging refers to advanced packaging systems that incorporate sensors, indicators, and data carriers to monitor the condition of food products throughout the supply chain. Unlike conventional packaging, smart packaging provides real-time information regarding food quality, freshness, and safety. This technology plays a crucial role in minimizing food waste caused by spoilage, improper storage, and misinterpretation of expiry dates. Time–temperature indicators (TTIs) are among the most widely used smart packaging tools. These indicators monitor temperature variations during storage and transportation and provide visual signals when products are exposed to unfavorable conditions. Similarly, freshness indicators detect chemical changes such as pH shifts or gas formation, which are associated with microbial growth in perishable foods. By providing accurate quality status, smart packaging prevents premature disposal of safe food products. Biosensors embedded in packaging materials can detect pathogens, toxins, or spoilage metabolites. The application of these sensors enables early detection of contamination, reducing large-scale recalls and associated waste. Smart packaging also supports improved inventory management by enabling retailers to prioritize the sale of products nearing the end of their shelf life, thus significantly lowering food losses at the retail level (Liegeard & Manning, 2020).

11.5.2 Artificial Intelligence and Digital Monitoring

Artificial intelligence has emerged as a powerful tool for analyzing complex datasets across food supply chains. AI-based digital monitoring systems help predict food demand, optimize production schedules, and reduce overproduction, which is a major contributor to food waste. Machine learning algorithms analyze historical sales data, weather patterns, and consumer behavior to generate accurate demand forecasts. In food processing and retail environments, AI-powered computer vision systems are increasingly used to identify defective, spoiled, or misshapen food products. Instead of discarding these items, they can be redirected for processing or value addition. Digital monitoring tools also assist in dynamic pricing strategies, where prices of products nearing expiration are adjusted automatically to encourage quicker sales and reduce disposal. AI applications in food waste management extend to waste auditing and analysis. Automated waste tracking systems identify critical loss points within production and distribution processes, allowing food technologists to implement targeted interventions. The integration of AI with predictive analytics improves operational efficiency and supports sustainable food production systems (da Costa et al., 2022).

11.5.3 Internet of Things (IoT) in Food Supply Chains

The Internet of Things (IoT) involves interconnected devices equipped with sensors that collect and transmit real-time data across supply chains. IoT technology enhances visibility, traceability, and control over food products from farm to consumer. This continuous flow of information plays a vital role in reducing food waste caused by improper handling and delayed response to quality deterioration. IoT-enabled temperature and humidity sensors are extensively used in cold-chain logistics to ensure optimal storage conditions for perishable foods. Real-time alerts notify stakeholders when deviations occur, enabling corrective actions before spoilage happens. Additionally, GPS-enabled tracking systems help reduce delays during transportation, minimizing waste caused by prolonged transit times. In food processing facilities, IoT systems monitor equipment performance and process parameters, reducing production errors and batch rejections. Smart warehouses use IoT-based inventory management systems that track stock movement and shelf life, ensuring efficient stock rotation and reduced expiration-related losses. Overall, IoT technologies significantly enhance coordination among supply chain actors and promote data-driven waste reduction strategies (da Costa et al., 2022).

11.5.4 Blockchain for Traceability and Waste Reduction

Blockchain technology provides a decentralized and tamper-proof digital ledger for recording transactions and product movements across the food supply chain. One of the primary causes of food waste is the lack of transparency and traceability, which leads to delayed recalls, quality disputes, and unnecessary disposal of food products. Blockchain addresses these challenges by enabling end-to-end traceability and trust among stakeholders. Each transaction recorded on a blockchain contains time-stamped data related to production, processing, storage, and distribution. This allows rapid identification of contamination sources, reducing the need for broad product recalls that often result in massive food waste. Blockchain also enhances accountability by ensuring that quality and safety standards are followed at every stage of the supply chain. Furthermore, blockchain-enabled smart contracts automate transactions and compliance verification, reducing administrative delays and inefficiencies. Consumers can access detailed product information such as origin, freshness, and handling history, enabling informed purchasing decisions and reducing household food waste. By improving transparency and collaboration, blockchain technology contributes significantly to sustainable food waste reduction practices (Ahmadzadeh et al., 2023).

11.6 Application and Advantages

11.6.1 Applications of Food Waste Reduction Technologies

11.6.1.1 Recovery of Bioactive Compounds and Functional Ingredients

Advanced extraction technologies such as microwave-assisted extraction (MAE), supercritical fluid extraction (SFE), ultrasonic-assisted extraction (UAE), pressurized liquid extraction (PLE), enzyme-assisted extraction (EAE), and ionic liquid extraction (ILE) are widely applied for the recovery of bioactive compounds from food waste. These include polyphenols, flavonoids, carotenoids, pectin, dietary fibers, antioxidants, and essential oils from fruit peels, vegetable pomace, cereal bran, and oilseed residues (Marić et al., 2018; Liu et al., 2023). Extracted compounds are utilized in functional foods, nutraceuticals, pharmaceuticals, and active food packaging materials.

11.6.1.2 Bioenergy and Biofuel Production

Food waste rich in carbohydrates, proteins, and lipids is extensively applied in anaerobic digestion, fermentation, pyrolysis, gasification, and hydrothermal carbonization (HTC) processes. Anaerobic digestion converts biodegradable waste into biogas and biomethane, while fermentation produces bioethanol, organic acids, and biohydrogen (Hunter et al., 2021; El Sheikha & Ray, 2022). Thermochemical processes such as pyrolysis and gasification generate syngas, bio-oil, and biochar, contributing to renewable energy production and waste volume reduction (Czerski, 2022).

11.6.1.3 Development of Sustainable Packaging and Materials

Recovered biopolymers such as pectin, starch, proteins, and cellulose from food waste are applied in the development of edible coatings, biodegradable films, and bio-nanocomposites. Technologies like PLE, UAE, and HHP enable the extraction of high-quality polymers suitable for food packaging applications (Perez-Vazquez et al., 2023). These materials enhance shelf life and reduce dependency on petroleum-based plastics.

11.6.1.4 Shelf-Life Extension and Waste Prevention

Emerging digital technologies, including smart packaging, AI, IoT, and blockchain, are applied across the food supply chain to prevent waste before it occurs. Smart packaging with sensors and indicators monitors freshness and storage conditions, reducing spoilage-related losses (Liegeard & Manning, 2020). AI-driven demand forecasting and IoT-enabled cold chain monitoring optimize inventory management and reduce overproduction and expiration losses (da Costa et al., 2022).

11.6.2 Advantages of Food Waste Reduction Technologies

1. *Environmental Sustainability*

Food waste reduction technologies significantly lower environmental impacts by reducing landfill disposal, greenhouse gas emissions, and resource wastage. Green extraction techniques such as SFE, MAE, UAE, and PLE minimize solvent use and energy consumption, aligning with green chemistry principles (Afraz et al., 2023). Anaerobic digestion and HTC further contribute to carbon neutrality through energy recovery.

2. *Economic Value Addition*

Valorization of food waste transforms low-value residues into high-value products such as antioxidants, functional ingredients, enzymes, biofuels, and bioplastics. This creates additional revenue streams for food processors and supports sustainable business models. Technologies like SFE and EAE have demonstrated commercial viability in applications such as decaffeination and oil extraction (Marić et al., 2018).

3. *Preservation of Nutritional and Functional Quality*

Non-thermal technologies such as HHP, PEF, and UAE preserve thermolabile compounds and maintain the structural integrity of bioactives. These methods enhance extraction efficiency while retaining antioxidant activity, gelling properties, and bioavailability of recovered compounds (Grassino et al., 2020; Tomas et al., 2024).

4. *Improved Food Safety and Traceability*

Blockchain, AI, and IoT technologies improve transparency, traceability, and accountability within food supply chains. Rapid identification of contamination sources reduces large-scale recalls and associated food waste. Smart contracts and real-time monitoring enhance compliance with food safety standards (Ahmadzadeh et al., 2023).

5. *Alignment with Circular Economy Principles*

Food waste reduction technologies support circular food systems by closing material loops and promoting resource efficiency. Waste streams are reintroduced into the value chain as energy, materials, or functional ingredients, reducing dependence on virgin resources and fostering sustainable food system innovation (Hunter et al., 2021).

11.7 Conclusions

Food waste has emerged as one of the most pressing challenges facing modern food systems, with significant environmental, economic, and social implications. Large quantities of waste generated at household, processing, retail, and food service levels not only represent financial losses but also signify the underutilization of valuable bioresources embedded within food by-products. These wastes are rich in bioactive compounds such as polyphenols, antioxidants, carotenoids, vitamins, minerals, peptides, fatty acids, pectin, prebiotics, and enzymes, all of which hold considerable potential for applications in functional foods, nutraceuticals, pharmaceuticals, cosmetics, packaging materials, and bioenergy production. Harnessing this potential requires the adoption of advanced and sustainable food waste reduction and valorization technologies.

A wide range of innovative processing technologies including microwave-assisted extraction, supercritical fluid extraction, ultrasonic-assisted extraction, pressurized liquid extraction, high hydrostatic pressure, pulse electric field, enzyme-assisted extraction, ionic liquid extraction, anaerobic digestion, fermentation, pyrolysis, gasification, and hydrothermal carbonization have demonstrated strong capability in recovering valuable compounds from diverse food waste streams. These approaches offer clear advantages over conventional methods by reducing solvent use, energy consumption, and processing time while preserving the functional and nutritional integrity of bioactive compounds. As such, they align closely with green chemistry principles and circular economy frameworks, transforming food waste from an environmental burden into a valuable resource.

Despite these advances, several challenges remain that must be addressed to enable large-scale and economically viable implementation. High initial investment costs, technological complexity, limited scalability, regulatory constraints, and the need for standardized quality and safety guidelines for waste-derived products continue to hinder widespread adoption. Additionally, variability in food waste composition and the lack of integrated waste collection and processing infrastructure pose technical and logistical challenges. Addressing these issues will require continued research, process optimization, supportive policy frameworks, and cross-sector collaboration among industry, government, and academia.

Future trends in food waste management point toward the integration of digital and smart technologies across the food supply chain. The increasing challenge of food waste has accelerated the adoption of emerging technologies such as smart packaging and sensors, artificial intelligence and digital monitoring, Internet of Things (IoT) applications, and blockchain-based traceability systems. These technologies enable real-time monitoring of food quality, storage conditions, and product movement from production to consumption, thereby preventing spoilage, optimizing inventory management, and enhancing transparency. AI-driven forecasting and decision-support tools help reduce overproduction and identify waste hotspots, while IoT-enabled cold chain monitoring minimizes losses during storage and transportation. Blockchain technology further strengthens traceability and accountability, allowing rapid response to quality or safety issues and reducing unnecessary product recalls.

Looking ahead, the convergence of advanced extraction technologies with digital monitoring systems is expected to play a transformative role in building sustainable, resilient, and circular food systems. Policy reforms, standardized data labeling, economic incentives, and increased consumer awareness will be equally important in supporting technological adoption and behavioral change. Ultimately, embracing these emerging technologies and integrated strategies offers a pathway toward reducing food waste, improving resource efficiency, generating new economic opportunities, and enhancing food security and environmental sustainability.

References

Abdolshahi, A., Majd, M. H., Rad, J. S., Taheri, M., Shabani, A., & Da Silva, J. A. T. (2015). Choice of solvent extraction technique affects fatty acid composition of pistachio (Pistacia vera L.) oil. *Journal of Food Science and Technology, 52*, 2422–2427.

Adhikari, B. B., Chae, M., & Bressler, D. C. (2018). Utilization of slaughterhouse waste in value-added applications: Recent advances in the development of wood adhesives. *Polymers, 10*, 176.

Aditya, S., Stephen, J., & Radhakrishnan, M. (2021). Utilization of eggshell waste in calcium-fortified foods and other industrial applications: A review. *Trends in Food Science & Technology, 115*, 422–432.

Afraz, M. T., Xu, X., Adil, M., Manzoor, M. F., Zeng, X. A., Han, Z., & Aadil, R. M. (2023). Subcritical and supercritical fluids to valorize industrial fruit and vegetable waste. *Food, 12*(12), 2417.

Ahmadzadeh, S., Ajmal, T., Ramanathan, R., & Duan, Y. (2023). A comprehensive review on food waste reduction based on IoT and big data technologies. *Sustainability, 15*(4), 3482.

Akhila, D. S., Ashwath, P., Manjunatha, K. G., Akshay, S. D., Surasani, V. K. R., Sofi, F. R., et al. (2024). Seafood processing waste as a source of functional components: Extraction and applications for various food and non-food systems. *Trends in Food Science & Technology, 145*, 104348.

Al-Tayawi, A. N., Sisay, E. J., Beszédes, S., & Kertész, S. (2023). Wastewater treatment in the dairy industry from classical treatment to promising technologies: An overview. *Processes, 11*(7), 2133.

Amasuomo, E., & Baird, J. (2016). The concept of waste and waste management. *Journal of Management and Sustainability, 6*, 88.

Andrés, A., Petrón, M., Adámez, J., López, M., & Timón, M. (2017). Food by-products as potential antioxidant and antimicrobial additives in chill stored raw lamb patties. *Meat Science, 129*, 62–70.

Ashayerizadeh, O., Dastar, B., Samadi, F., Khomeiri, M., Yamchi, A., & Zerehdaran, S. (2017). Study on the chemical and microbial composition and probiotic characteristics of dominant lactic acid bacteria in fermented poultry slaughterhouse waste. *Waste Management, 65*, 178–185.

Augustin, M. A., Sanguansri, L., Fox, E. M., Cobiac, L., & Cole, M. B. (2020). Recovery of wasted fruit and vegetables for improving sustainable diets. *Trends in Food Science & Technology, 95*, 75–85.

Balandrán-Quintana, R. R., Mercado-Ruiz, J. N., & Mendoza-Wilson, A. M. (2015). Wheat bran proteins: A review of their uses and potential. *Food Review International, 31*, 279–293.

Barp, L., Višnjevec, A. M., & Moret, S. (2023). Pressurized liquid extraction: A powerful tool to implement extraction and purification of food contaminants. *Food, 12*(10), 2017.

Bellur Nagarajaiah, S., & Prakash, J. (2015). Nutritional composition, acceptability, and shelf stability of carrot pomace-incorporated cookies with special reference to total and β-carotene retention. *Cogent Food & Agriculture, 1*, 1039886.

Ben-Othman, S., Jõudu, I., & Bhat, R. (2020). Bioactives from agri-food wastes: Present insights and future challenges. *Molecules, 25*, 510.

Boateng, I. D. (2022). Recent processing of fruits and vegetables using emerging thermal and non-thermal technologies. A critical review of their potentialities and limitations on bioactives, structure, and drying performance. *Critical Reviews in Food Science and Nutrition., 64*, 4240.

Boulila, A., Hassen, I., Haouari, L., Mejri, F., Amor, I. B., Casabianca, H., & Hosni, K. (2015). Enzyme-assisted extraction of bioactive compounds from bay leaves (Laurus nobilis L.). *Industrial Crops and Products, 74*, 485–493.

Carmona-Cabello, M., García, I. L., Sáez-Bastante, J., Pinzi, S., Koutinas, A. A., & Dorado, M. P. (2020). Food waste from restaurant sector–characterization for biorefinery approach. *Bioresource Technology, 301*, 122779.

Cassano, A., & Galanakis, C. M. (2018). Membrane technologies for the fractionation of compounds recovered from cereal processing by-products. In C. M. Galanakis (Ed.), *Sustainable recovery and reutilization of cereal processing by-products* (pp. 159–187). Woodhead Publishing.

Chatzimitakos, T., Athanasiadis, V., Kalompatsios, D., Mantiniotou, M., Bozinou, E., & Lalas, S. I. (2023). Pulsed electric field applications for the extraction of bioactive compounds from food waste and by-products: A critical review. *Biomass, 3*(4), 367–401.

Chaudhary, V., Kajla, P., Verma, D., Singh, T. P., Kothakota, A., Prasath, V. A., et al. (2023). Valorization of dairy wastes into wonder products by the novel use of microbial cell factories. *Trends in Food Science & Technology, 142*, 104221.

Cheng, C., Chen, L., Zhang, D., Yu, J., Zhu, M., Li, C., et al. (2024). Value-added utilization of hemoglobin and its hydrolysis products from livestock and poultry blood processing by-products: A review. *Trends in Food Science & Technology, 151*, 104645.

Czerski, G. (2022). Pyrolysis and gasification of biomass and waste. *Energies, 15*(19), 7299.

da Costa, T. P., Gillespie, J., Cama-Moncunill, X., Ward, S., Condell, J., Ramanathan, R., & Murphy, F. (2022). A systematic review of real-time monitoring technologies and its potential application to reduce food loss and waste: Key elements of food supply chains and IoT technologies. *Sustainability, 15*(1), 614.

de Andrade Lima, M., Kestekoglou, I., Charalampopoulos, D., & Chatzifragkou, A. (2019). Supercritical fluid extraction of carotenoids from vegetable waste matrices. *Molecules, 24*(3), 466.

Defra. (2009). UK food security assessment: our approach. <http://archive.defra.gov.uk/foodfarm/food/pdf/foodassess-approach-0908.pdf>. Accessed 25 Jan 12.

Diwan, B., Parkhey, P., & Gupta, P. (2018). From agro-industrial wastes to single cell oils: A step towards prospective biorefinery. *Folia Microbiologica, 63*, 547–568.
Dorta, E., Lobo, M. G., & González, M. (2013). Improving the efficiency of antioxidant extraction from mango peel by using microwaveassisted extraction. *Plant Foods for Human Nutrition, 68*, 190–199.
El Sheikha, A. F., & Ray, R. C. (2022). Bioprocessing of horticultural wastes by solid-state fermentation into value-added/innovativebioproducts: A review. *Food Reviews International, 39*, 3009.
Elbadrawy, E., & Sello, A. (2016). Evaluation of nutritional value and antioxidant activity of tomato peel extracts. *Arabian Journal of Chemistry, 9*, S1010–S1018.
El-Safy, F. S., Salem, R. H., & Abd El-Ghany, M. (2012). Chemical and nutritional evaluation of different seed flours as novel sources of protein. *World Journal of Dairy & Food Sciences, 7*, 59–65.
European Commission. (2014). Food waste and its impacts: European week for waste reduction.
European Parliament Council. (1975). Council directive of 15 July 1975 on waste 75/442/EEC. Off. J. L 194, 3941.
FAO. (2023). State of food security and nutrition in the world report.
Faria, D. J., Carvalho, A. P. A. D., & Conte-Junior, C. A. (2023). Valorization of fermented food wastes and byproducts: Bioactive and valuable compounds, bioproduct synthesis, and applications. *Fermentation, 9*(10), 920.
Fishman, M. L., Chau, H. K., Cooke, P. H., & Hotchkiss, A. T., Jr. (2008). Global structure of microwave-assisted flash-extracted sugar beet pectin. *Journal of Agricultural and Food Chemistry, 56*, 1471–1478. Germany, 2010; pp. 25–43.
Foresight. (2010). How can waste reduction help to healthily and sustainably feed a future global population of nine billion people? London. Retrieved from: http://www.bis.gov.uk/assets/foresight/docs/food-and-farming/workshops/11-608-w4-expert-forum-reduction-of-foodwast.
Giordano, C., & Franco, S. (2021). Household food waste from an international perspective. *Sustainability, 13*(9), 5122.
Grassino, A. N., Ostojić, J., Miletić, V., Djaković, S., Bosiljkov, T., Zorić, Z., et al. (2020). Application of high hydrostatic pressure and ultrasound-assisted extractions as a novel approach for pectin and polyphenols recovery from tomato peel waste. *Innovative Food Science & Emerging Technologies, 64*, 102424.
Guo, X., Han, D., Xi, H., Rao, L., Liao, X., Hu, X., & Wu, J. (2012). Extraction of pectin from navel orange peel assisted by ultra-high pressure, microwave or traditional heating: A comparison. *Carbohydrate Polymers, 88*, 441–448.
Guo, X., Zhao, W., Liao, X., Hu, X., Wu, J., & Wang, X. (2017). Extraction of pectin from the peels of pomelo by high-speed shearing homogenization and its characteristics. *LWT-Food Science and Technology, 79*, 640–646.
Gustavsson, J., Cederberg, C., Sonesson, U., van Otterdijk, R., & Meybeck, A. (2011). *Global food losses and food waste: Extent, causes and prevention*. Food and Agriculture Organisation of the United Nations (FAO).
Hassas-Roudsari, M., Chang, P. R., Pegg, R. B., & Tyler, R. T. (2009). Antioxidant capacity of bioactives extracted from canola meal by subcritical water, ethanolic and hot water extraction. *Food Chemistry, 114*, 717–726.
Hua, X., Yang, R., Zhang, W., Fei, Y., Jin, Z., & Jiang, B. (2010). Dual-enzymatic synthesis of lactulose in organic-aqueous two-phase media. *Food Research International, 43*, 716–722.
Hunter, S. M., Blanco, E., & Borrion, A. (2021). Expanding the anaerobic digestion map: A review of intermediates in the digestion of food waste. *Science of the Total Environment, 767*, 144265.
Hwang, I.-W., Jeong, M.-C., & Chung, S.-K. (2011). The physicochemical properties and the antioxidant activities of persimmon peel powders with different particle sizes. *Journal of Korean Society for Applied Biological Chemistry, 54*, 442–446.

Kader, A.A., (2009). Handling of horticultural perishables in developing vs. developed countries. In: *VI International Post Harvest Symposium* (Vol. 877, pp. 121–126). International Society for Horticultural Science (ISHS).

Katsinas, N., Bento da Silva, A., Enríquez-de-Salamanca, A., Fernández, N., Bronze, M. R., & Rodríguez-Rojo, S. (2021). Pressurized liquid extraction optimization from supercritical defatted olive pomace: A green and selective phenolic extraction process. *ACS Sustainable Chemistry & Engineering, 9*(16), 5590–5602.

Kavitha, S., Kannah, R. Y., Kumar, G., Gunasekaran, M., & Banu, J. R. (2020). Introduction: Sources and characterization of food waste and food industry wastes. In *Food waste to valuable resources* (pp. 1–13). Academic.

Kiran, E. U., Trzcinski, A. P., Ng, W. J., & Liu, Y. (2014). Bioconversion of food waste to energy: A review. *Fuel, 134*, 389–399.

Kothari, V., & Seshadri, S. (2010). In vitro antibacterial activity in seed extracts of Manilkara zapota, Anona squamosa, and Tamarindus indica. *Biological Research, 43*, 165–168.

Kushwaha, J. P., Srivastava, V. C., & Mall, I. D. (2011). An overview of various technologies for the treatment of dairy wastewaters. *Critical Reviews in Food Science and Nutrition, 51*, 442–452.

Lappa, I. K., Papadaki, A., Kachrimanidou, V., Terpou, A., Koulougliotis, D., Eriotou, E., & Kopsahelis, N. (2019). Cheese whey processing: Integrated biorefinery concepts and emerging food applications. *Food, 8*, 347.

Liegeard, J., & Manning, L. (2020). Use of intelligent applications to reduce household food waste. *Critical Reviews in Food Science and Nutrition, 60*(6), 1048–1061.

Lipinski, B., Hanson, C., Lomax, J., Kitinoja, L., Waite, R., & Searchinger, T. (2013). *Reducing food loss and waste* (p. 140). World Resource Institute.

Liu, Z., de Souza, T. S., Holland, B., Dunshea, F., Barrow, C., & Suleria, H. A. (2023). Valorization of food waste to produce value-added products based on its bioactive compounds. *PRO, 11*(3), 840.

Luo, W., Zhang, J., Ahmmed, M. K., Sakai, K., Shahidi, F., Zhi, Z., & Wu, H. (2025). Valorization of animal by-product enzymes: Advancing sustainable food processing through innovative extraction, purification, and application strategies. *Trends in Food Science & Technology, 156*, 104870.

Manasa, D., Srinivas, P., & Sowbhagya, H. (2013). Enzyme-assisted extraction of bioactive compounds from ginger (Zingiber officinale Roscoe). *Food Chemistry, 139*, 509–514.

Marciniak, A., Suwal, S., Naderi, N., Pouliot, Y., & Doyen, A. (2018). Enhancing enzymatic hydrolysis of food proteins and production of bioactive peptides using high hydrostatic pressure technology. *Trends in Food Science & Technology, 80*, 187–198.

Marić, M., Grassino, A. N., Zhu, Z., Barba, F. J., Brnčić, M., & Brnčić, S. R. (2018). An overview of the traditional and innovative approaches for pectin extraction from plant food wastes and by-products: Ultrasound-, microwaves-, and enzyme-assisted extraction. *Trends in Food Science & Technology, 76*, 28–37.

Marques, R. V., Paz, M. F. D., Duval, E. H., Corrêa, L. B., & Corrêa, É. K. (2016). Staphylococcus xylosus fermentation of pork fatty waste: Raw material for biodiesel production. *Brazilian Journal of Microbiology, 47*, 675–679.

Milner, J. A., & Romagnolo, D. F. (2010). Cancer biology and nutrigenomics. In *Bioactive compounds and cancer*. Springer.

Monrad, J. K., Howard, L. R., King, J. W., Srinivas, K., & Mauromoustakos, A. (2010). Subcritical solvent extraction of anthocyanins from dried red grape pomace. *Journal of Agricultural and Food Chemistry, 58*, 2862–2868.

Niksiar, A., & Nasernejad, B. (2017). Activated carbon preparation from pistachio shell pyrolysis and gasification in a spouted bed reactor. *Biomass Bioenergy, 106*, 43–50.

Pan, X., Niu, G., & Liu, H. (2003). Microwave-assisted extraction of tea polyphenols and tea caffeine from green tea leaves. *Chemical Engineering and Processing Process Intensification, 42*, 129–133.

Papageorgiou, M., & Skendi, A. (2018). Introduction to cereal processing and by-products. In *Sustainable recovery and reutilization of cereal processing by-products* (pp. 1–25). Elsevier.

Pappas, V. M., Athanasiadis, V., Palaiogiannis, D., Poulianiti, K., Bozinou, E., Lalas, S. I., & Makris, D. P. (2021). Pressurized liquid extraction of polyphenols and anthocyanins from saffron processing waste with aqueous organic acid solutions: Comparison with stirred-tank and ultrasound-assisted techniques. *Sustainability, 13*(22), 12578.

Parashar, A., Jin, Y., Mason, B., Chae, M., & Bressler, D. C. (2016). Incorporation of whey permeate, a dairy effluent, in ethanol fermentation,to provide a zero waste solution for the dairy industry. *Journal of Dairy Science, 99*, 1859–1867.

Parfitt, J., Barthel, M., & Macnaughton, S. (2010). Food waste within food supply chains: Quantification and potential for change to 2050. *Philosophical Transactions of the Royal Society B, 365*, 30653081.

Perez-Vazquez, A., Carpena, M., Barciela, P., Cassani, L., Simal-Gandara, J., & Prieto, M. A. (2023). Pressurized liquid extraction for the recovery of bioactive compounds from seaweeds for food industry application: A review. *Antioxidants, 12*(3), 612.

Peron, G., Ferrarese, I., Carmo Dos Santos, N., Rizzo, F., Gargari, G., Bertoli, N., et al. (2024). Sustainable extraction of bioactive compounds and nutrients from agri-food wastes: Potential reutilization of berry, honey, and chicory byproducts. *Applied Sciences, 14*(23), 10785.

Petrović, J., Ercegović, M., Simić, M., Koprivica, M., Dimitrijević, J., Jovanović, A., & Janković Pantić, J. (2024). Hydrothermal carbonization of waste biomass: A review of hydrochar preparation and environmental application. *Processes, 12*(1), 207.

Pilarska, A. A., Kulupa, T., Kubiak, A., Wolna-Maruwka, A., Pilarski, K., & Niewiadomska, A. (2023). Anaerobic digestion of food waste—A short review. *Energies, 16*(15), 5742.

Pontonio, E., Stringari, A., Di Cagno, R., Filannino, P., Rizzello, C. G., Polo, A., et al. (2024). Plant-derived food waste management, valorization, and recycling through sourdough fermentation. *Trends in Food Science & Technology, 150*, 104589.

Prameela, K., Venkatesh, K., Immandi, S. B., Kasturi, A. P. K., Krishna, C. R., & Mohan, C. M. (2017). Next generation nutraceutical from shrimp waste: The convergence of applications with extraction methods. *Food Chemistry, 237*, 121–132.

Ren, Y., Yu, M., Wu, C., Wang, Q., Gao, M., Huang, Q., & Liu, Y. (2018). A comprehensive review on food waste anaerobic digestion: Research updates and tendencies. *Bioresource Technology, 247*, 1069–1076.

Sabio, E., Lozano, M., Montero de Espinosa, V., Mendes, R., Pereira, A., Palavra, A., & Coelho, J. (2003). Lycopene and _-carotene extraction from tomato processing waste using supercritical CO2. *Industrial and Engineering Chemistry Research, 42*, 6641–6646.

Sherazi, S. T. H., & Mahesar, S. A. (2016). Vegetable oil deodorizer distillate: A rich source of the natural bioactive components. *Journal of Oleo Science, 65*, 957–966.

Shirsath, A. P., & Henchion, M. M. (2021). Bovine and ovine meat co-products valorisation opportunities: A systematic literature review. *Trends in Food Science & Technology, 118*, 57–70.

Shukla, S., Sondhi, A., Tripathi, A. D., Lee, J. K., Patel, S. K., & Agarwal, A. (2024). Valorisation of fruit waste for harnessing the bioactive compounds and its therapeutic application. *Trends in Food Science & Technology, 144*, 104302.

Sinha, S., & Tripathi, P. (2021). Trends and challenges in valorisation of food waste in developing economies: A case study of India. *Case Studies in Chemical and Environmental Engineering, 4*, 100162.

Skendi, A., Zinoviadou, K. G., Papageorgiou, M., & Rocha, J. M. (2020). Advances on the valorisation and functionalization of by-products and wastes from cereal-based processing industry. *Food, 9*(9), 1243.

Tomas, M., Kamiloglu, S., Nemli, E., Ozdal, T., Haque, S., Apak, R., & Capanoglu, E. (2024). From carotene-rich waste-to-food: Extraction, food applications, challenges and opportunities. *Trends in Food Science & Technology, 155*, 104756.

Valcarcel, J., Novoa-Carballal, R., Pérez-Martín, R. I., Reis, R. L., & Vázquez, J. A. (2017). Glycosaminoglycans from marine sources as therapeutic agents. *Biotechnology Advances, 35*, 711–725.

Viganó, J., Machado, A. P. D. F., & Martínez, J. (2015). Sub- and supercritical fluid technology applied to food waste processing. *Journal of Supercritical Fluids, 96*, 272–286.

Wang, Y., Jiang, L., Dai, L., Yu, Z., Liu, Y., Ruan, R., Fu, G., Zhou, Y., Fan, L., & Duan, D. (2018). Microwave-assisted catalytic co-pyrolysis of soybean straw and soapstock for bio-oil production using SiC ceramic foam catalyst. *Journal of Analytical and Applied Pyrolysis, 133*, 76–81.

Wang, Y., Pan, S., Yin, J., Feng, H., Wang, M., & Chen, T. (2022). Resource potential and global warming potential of fruit and vegetable waste in China based on different treatment strategies. *Waste Management, 140*, 225–232.

Wanyo, P., Meeso, N., & Siriamornpun, S. (2014). Effects of different treatments on the antioxidant properties and phenolic compounds of rice bran and rice husk. *Food Chemistry, 157*, 457–463.

Wijngaard, H., Hossain, M. B., Rai, D. K., & Brunton, N. (2012). Techniques to extract bioactive compounds from food by-products of plant origin. *Food Research International, 46*, 505–513.

Xing, J.-J., Cheng, Y.-L., Chen, P., Shan, L., Ruan, R., Li, D., & Wang, L.-J. (2018). Effect of high-pressure homogenization on the extraction of sulforaphane from broccoli (Brassica oleracea) seeds. *Powder Technology, 358*, 103–109.

Xing, T., Yu, S., Tang, J., Liu, H., Zhen, F., Sun, Y., & Kong, X. (2023). Liquid–liquid extraction of volatile fatty acids from anaerobic acidification broth using ionic liquids and cosolvent. *Energies, 16*(2), 785.

Yaakob, M. A., Mohamed, R. M. S. R., Al-Gheethi, A., Tiey, A., & Kassim, A. H. M. (2019). Optimising of Scenedesmus sp. biomass production in chicken slaughterhouse wastewater using response surface methodology and potential utilisation as fish feeds. *Environmental Science and Pollution Research, 26*, 12089–12108.

Zhu, Y., Luan, Y., Zhao, Y., Liu, J., Duan, Z., & Ruan, R. (2023). Current technologies and uses for fruit and vegetable wastes in a sustainable system: A review. *Foods, 12*(10), 1949.

Zuorro, A., Fidaleo, M., & Lavecchia, R. (2011). Enzyme-assisted extraction of lycopene from tomato processing waste. *Enzyme and Microbial Technology, 49*, 567–573.

Index

Y. Kumar, S. Dey, *Emerging Technology in Food Processing*,
https://doi.org/10.1007/978-3-032-24505-2

The manufacturer's authorised representative in the EU is Springer Nature Customer Service Centre GmbH, Europaplatz 3, 69115 Heidelberg, Germany. If you have any concerns regarding our products, please contact ProductSafety@springernature.com

Printed and bound by CPI Group (UK) Ltd, Croydon, CR0 4YY
07/10/2026
02229349-0016